Paula von Gleich
The Black Border and Fugitive Narration in Black American Literature

American Frictions

Editors
Carsten Junker
Julia Roth
Darieck Scott

Editorial Board
Arjun Appadurai, New York University
Mita Banerjee, University of Mainz
Tomasz Basiuk, University of Warsaw
Isabel Caldeira, University of Coimbra

Volume 4

Paula von Gleich

The Black Border and Fugitive Narration in Black American Literature

—

DE GRUYTER

The open access publication of this title was supported by the publication fund of the Staats- und Universitätsbibliothek Bremen.

ISBN 978-3-11-135820-8
e-ISBN (PDF) 978-3-11-076103-0
e-ISBN (EPUB) 978-3-11-076128-3
ISSN 2698-5349
DOI https://doi.org/10.1515/9783110761030

This work is licensed under a Creative Commons Attribution-NonCommercial-NoDerivatives 4.0 International License. For details go to https://creativecommons.org/licenses/by-nc-nd/4.0/

Library of Congress Control Number: 2021950661

Bibliographic Information published by the Deutsche Nationalbibliothek
The Deutsche Nationalbibliothek lists this publication in the Deutsche Nationalbibliografie; detailed bibliographic data are available on the Internet at http://dnb.dnb.de.

© 2023 Paula von Gleich, published by Walter de Gruyter GmbH, Berlin/Boston
This volume is text- and page-identical with the hardback published in 2022.
The book is published with open access at www.degruyter.com
Printing and binding: CPI books GmbH, Leck

www.degruyter.com

Contents

Introduction or Looking for the Fugitive Life in Social Death —— 1

Part 1 Fugitivity against the Border: Afro-Pessimism, Black Feminist Fugitive Thought, and the Border to Social Death —— 12

1.1 A First Note on Method: Concept-Driven Cultural Analysis —— 13
1.2 Travelling Border Concepts —— 14
1.3 Afro-Pessimism and the Border(s) of "the Human" —— 19
1.4 "Black Feminist Fugitivity" against the Border —— 32
1.5 A Second Note on Method: The Texts, the Tools, and Positionalities —— 54

Part 2 Practices of Flight: Captivity and Fugitivity in Black American Literature —— 64

2.1 Fugitive Forbears —— 64
Reiterations of life and "comparative freedom" against social death in Frederick Douglass's *Narrative* and *My Bondage* —— 72
"Loophole[s] of retreat" from the garret to the North: Harriet Jacobs's *Incidents in the Life of a Slave Girl* —— 79

2.2 Fugitive Black Power —— 91
"A slave in a captive society": George Jackson's prison letters —— 109
Conjuring captive and fugitive communities in Angela Davis's autobiography —— 120
"You'll be in jail wherever you go": Confinement and flight in *Assata: An Autobiography* —— 135
From Black Power to Black Lives Matter —— 151

2.3 Fugitive Fictions of Slavery and Flight —— 154
"Do you think it'll be any different in Canada?": Fugitive narration from Josiah Henson's slave narrative to Ishmael Reed's *Flight to Canada* —— 167
"Rememories" of captivity and flight in Toni Morrison's *Beloved* —— 187
Futile homemaking, impossible homecoming, and "perpetual migration" in Lawrence Hill's *Someone Knows My Name* —— 204
The "fugitive notes" of Teju Cole's *Open City* —— 221
Fugitive insights of narratives of slavery and flight —— 244

Fugitive Conclusions or the Inescapability of Captivity, Flight, and Fugitive Narration —— 248

Works Cited —— 258

Acknowledgements —— 286

Index —— 288

Introduction or Looking for the Fugitive Life in Social Death

Flight generally entails the crossing of borders – may they be prison walls, estate boundaries, or state borders, being fugitive implies that borders have been and/or still need to be overcome. One might assume that flight ends when captives have successfully crossed the borders that stood between them and their freedom. Enslaved Black people frequently fled their enslavers and legal owners in North America since the commencement of the transatlantic slave trade to gain freedom, crossing, for instance, the demarcating lines between plantations and the 'wilderness' or the Mason-Dixon Line, the Ohio River, and the borders to Canada and Mexico into what they hoped would be free territory.[1] However, with legislature such as the Fugitive Slave Acts (1793 and 1850) a fugitive remained retrievable property even in the supposedly 'free North' so that freedom for fugitive enslaved people in nineteenth-century North America was only a constrained form of freedom, if the term applies at all. Even after the Emancipation Proclamation, many Black Americans enjoyed only few freedoms because the criminalizing Black Codes, convict leasing, and imprisonment quickly superseded slavery followed by Jim Crow segregation as mechanisms to control the formerly enslaved and their descendants in the United States of America (Alexander, ch. 1; Blackmon). For these reasons African Americans appropriated "the biblical story" and "American myth" of the "Promised Land" as a "floating signifier" with which they "locate[d] freedom from oppression in a Promised Land that is always elsewhere, so to speak, and often outside of the US" and Canada – from fugitive enslaved people escaping slavery, to African Americans leaving the US South for the urban North during the Great Migration in the late nineteenth and early twentieth centuries to migration between North America and the African continent since the Middle Passage until today (Paul, *The Myths* 170–71). In other words, Black people in North America have had to escape from anti-black violence and captivity continuously since the transatlantic slave trade. So what if the "social death" (Patterson) that the transatlantic slave trade and chattel slav-

---

[1] In this study, I use variations of the terms *Canada* and the *United States* to refer to the nation states, while I use the term *North America* to refer to the geographical region on the continent in the northern hemisphere. I acknowledge that Canada and the United States were founded as and continue to be not only enslavist (Broeck, *Gender*) but also settler colonial states (Simpson) on lands that were stolen from Indigenous communities. On the relations between anti-blackness and settler colonialism as well as Indigenous and Black forms of refusal, see, e.g., Day; Maynard; T. L. King; T. L. King, Navarro, and Smith.

ⓐ OpenAccess. © 2022 von Gleich, published by De Gruyter. (cc) BY-NC-ND This work is licensed under the Creative Commons Attribution-NonCommercial-NoDerivatives 4.0 International License.
https://doi.org/10.1515/9783110761030-001

ery brought over "people racialised as Black" (~~Coleman~~) has been never-ending in the United States and elsewhere as the Afro-pessimist Frank B. Wilderson III suggests (*Red, White, and Black*)? And if so, how has Black North American literature and theory responded to this perpetual captivity and flight? How have they accounted for Black social life that has undoubtedly endured despite social death in such a framework?

In order to pursue these questions, this study examines concepts of captivity and fugitivity in contemporary North American Black studies theory and in African American and Black diasporic autobiographical and fictional writing traditions. Considering the briefly sketched Black North American history of mobility and escape,[2] it seems unsurprising that fugitivity has become not only a buzzword as recent publication titles in African American studies, Black studies, and Black Diaspora studies show,[3] but also a preferred vantage point from which to examine social life in social death. One major appeal of the term, it seems, is its conceptual ambiguity, its capacity to evade clear definition, just as fugitives might flee and hide, so as not to be captured and pinned down, while at the same time pointing to very specific experiences and concrete historical movements of Black people in North America through the ages.

The definitions and examples given for the term *fugitive* in the *Merriam-Webster Dictionary* immediately establish, among others, a clear relation to slavery, the prison, and refuge that the theory and literature under examination also invoke ("Fugitive").[4] As the entry shows, the connotations are strongly determined by the context in which the term is used and the speakers' or readers' judgment about the fugitive subject. As Alexandra Ganser, Katharina Gerund, and Heike Paul note, "[t]he figure of the fugitive is readily associated with both dangerous criminals on the run from the law and heroic individuals fleeing oppression and injustice. It has often been simplistically determined by discourses of illegality and liberation" (22). The Black fugitive, too, has been a "contested figure," cast as "foundational hero, helpless victim, escapist adventurer, dissenting rebel, or dangerous outlaw" who "defies surveillance, control, and containment (at least to a certain degree)" (22). They prove "essential to the imagination of the

---

[2] For a brief summary of the African American history of flight and migration from the era of slavery to the post-Civil Rights period, see Commander, *Afro-Atlantic Flight* 7–13.

[3] Best, *The Fugitive's Properties* (2004); Bey, *Them Goon Rules* (2019), "Black Fugitivity Un/Gendered" (2019); Commander, *Afro-Atlantic Flight* (2017); T. Davis, "Recovering Fugitive Freedoms" (2015); Dillon, *Fugitive Life* (2018); Goyal, *Runaway Genres* (2019); McCormick, *Staging Black Fugitivity* (2019); Sawallisch, *Fugitive Borders* (2019), to name but a few.

[4] The term *fugitivity* has no single dictionary entry, but the *Merriam-Webster Dictionary* refers to the term fugitive instead ("Fugitive").

nation," reaffirming the legal basis of the United States and its limitations to and freedoms of mobility (Gerund 293). As Mieke Bal reminds us, concepts represent "miniature theories" and provide a "common ground" for "cultural analysis" through their "processes of differing" (*Travelling Concepts* 22, 8, 24; see also ch. 1.1. in this study). So the ambiguity of the term fugitivity can be understood as ultimately defining and its "processes of differing" (24) shall lead us through this study.

I first began pondering confinement and flight in Black American literature and theory for this study during the second presidential term of Barack Obama.[5] While many celebrated the presumed arrival of a 'post-racial' era in the wake of the first Black US president,[6] the queer Black women[7] Alicia Garza, Patrisse Khan-Cullors, and Opal Tometi founded the Black Lives Matter movement (BLM) in the United States as a reaction to numerous cases of fatal police violence against young unarmed African Americans, such as Trayvon Martin in 2012 and Michael Brown in 2014.[8] The formation of Afro-pessimism[9] as a contem-

---

[5] For a more elaborate discussion of the early twenty-first century United States and its study caught between the optimism of African American progress supposedly symbolic of the Obama presidency and the critique of pervasive anti-black racism, see, e. g., Obenland, Sawallisch, and West.

[6] The term *post-racialism* describes the assumption that race and racism have dwindled in societal importance due to "racial progress" in the United States (Cho 1594). As Erica R. Edwards argues, post-racialism went hand in hand with "[n]eoliberal multiculturalism [that] demanded the incorporation of brown and black peoples into higher and higher realms of achievement in culture, the military, the government, the church, and other sites of institutional culture [...] against the backdrop of the late-twentieth-century racialization of immigrants from Mexico, Central America, [...] the Caribbean, the Middle East, [...] India[,] and Pakistan" in "the War on Terror as a racial-spatial regime" (668, 665). "In this context, black people, Latinas/os, Latin American immigrants, Middle Easterners, and South Asians could at once ignite the public imagination of the ideal patriot – and, by extension, a post-racial future – and call up fears of treason, invasion, and mass violence" (668).

[7] In this study, I use variations of the terms *man/men/male* and *woman/women/female* to refer to the dominant Euro-American heteronormative conceptions of sex and gender, if not stated otherwise. I acknowledge that gender is not and never has been binary but includes trans, gender-nonconforming, and queer positionalities. For a recent discussion of trans studies, Black feminism, and fugitivity, see, e. g., Bey, "Black Fugitivity Un/Gendered."

[8] Since 2013, activists and community organizers have joined forces under Garza's hashtag #BlackLivesMatter to draw attention to case after case of fatal violence, often at the hands of the police, against unarmed Black children, women, trans people, and men. The cases are too many to name, but I include some of the victims whose names became emblematic of the movement in the United States over the last decade: Ahmaud Arbery, Sandra Bland, Michael Brown, Philando Castile, George Floyd, Eric Garner, Freddie Gray, Travon Martin, Tamir Rice, Breonna Taylor, Walter Scott, and Alton Sterling. For a personal account of the emergence of the move-

porary theoretical trajectory in US Black studies developed almost at the same historical moment as BLM and the publication of the most recent narratives included in this study (Cole, *Open City*; Hill, *Someone*). Afro-pessimists, such as Wilderson and Jared Sexton started to analyze the epistemic and ontic structures that allowed the continued gratuitous oppression, incarceration, and killing of Black people, such as Martin and Brown, in an anti-black, white supremacist,[10] and what Sexton has described as "people-of-color-blind" ("People-of-Color-Blindness")[11] society about 150 years after formal emancipation of the enslaved.

In 2016, when Donald J. Trump was elected as the 45th US president, with support from both openly racist and white supremacist groups and less radical conservative, overwhelmingly white voters, Colson Whitehead's *Underground Railroad* and Yaa Gyasi's *Homegoing* were published. While not the focus of this study, these two acclaimed novels belong to the literary archive of "African and African-descent slaves and their progeny in the Americas" that this study focusses on and which Michael Hames-García considers "[p]erhaps one of the richest and yet intellectually undervalued traditions in the West" (95). Like the narratives examined in this study, they tackle questions about Blackness,[12] con-

---

ment, see Khan-Cullors and bandele, esp. chapters 11, 13, and 14. On state violence and surveillance in a "New Jim Crow" era in the United States, see, e.g., Alexander; Martinot and Sexton; Sexton, "Racial Profiling." In the last decade, BLM developed into an international movement with chapters across the Americas and Europe. For a self-representation of the Black Lives Matter Global network that emerged from the hashtag and the Black Lives Matter Ride to Ferguson and St. Luis, Missouri, after the murder of Brown in 2014, see their mission statement on their website (Black Lives Matter Global Network). For a historical account, see Ransby.

9 The term *Afro-pessimism* (sometimes also spelled *Afropessimism*) describes a trajectory of North American Black studies that has theorized the position of people racialized as Black as incommensurable with the position of non-Black people (see ch. 1.3. below). It differs from the pessimist perspective on the future of Africa under the same name, while sharing some common features (see G. Thomas, "Afro-Blue Notes"). For a concise introduction to Afro-pessimism, see Douglass, Terrefe, and Wilderson. For a brief discussion of the challenges it poses especially to European and German American studies, see Weier.

10 For a definition and brief overview of the terms *whiteness* and *white supremacy* in the United States, see, e.g., Essi et al.; Perry.

11 In the article "People-of-Color-Blindness: Notes on the Afterlife of Slavery," Sexton describes the refusal to accept the different "structural position[s] born of discrepant histories between blacks and their political allies [of color]" as "people-of-color-blindness" (47).

12 For the purpose of this study, the term *Blackness* and the related adjectives *Black* and *Black American* refer to the structural positionality allocated to people racialized as Black in North America since its formation in the seventeenth and eighteenth centuries in the wake of the transatlantic slave trade and chattel slavery. It refers to "a non-cultural positionality, [i.e.,] not an identity" (Gordon et al. 124, see also ch. 1.3. below). In order to refer specifically to the positionality of descendants of the enslaved in the United States, I use the phrase *African American*.

finement, the (im)possibility of flight and unconfined freedom in North America since the historical era of slavery until today by drawing on autobiographical and fictional Black American writing traditions in North America, such as slave narratives and neo-slave narratives. In fact, in the 1970s and 80s, the themes and forms of the nineteenth-century slave narrative resurfaced – or "reincarnated" in Chinosole's words (*The African Diaspora* 116) – with intensity in both fictional neo-slave narratives about life during and after chattel slavery in North America and autobiographies of (formerly) incarcerated and fugitive Black Power activists. And the same seems to be true of our current moment in which neo-slave narratives and autobiographies of the Black Power movement as well as the Black Lives Matter movement continue to be published to great acclaim year after year (see, e.g., Edugyan, Gyasi, Khan-Cullors and bandele, Moore, Whitehead, Woodfox). This study thus offers not only an in-depth analysis of the ways in which narratives of Black American writing traditions from the antebellum nineteenth, the late twentieth, and early twenty-first centuries have addressed the fugitive lives of Black Americans who the dominant white society has held captive physically and in many other ways. This study is also concerned with how these narratives continue to speak to our present moment by (re-)imagining the transatlantic world in the time of slavery and its captive and fugitive "afterli[ves]" (Hartman, *Lose Your Mother* 6). In what Christina Sharpe describes as "a lived and undeclared state of emergency" for Black people around the world (*In the Wake* 100), examining this autobiographical and fictional archive for its conceptualizations of captivity and flight seems more pressing than ever, or, as many of the writers and theorists included in this study would have it, as urgent as ever.

This study focusses on fugitivity as a concept that has emerged in recent Black feminist theory in North America and has been ever-present in Black American literature through the ages. It argues that the concept can both account for the rigorous Afro-pessimist analysis of the "structural positionality" (Wilderson, *Red, White, and Black* 31) of Black people as socially dead outside of civil society and testify to the fact of their refusing, enduring, resisting, and thinking beyond that status towards future possibilities of unconfined freedom. In this way, fugitivity as a concept takes the groundbreaking insights of Afro-pessimism in recent years seriously while also reacting through indirection to some of its critique. Critics of Afro-pessimism have, in particular, bemoaned that its primary

---

Black North American emphasizes the inclusion of Black Canadian culture and history explicitly, and – like *Black diaspora* – refers to Black presences outside of the African continent that may stem from the history of slavery and colonialism as well as migration through the ages.

focus on the "grammar of suffering" (11, 37) risks pathologizing Blackness as it fails to also analyze past, present, and future escapes, survival, and resistance (see, e.g., Gordon et al.; G. Thomas, "Afro-Blue Notes").

In the first part of this study that builds its methodological and theoretical framework I suggest to understand Afro-pessimism and to a lesser degree Black feminist theory as identifying a challenging border concept between 'the Human' and the structural position of the 'non-human Black.' Afro-pessimism and Black feminist theory, not least as expressed in Sylvia Wynter's work, expose the definition of the modern 'Human' to have been continuously formed through a structurally absolute delimitation between Blackness and being 'Human,' or the dominant "genre of Man" and the structural position of the "non-human" (Thomas and Wynter 24). This *Black border* demarcates Blackness and non-Blackness, enslavement (or imprisonment) and unconfined freedom, premised on the history of colonialism and enslavement in the Americas and its continuing legacy in today's "afterlife of slavery" (Hartman, *Lose Your Mother* 6). It renders relations between its two sides, including its crossing, impossible due to its absolute, ultimate, and antagonistic manifestation as part and parcel of an "antiblack world" (Gordon, *Bad Faith*).[13] Especially Wilderson's and Sexton's work focusses on the analysis of the epistemic and structural genesis of the 'free white subject' as 'Human' in dissociation from Blackness as "Slaveness" in the United States (see, e.g., Wilderson, *Red, White, and Black* 52; Sexton, "Racial Profiling" 213). In doing so, they not only question narratives of recuperation and redemption in the "afterlife of slavery" in which Black life continues to be fundamentally threatened and precarious (Hartman, *Lose Your Mother* 6). They also challenge the majority of cultural and literary studies by showing how comparative, inter-cultural, and egalitarian approaches are stretched to their limits when confronted with the structural positionality of Blackness and the role anti-blackness plays in US American civil society and elsewhere. As we will see in my engagement with Wilderson's work in particular, the Black border differs considerably from dominant border concepts of US American Border studies that originated in Chicanx and US Latinx studies, not least because, in an Afro-pessimists framework, the Black border can only be dismantled if the anti-black world ended. Until then structural change seems impossible.

Fugitivity as a central concept elaborated in recent Black feminist work, especially by Saidiya Hartman, Tina Campt, and Christina Sharpe, has anticipated

---

[13] Gordon describes the world as "an antiblack racist project" and distances his work from Afro-pessimist thought because, according to him, it conceptualizes anti-blackness not as a project or process but as an irrevocable *"achievement"* (Gordon et al. 106). For a more detailed discussion of Afro-pessimism, see ch. 1.3. and 1.4.

and continues to engage with Afro-pessimism without pathologizing Blackness in the anti-black environment of the US society and beyond, but also without premature moves into discourses of hope and liberation in the face of ongoing anti-black violence. Campt, Hartman, and Sharpe have been covering ground between the seeming extremes of Afro-pessimism and other trajectories in cultural and literary studies with their approaches and analytic vocabulary, such as "refusal," "fabulation," and "wake work," that enable them to address both social death and social life facing anti-blackness in nuanced ways. Thus, the initial juxtaposition of Afro-pessimism with Gloria Anzaldúa's "borderlands," Mary Louise Pratt's "contact zones," and Walter Mignolo's "border thinking" in chapters 1.2. and 1.3. serves as a basis to fathom the possibilities and challenges of the alternative concept of fugitivity to conceptualize no more and no less than the capacity of the captives and fugitives of anti-blackness to create possibility in the "afterlife of slavery" (Hartman, *Lose Your Mother* 6) and to imagine futures of Black life and unconfined freedom. In fact, in recent Black feminist theory, fugitivity and fugitive refusal seem to have replaced the concept of resistance that has been increasingly questioned in relation to Blackness due to its implication of agency where, as Afro-pessimists would argue, there is none to speak of on a structural level. Chapter 1.4. will look at what I call *Black feminist fugitive thought* – based on Alexis Pauline Gumbs's notion of "Black Feminist Fugitivity" – and examine the shifts in meaning when we speak of fugitivity, flight, and fugitive refusal instead of resistance. After all, as Bal suggests, choosing to use a concept influences the analysis and its potential outcomes in important ways (see ch. 1.1. below).

As we will see in chapter 1.4., Black feminist fugitive thought frequently turns to the arts and literature as interlocutors in order to contemplate the impossible position of social life in social death. Sharpe, Campt, Hartman, and others not only analyze literature, music, and photography, but also make use of artistic techniques, registers, and genres for their methodological and theoretical approaches (see Campt, *Image Matters, Listening to Images*; Hartman, *Lose Your Mother*; Sharpe, *In the Wake*).[14] Cultural forms of expression, such as photography and literature, figure prominently in their conceptualizations since they seem predestined to not only testify to the continued "abjection" of Blackness (Broeck, *Gender* 13)[15] but also imagine "otherwise worlds" (King, Navarro, and

---

[14] Note that Wilderson has also used literature and film in relation to his theoretical work. Apart from *Red, White, and Black*, he also published the memoir *Incognegro* and most recently *Afropessimism*, a book that mixes memoir with Afro-pessimist theory.

[15] Sabine Broeck traces anti-blackness back to *abjection*, a term she lends from feminist theorist Julia Kristeva and fundamentally reinterprets in what she calls a "post-Fanonion vein in

Smith). While autobiographical and even more so fictional writing are relatively free of the constraints of rigorous scholarly analysis to imagine 'otherwise,' they are nonetheless always already inscribed into and a product of the world they are produced in. African American captivity and fugitivity narratives have dealt with experiences of enslavement, confinement, and flight ever since the transatlantic slave trade until today. In doing so, they have grappled with the histories of the Middle Passage, slavery, emancipation, segregation, and the Civil Rights and Black liberation movements. This study therefore follows the lead of Black feminist scholars and examines Black North American literature as a vital space where the contradictions, tensions, and paradoxes of the social life of the socially dead are negotiated. Consequently, in the main part of this study an in-depth literary analysis of captivity and fugitivity in Black American literature follows. The corpus, which engages African American discourses of slavery, Black liberation movements, and incarceration, consists of slave narratives from the nineteenth century by Frederick Douglass, Harriet Jacobs, and Josiah Henson; prison narratives of the Black Power movement of the twentieth century by Angela Davis, George Jackson, and Assata Shakur; and neo-slave narratives as well as novels of slavery and migration of the late twentieth and early twenty-first centuries by Teju Cole, Lawrence Hill, Toni Morrison, and Ishmael Reed. With the help of Afro-pessimist and Black feminist fugitive thought, the narratives' capacity to tell complex stories of anti-blackness, (social) death, flight, endurance, and survival will be scrutinized. As I argue, analyzing narratives of captivity and flight with an Afro-pessimist and Black feminist theoretical framework highlights the texts' conceptualizations of Blackness, anti-blackness, and (un)freedom as well as their long-standing nuanced negotiation of Black life and endurance in the face of anti-blackness as contributions to Afro-pessimism and Black feminist theory.

Slave narratives represent one of the founding writing traditions of an ever-growing autobiographical and fictional archive of Black captivity and flight. The first chapter conducts a thematic and formal analysis of Jacobs's and Douglass's slave narratives as central intertexts for the texts discussed in the other two chapters. Chapter 2.1. thus investigates how slave narrative narrators escape to

---

which it appears in Saidiya Hartman's and Hortense Spiller[s]'s work" (*Gender* 13, 16–17). She defines abjection as a white praxis in which people racialized as Black have been "structurally, not contingently, cut off from the human, from the self-possessed possessor of the world and its things," i.e., from the Enlightenment binary of subject and object positions ever since the transatlantic slave trade and made into what Spillers coined "flesh" (*Gender* 13). See also Broeck, "The Challenge." Broeck calls this genealogy the "regime of modern enslavement" ("Lessons for A-Disciplinarity" 351).

seek not only legal freedom but unconfined forms of life. Even though the flight plots remove the autobiographical subjects from geographies of slavery, Douglass and Jacobs indirectly show how the fugitive is still situated temporally, ontologically, and epistemically in the historical time of slavery. Published more than one hundred years later, writing by formerly incarcerated Black Power activists heavily draws on slave narratives as their literary and autobiographical predecessors. In chapter 2.2., I analyze how autobiographical texts by Jackson, Davis, and Shakur raise and revise fugitivity and captivity tropes from the slave narrative tradition. In his prison letters published in *Soledad Brother* in 1970, Jackson explores fugitivity in the movement era as a form of refusal and self-defense from the position of a "slave" trapped in a prison that resembles a plantation and extends beyond the carceral system into US society, anticipating the racialized mass incarceration that would follow in the 1980s and 90s. *Angela Davis: An Autobiography* (1974) relies on the long history of fugitivity of the enslaved to create a community of prisoners who refuse state violence while incidentally recognizing the violence's inescapability. The analysis of fugitivity in *Assata: An Autobiography* (1987) leads from adolescent runaway and "astro-travel" via prison escape to lasting exile and shows how Shakur's flight performs radical Black feminist futurity. As a glance at Khan-Cullors's recent Black Lives Matter memoir at the end of this chapter illustrates, especially the writing of the Black feminists Shakur and Davis who highlight communal forms of care "in the wake" of death, violence, and imprisonment (Sharpe, *In the Wake*) proves inspirational for today's queer and intersectional Black Lives Matter movement.

While perpetual confinement and flight, the question of refuge at risk of invasion, and the circular, non-linear, and expanding time and place of slavery and its anti-black afterlives play a central role in slave narratives and Black Power autobiographies, they also reappear in neo-slave novels of the late twentieth and early twenty-first century. Yet, neo-slave narratives, such as Toni Morrison's *Beloved* (1987), more explicitly navigate the ethics and poetics of remembering the past from post-Civil Rights perspectives. They search for ways to tell of slavery's atrocities, while emphasizing the enslaved peoples' humanity and questioning the narrative drive for closure and transformation as strategies to overcome slavery's legacy. Ishmael Reed's *Flight to Canada* (1976) rewrites the tradition of slave narratives and the history of enslavement in North America with the postmodern techniques of non-linear narration, intertextuality, and satire. It draws not only on slave narratives, such as Henson's, but also critically comments on US Black Power discourses, whereas Morrison's *Beloved*, probably the most well-known neo-slave narrative, tells stories of slavery and flight between and beyond the conventions of the historical novel, the gothic, and romance. Its focus on motherhood in captivity and flight and the spread of confinement from the pre- into the postbellum period speaks not only to Jacobs's slave narrative but also to Black Power autobiographies and their conceptualization of a 'neo-slavery.' Whereas Reed's *Flight to Canada* directs its view towards Canada, Hill's novel offers a transnational angle that covers the Black Atlantic (Gilroy) between North America and West Africa. With a resilient Black female storyteller at its center *Someone Knows My Name*, originally published in Canada in 2007 under the title *The Book of Negroes*, stylistically revisits nineteenth-century realist and sentimental slavery literature. But loss and reiterative escapes frequently undermine the plot's hopeful push towards resolution and coherence in the form of homemaking, resettlement, and return.

Finally, *Open City* (2011) by Nigerian American writer Teju Cole expands the perspective of the neo-slave narrative chapter beyond historical novels of slavery. It thereby returns us to where chapter 2.2. on Black Power autobiographies left off with respect to their engagement of the late-twentieth-century carceral afterlives of slavery in the United States. Set in twenty-first century New York City, the novel narrated from a Nigerian German immigrant's viewpoint places neo-slave narrative elements side by side with vignettes of contemporary migration, incarceration, war, and the histories of settler colonialism and genocide as they have marked the lives of Black people and people of color since the early colonial period. Cole's narrative exposes the continued inescapability of confinement and flight for the 'Afropolitan' narrator and other less privileged characters in the post-9/11 era, ultimately adopting flight as a narrative strategy of evasion and overexposure. As the analytic chapters show, Black North American literature can be better understood through Afro-pessimist and Black feminist concepts of confinement and flight, while they also actively contribute to their theorization by having developed their own concepts of captivity and flight across different periods and writing traditions – from literal or physical, through ambiguous or unconscious, to mental or spiritual confinement and escape. Black American narratives of captivity and flight ultimately shape an intertextual archive that approaches in literary form and with specific narrative strategies, such as strategic ellipses, the de-emphasizing and fragmentation of the narrating I, and non-linear or circular narration, the seeming paradox of social life of the socially dead in an anti-black world, subtly narrating life that should not be but still is from the historical time of slavery until today.

Part 1 provides the theoretical and methodological framework for this study. Chapter 1.1. discusses concept-driven cultural analysis as developed by Bal as the methodological approach which is further refined in chapter 1.5. with respect to the corpus selection and the theoretical framework. The latter is developed out of Black feminist and Afro-pessimist theory in chapters 1.3. and 1.4. The majority of the study is made up of three analytic chapters in part 2 on the three literary writing traditions mentioned above. The chapters in part 2 are each dedicated to one of the aforementioned writing traditions at a time: Harriet Jacobs's and Frederick Douglass's slave narratives in chapter 2.1.; autobiographical texts by the Black Power activists Jackson, Davis, and Shakur in chapter 2.2.; and the neo-slave novels *Flight to Canada*, *Beloved*, *Someone Knows My Name*, and *Open City* in chapter 2.3.[16] The diachronic analysis of these narratives is conducted with a double focus on the travel of the conceptual tropes of captivity and flight in the narratives and on the narrative techniques used to convey these notions. The study closes with a summary and systematization of the forms of captivity and fugitivity identified during the literary analyses and an outlook for further study.

The Black Border and Fugitive Narration in Black American Literature tests the limits of fugitivity as a concept expressed not only in Black feminist fugitive thought but also developed and continuously revised in Black American autobiographical and fictional writing from the antebellum period until today to account for the paradox of social life of the socially dead in an anti-black Atlantic world. It explores fugitivity's conceptual possibilities both to take seriously the Afro-pessimist analysis of the structural positionality of Blackness as social death outside of civil society and at the same time to appreciate the praxis of flight from and endurance of the structural abjection – without, however, resolving or overcoming the tensions between social death and fugitive life. Ultimately, the study asks whether, and if so, how fugitivity may pose a differently accentuated perspective that continues to focus 'Afro-pessimistically' on the structural impossibility of Black life across the Black border but adds further layers to the rigid equation of Blackness as "Slaveness" (Wilderson, *Red, White, and Black* 52) by acknowledging the performances of Black life that have occurred and been reflected upon outside of civil society for centuries and by thinking beyond the current 'afterlife of slavery' towards possibilities of Black life and freedom elsewhere and 'else when.'

---

16 In this study, the term *text* refers only to written fictional and autobiographical narratives if not stated otherwise. Of course, captivity and flight also play a central role in other North American writing traditions, such as Jewish and Indigenous literatures, literatures of migrant communities of color, as well as literatures concerned with white American outlaws, such as 'Billy the Kid.' While the different genres, including drama and poetry, indeed share some central tropes, the study at hand focusses exclusively on a selection of Black North American narrative genres. For discussions that touch on the tropes of captivity and flight across different writing traditions in the United States, Canada, and beyond, see, e.g., Adams; Beverly; H. B. Franklin, *The Victim as Criminal*; Ganser, Gerund, and Paul; Goyal, *Runaway Genres*; Irr; Jay; Manzanas Calvo; McCormick, "Fugitivity and Neo-Slave Performance," *Staging Black Fugitivity*; Sadowski-Smith.

Part 1 Fugitivity against the Border: Afro-Pessimism, Black Feminist Fugitive Thought, and the Border to Social Death

This part develops the methodological and theoretical framework for the study by way of a cultural analysis of the travelling of border and fugitivity concepts. After a brief introduction into concept-driven cultural analysis and travelling concepts by Mieke Bal, I discuss Afro-pessimism as one of the most groundbreaking trajectories of contemporary Black studies in North America. Frank B. Wilderson's Afro-pessimist concept of "Slaveness," the "ruse of analogy," and his notion of the "end of the world" will be contrasted with a selection of Chicanx and Latinx Border studies concepts (Gloria Anzaldúa's "borderlands," Mary Louise Pratt's "contact zone," and Walter Mignolo's "border thinking") as popular cultural and literary theory concepts. In juxtaposing concepts of Afro-pessimism and Border studies, the underlying logics that govern knowledge produced in these scholarly trajectories will become apparent. Summarizing central Afro-pessimist insights, I describe the antagonistic relation that Afro-pessimism proposes between Blackness and 'the Human' as a Black border, the ultimate border between Blackness and the position of 'the Human.' I then look at how scholars have developed concepts to account for this Black border and the Afro-pessimist ontology of Blackness while also focusing on performances and experiences of social life that might otherwise escape an Afro-pessimist analytic lens. By drawing on work by Black feminist scholars, such as Saidiya Hartman, Tina Campt, and Christina Sharpe, the second section of part 1 inquires, how we may account for the social life of the socially dead while taking the Afro-pessimist antagonism seriously. Focusing on their notions of flight and refusal, I propose to think of fugitivity as a constant struggle against the 'Black border' without, however, dismantling it or arriving at the other side that bodes civil life inside civil society only for the 'non-Black.' In this way, the concept of fugitivity successfully links analyses of fugitive experiences and performances with an Afro-pessimist structural focus on the position of Blackness. After developing the theoretical framework in this manner, the chapter closes with a second note on methodology, explaining the corpus selection and refining cultural analysis as an approach for the following literary analyses of the novels and autobiographical texts.

1.1 A First Note on Method: Concept-Driven Cultural Analysis

The term "travelling concept" was coined by Dutch narratologist and cultural theorist Mieke Bal in her book *Travelling Concepts in the Humanities* (2002) in order to account for the movement of concepts "between disciplines, between individual scholars, between historical periods, and between geographically dispersed academic communities" (24).[17] Bal proposes "cultural analysis" as a "methodological base" (5) that answers to the demands of interdisciplinary work in the humanities and its need for "methodological common ground" in the form of concepts (8). Cultural analysis, Bal submits, supplies the necessary *"sensitivity to the provisional nature of concepts"* (55) so that they can act as "counterparts" to the cultural object of analysis (8) and therefore operate as a "third partner in the otherwise totally unverifiable and symbiotic interaction between critic and object" (23). I follow Bal in my use of the term 'concept.'[18] Her "cultural analysis" as a "concept-based methodology" (5) will serve as the overarching methodological approach for both the development of the theoretical framework of Afro-pessimism and the concept of fugitivity in this chapter and the literary analyses that follow in the main part of the study (part 2).

Bal sums up the "priorities" that guide the practice of cultural analysis as first, "cultural processes over objects," second, "intersubjectivity over objectivity," and third, "concepts over theories" (44). This study takes its methodological cue especially from the last point. As Bal points out, concepts do not present themselves as exhaustive theories. They are ever-changing "miniature theories" (22), and as complex points of "accumulation of [their] own components" (51–52), they can never be used in precisely the same way. Thus, concepts are less than elaborate theories and much more than mere "tools" (22). They "focus interest," "organize a group of phenomena, define the relevant questions to be addressed to them, and determine the meanings that can be given to observations regarding phenomena" (31). Therefore, choosing to work with specific concepts, such as the border, fugitivity, social death, and Blackness, shapes the knowledge produced and disseminated in many important ways. Not least, they supply this study with contexts and registers of space, place, confinement,

---

[17] Of course, Bal was not the first to address the development of concepts, discourses, and theories as a form of travel. See, e.g., Bloch; Clifford, "Notes on Travel," *Routes*; Said.

[18] For further considerations of the term 'travelling concept,' see, e.g., Bachmann-Medick; Baumbach, Michaelis, and Nünning; Neumann and Nünning; Teller. For discussions of the term's appropriateness and a problematization of the metaphorical use of the notion of 'travel,' see, e.g., Bachmann-Medick; Clifford, "Notes on Travel."

movement, and territorial demarcation one the one hand and of enslavement, anti-blackness, and North American race relations on the other.

According to Bal, examining the concepts' "processes of differing" (24) makes their travels, their "shorthand theories" (23), and their contexts of development accountable (40). Defining a concept is a central part of this assessment. Provisional definitions reveal what concepts do rather than what they denote. For Bal, "the valuable work lies" in the "groping to define, provisionally and partly, what a particular concept may *mean*" (11). The parameter according to which she measures 'proper' uses of concepts that travel is therefore not correctness or precision but meaningfulness (16–17). When we discuss and use them to practice "detailed analysis from a theoretical perspective" (44), concepts yield "analytical insight" by enabling the analyst to ask meaningful questions with respect to both the concept used and the object of analysis (17). This leads Bal to infer that "a good concept founds a scientific discipline or field" (33; cf. Nünning 42), which makes looking at concepts like the border and fugitivity at the intersection of (German) North American studies and Black studies all the more pertinent.

Clearly, Bal's delineation of cultural analysis and travelling concepts is far from univocal. The differentiation between "ordinary words" (23), concepts, and elaborate theories remains to be determined as the case arises, just like the ways in which a concept may act as a "common language" (22) even though it is ever-changing, flexible, and never the same during its travels. Instead of providing a full-fledged method with clear instructions, Bal supplies what she calls a "rough guide," offering a very basic "common ground" (8) in order to remain flexible and cater to the various needs different concepts, cultural objects, and disciplines bring to the interdisciplinary approach. I use this concept-driven 'rough guide' to navigate through the theoretical deliberations of Afro-pessimism and the travelling route of the concept of fugitivity and the border in the following. Bal's deliberations on the relevance of concepts and their travels guide the way through the concepts' "processes of differing," their provisional definitions, and, most importantly, their varying "miniature theories" and underlying logics that require close reading. They will also steer us through the literary analyses in the main part of this study that will perform close and wide readings of the literary corpus.

1.2 Travelling Border Concepts

Geographical borders are sites of intensified cultural contact and conflict where people, languages, and cultures meet, mix, and clash. As such, they are central sites of knowledge production and dissemination. Theoretical conceptualizations of borders have emerged in different localities, across various periods, and in numerous disciplines and fields of inquiry, e. g., in the political and social sciences, anthropology, and cultural and literary studies.[19] Thus, border concepts may be understood with Bal as prime examples of "travelling concepts." Research on the border between the United States and Mexico represents a key point of departure for the travels of border concepts in North American studies. As Scott Michaelsen and David E. Johnson observe, Chicanx studies were at the fore, before ethnic and postcolonial studies, in making "the idea of the border available, indeed necessary, to the larger discourses of American literary studies, US history, and cultural studies in general" (22). Not least with the publication of *Borderlands/La Frontera* by Chicana feminist writer Gloria Anzaldúa in 1987, the border emerged from Chicanx studies as a concept to describe and criticize (cultural) contact and exchange informed by asymmetrical power relations in the Americas. Chicanx and Latinx experiences with the US-Mexican border as a border not only between the United States and Mexico but also between North and South America have been in many cases a source of inspiration for influential border conceptualizations.[20]

With her mix of autoethnography, autobiography, poetry, and prose in *Borderlands/La Frontera*, Anzaldúa developed her border concept both as referring to a contested, historically and culturally specific border region in the US Southwest and as addressing social boundaries in interpersonal, intracultural, and intercultural relations.[21] She explains in the preface to *Borderlands:* "The actual physical borderland that I'm dealing with [...] is the Texas-US Southwest/Mexican border. The psychological borderlands, the sexual borderlands[,] and the spiritual borderlands are not particular to the Southwest." In their introduction to *Border Women: Writing from la Frontera* (2002), Debra Castillo and María Soccoro Tabuenca Córdoba describe Anzaldúa's border concept as evoking "the intellectual project of a discursively based alternative national culture while gesturing toward a more transnational space of identity formation" (3). Anzaldúa

---

[19] For a broad overview of border concepts in anthropology and literary studies, see, e.g., Donnan and Wilson, Manzanas Calvo.
[20] See, e. g., Castillo and Tabuenca Córdoba, Manzanas Calvo, Michaelsen and Johnson, Mae G. Henderson, Saldívar. Of course, the US-Mexican border is not the first and only border on the North American continent that has instigated border concepts. Scholarship on the 'American frontier' precedes work in Chicanx studies (Durán 124–27), while research on the Canadian-US border and the inner-national and cross-national borders of Native American communities and First Nations have played an increasing role in Border studies as well (Sadowski-Smith).
[21] For an in-depth analysis of Anzaldúa's *Borderlands*, see, e.g., Saldívar-Hull.

makes such an extension of the concept of the "borderlands" explicit by including varied border experiences. She adds to the above that "[i]n fact, the Borderlands are physically present wherever two or more cultures edge each other, where people of different races occupy the same territory, where under, lower, middle and upper classes touch, where the space between two individuals shrinks with intimacy."

Anzaldúa carefully combines this referencing beyond the specific cultural frame of Chicanx experiences in the US Southwest, toward an alternative, potentially idealist "contact zone" (Pratt), with a clear focus on structural violence – as expressed in a drastic, albeit very poetic way through the oft-cited metaphor of the US-Mexican border as an "open wound" (2). This combination of the role of structural violence and the specific US-Mexican context as well as the concept's inherent potential to pertain to other (national, cultural, social, and inter-personal) contexts seem to have made Anzaldúa's "borderlands" "[o]ne of the most widely used critical concepts in Latino/a[/x] studies [...] and in border theory more generally" (Allatson 39). As Richard T. Rodríguez claims, "[i]n many ways, *Borderlands* set the stage for scholars who [...] would begin identifying their work under the rubric of 'Border studies' (or 'Border Theory')" (202).

The notion of a contact zone that may include Anzaldúa's borderlands as one possible form is well-known in and beyond cultural and literary studies for its more general conceptualization of a space of cultural contact across asymmetrical power relations in the long aftermaths of colonialism, the transatlantic slave trade, and slavery in the Americas and the Caribbean. First coined in her essay "Arts of the Contact Zone" and further developed in her study of European eighteenth and nineteenth century travel writing *Imperial Eyes: Travel Writing and Transculturation*, Pratt defines contact zones as "social spaces where disparate cultures meet, clash, and grapple with each other, often in highly asymmetrical relations of domination and subordination – like colonialism, slavery, or their aftermaths as they are lived out across the globe today" (4). Contact zones conceptualize (post)colonial cultural contact and communication between the (former) colonizers and the (former) colonized and enslaved (6). As she shows in her analysis of Guáman Poma's writing,[22] Pratt understands this contact as a form of forced conversation on unequal grounds in which "the subordinate peoples" find ways to talk back and self-represent through "transculturation" and "autoethnography" ("Arts" 36). In this way, the contact zone takes on the issue of resistance to subjugation and the role knowledge production

---

[22] Felipe Guáman Poma de Ayala (ca. 1535–1615) was a Peruvian writer and illustrator during the early years of the Spanish conquest (Encyclopedia Britannica editors, "Felipe").

and dissemination plays in this context. It therefore refers less to a specific geographical location than to an improvised interpersonal and epistemic space for communication and interaction in the (post)colonial world. The space the two parties enter is hierarchically structured, but it still leaves room for 'the subordinate' to negotiate with 'the dominant' and therefore also presupposes (a limited form of) agency on the side of the former.

Besides *Borderlands*, Castillo and Tabuenca Córdoba also consider Walter Mignolo's *Local Histories/Global Designs* (2000) in their overview of border conceptualizations. They describe it as "one of the most complete and theoretically powerful surveys on recent discussions of the idea of the border in US, Latin American, Caribbean, European, and former British Commonwealth thought" (11). Interestingly, Mignolo references *Borderlands*' appropriation of the colonial languages of English and Spanish as an example of his concept of "border thinking" (*Local Histories* 222–23) suggesting that it describes a more general space, like Pratt's contact zones. He deems Anzaldúa's concept of the border a "powerful metaphor" that "establish[es] links with similar metaphors emerging from a diversity of colonial experiences," such as "double critique and une pensée autre, double consciousness, the Zapatista's double translation, creolité, transculturation, provincializing Europe, negative critique introduced by African philosophers, etc." (Delgado, Romero, and Mignolo 11). He understands these metaphors as a "conceptual arsenal making it possible to 'think otherwise,' from the interior exteriority of the border" (11), and aspires after a future of "pluri-versality" that is ruled by plurality as a universal, inter-epistemic, and dialogical concept connected through universal values ("Delinking" 452–53, 499). For this purpose, Mignolo makes broad connections between theoretical concepts localized and historicized in decidedly different contexts, assuming analogic relations to power between various groups of oppressed people around the globe, or in his words a "[c]ommon basis" in their "experience to have to come to terms with modernity/coloniality" (497).

Based on this assumption, Mignolo argues that "border thinking" occurs in various "border positions" (*Local Histories* 72), i.e., in the "geographical and epistemological location" of the border (309). According to him, the border offers a geopolitical position and a critical perspective from which remaining colonial structures, the "coloniality" in knowledge and knowledge production, can be decolonized. Yet, Mignolo not only considers postcolonial thinkers who live in and move between (formerly) colonizing and colonized countries as occupying these positions. He also includes as 'border thinkers' Native Americans and Chicanxs because not only did they or their ancestors move across borders, but "the world [read borders] around them moved," too (72–73). Moreover, Mignolo identifies "border thinking" more generally "in the borders lived by Afro-Americans, [...]

Arabs, [and] Jews" (157), thus theorizing a very diverse and hybrid concept of "border thinking" that encompasses people racialized as Black, Indigenous, and of color, as well as more generally people marginalized because of their ethnicity and/or religious beliefs.

At the same time, Mignolo clearly distances "border thinking" from optimistic theories of cultural hybridity when he points out that "border thinking [...] is not a new form of syncretism or hybridity, but an intense battlefield in the long history of colonial subalternization of knowledge and legitimation of the colonial difference" (12). In fact, he argues for a recognition that knowledge is situated geographically, politically, and bodily (37, 310, 327). Mignolo's approach thus also calls attention to the embodied construction of knowledge and the geographical contexts of its development, suggesting that the differences between the diverse "border positions" mentioned above should not be overlooked.

I understand Mignolo's focus on the "geopolitics of knowledge" through regionalizing "global designs" and recognizing, appreciating, or rather thinking from "local histories" as part of as well as a reaction to the transnational turn in American studies and literary and cultural studies more generally (cf. Jay). It apparently attempts to mediate between what Mae G. Henderson describes as "the dangers of particularism embodied in ideas of nationalism and ethnicity (or class and gender) on the one hand and the perils of universalism embodied in transnationalism and transculturalism" on the other (25–26). Unsurprisingly, Paul Jay includes Mignolo's "border thinking" (86–88), Pratt's "contact zone," and Anzaldúa's "borderlands" (76–78) in his study of *The Transnational Turn in Literary Studies* as "new approaches to location in American Studies" that may "reconfigure the spaces we study in ways that both predate and transcend traditional nation-state boundaries" (74). Through their predominantly hemispheric or global approaches, these concepts have indeed played a vital role in the necessary and ongoing project of reforming the (US) American literary canon and its underlying assumption of a homogenous American identity in the last three decades (see Sadowski-Smith, Saldívar).

In their travels within the interdisciplinary field of transnational American studies, however, the concepts were increasingly used to focus on the fluidity and hybridity of identities and the permeability of territorial, epistemological, and disciplinary borders in diaspora, minority, and Indigenous literatures and literatures of migration in North America (Jay 74–76). This often involved an increasing disregard of the cultural and historical contexts of the locations where the concepts emerged, risking the loss of their specificity during their travels. For example, Castillo and Tabuenca Córdoba have found fault with the use of border concepts that all too frequently overlook the precarious and violent realities of the actual state border between Mexico and the United States and privilege US over Mexican border perspectives (13). Thus, in Bal's words the concepts' "processes of differing" have not only foregrounded some of their "components" (*Travelling Concepts* 24, 52) but they have actually eclipsed others, such as the fact that Mignolo and Anzaldúa clearly historicize and localize their border concepts and address the tensions involved in their wider use with respect to interpersonal, intracultural, and intercultural relations.

While potentially challenging in itself, the concepts' oscillation between culturally and historically specific and all-too generalized definitions does not give cause to a conceptual conflict between these three border concepts and Afro-pessimist interrogations of the category of 'the Human.' As I show in what follows, the issue arises from the incompatible "assumptive logics" (Wilderson, "Gramsci's Black Marx" 225) that underlie the conceptualizations' "miniature theories" (Bal, *Travelling Concepts* 22) in the two different theoretical trajectories. The travelling border concepts discussed here rely on the basic notion that groups of people they locate in border positions, including African Americans and more generally people racialized as Black, are assumed to have a diversified but fundamentally common relation to what Mignolo calls with reference to Aníbal Quijano the "coloniality of power" (*Local Histories* 17). Border concepts such as Anzaldúa's, Pratt's, and Mignolo's fundamentally rely on the assumption that analogies and comparisons can and should be drawn between different groups of people subordinated to a dominant, often colonial and patriarchal white power based on their allegedly common experiences of oppression and similar positions toward that power. These theorists further construct borders as essentially relational, crossable, and hybridizing – no matter how violent, painful, and at times fatal, i.e., impossible, attempts of crossing national, cultural, and epistemic borders and the associated transformations may be.

1.3 Afro-Pessimism and the Border(s) of "the Human"

Arguing against the backdrop of a structural analysis of the role of Blackness for the prevalent category of 'the Human,' Afro-pessimist Frank Wilderson rejects the assumption of universal comparability in the majority of cultural and literary studies today. Instead, he suggests that the delimitation of 'the Human' from Blackness forms the "common ground" (Bal, *Travelling Concepts* 8) on which postcolonial, patriarchal white powers have fashioned their relation to people racialized as Black. In the following, I propose to envisage Wilderson's theorization of the delimitation between non-Black social life inside civil society and Black social death outside of it as pointing to an epistemic border that has defined 'being human' as 'not Black' at least since slavery. This 'Black border' appears utterly different from border concepts discussed so far because it renders problematic, if not almost impossible, relations between its two sides, its crossing, and any change or demolition of its border structure. From an Afro-pessimist perspective, it has formed the basis onto which other borders and boundaries, such as those involved in the categories of ethnicity, class, and gender, but also sexuality, age, and ability, have evolved in North America for more than two hundred years. To illustrate the central ways in which Afro-pessimism contrasts with the basic assumptions of Chicanx and Latinx studies, I first examine Wilderson's concepts of different positionalities and their "ruse of analogy" before putting his arguments in relation to the border concepts discussed above.

While Chicanx and Latinx studies have produced a large set of border concepts, African American and Black studies have been less often associated with border conceptualizations. Borders have manifested themselves in African American culture, literature, and theory from slavery until today, for instance, in the demarcating lines between enslaving and so-called free territory during the historical era of slavery, in Jim Crow segregation, and in today's "New Jim Crow" (Alexander). They have also played a central role in the analyses of social relations in slavery and of race relations in the United States more generally, such as by way of W. E. B. Du Bois's famous concept of the "color line,"[23] which resembles the border concepts discussed above at least with respect to their engagement with questions of mobility, access, and relationality. In contrast, the demarcation in Afro-pessimism, which I propose to understand as suggesting a border concept, emphasizes the sheer difficulty if not impossibility of passing through the 'border' between Blackness structurally understood as non-human and non-Blackness as 'Human.' In fact, protagonists of both Black feminist research on "racialization and the category of the human in Western modernity" (Weheliye 5) and Afro-pessimism – like Sylvia Wynter and Frank Wilderson respectively – appear to have refused the entry of border concepts, such as border thinking, into their work. Instead, by critically taking stock of the momentum of what Wynter calls the current "genre of Man" (Thomas and Wynter 24), they have the-

---

[23] Du Bois (1868–1963) was an African American sociologist. With his renowned essay collection *The Souls of Black Folk* (1903), he introduced the well-received concepts of the "color line" and "double consciousness," the latter of which Mignolo includes in his concept of border thinking (see ch. 1.2.). The concept of the "color line" points to socio-cultural and socio-political processes of exclusion of people racialized as Black or non-white in US civil society, processes that were part and parcel of what Du Bois describes as "being black here at the dawn of the twentieth century" in his "Forethought" to *The Souls*. As Katherine McKittrick points out, the color line is both "material" and "philosophical" in *The Souls*, "a line that separates, yet connects, what Du Bois calls 'two worlds,' in the United States" (*Demonic Grounds* 22).

orized in idiosyncratic ways a demarcation between 'the non-human' as Blackness and 'the Human' as non-Blackness that has not been interpreted as suggesting a border concept so far.

Black feminist scholar Wynter has contributed considerably to theorizing the structurally ostracized and confined position of people racialized as Black in the context of the legacies of colonialism and slavery through her decades-long engagement with the concept of 'the Human' as "Man" and its "self-instituted genres" (Scott and Wynter 206).[24] According to Wynter, "Man" describes a "Western bourgeois conception of what it is to be good man and woman of one's kind" (Thomas and Wynter 15) that has come to appear as the only understanding of what it means to be human by pushing people racialized as Black beyond 'the Human,' for instance through policing and incarceration (11).[25] Her aptly titled article "'No Humans Involved:' An Open Letter to My Colleagues" was published in 1994 as a reaction to the so-called L.A. Riots, an uprising that occurred after the acquittal of police officers who had been involved in a videotaped and broadcasted beating of the African American Rodney King in 1991. In it, Wynter criticizes an episteme which considers people racialized as Black as the "Lack of the human" (43) and the "least equal of all [minorities]" (42). While all other minorities are located *within* the social order, she maintains, people of African and "Afro-mixed descent" (42) are constructed as "*pariahs outside of* [it]" (45). Wynter thus implies a border here that encloses "Man" by excluding people racialized as Black, even though she never uses the term 'border.' Theorizing the position of people racialized as Black in this way, Wynter also acknowledges that other oppressed groups such as Latinx and "non-owning jobless young [people] of the inner cities" (45) have increasingly been associated with the status of the outcast that Blackness entails, or have been "blackened" to use Christina Sharpe's words

---

24 For Wynter, "Man" is that "which over-represents itself as if it were 'the human'" (Thomas and Wynter 11). She identifies two dominant "genres of Man": first, "Man(1)" which represented itself as "a rational political subject of the state" in Renaissance Europe and, second, "Man(2)" which represented itself, in contrast to "the dysselected," as "*selected by evolution*" and emerged from eighteenth-century secularism and the developing European sciences (Scott and Wynter 202, 204). For an in-depth study of Wynter's large, complex, interdisciplinary oeuvre, see McKittrick's edited volume *Sylvia Wynter*.

25 Of course, a multitude of fields has interrogated the question of what it means to be a human for centuries and remains disputed. For a brief discussion of poststructuralist and posthumanist conceptualizations of the free modern 'Human' and their relation to the categories of race and Blackness, see Weheliye, esp. 8–11. For further Black feminist interrogations, see also, e.g., Hartman, *Scenes of Subjection* and *Lose Your Mother*; Sharpe, *In the Wake*; and Spillers, "Mama's Baby, Papa's Maybe."

(*Monstrous Intimacies* 190–91).[26] Thus, Wynter's theoretical argument takes into consideration how close or far away groups of people racialized as non-Black are structurally positioned to or from Blackness. Ultimately, a "genre of the Human" (Thomas and Wynter 25) that is all-encompassing seems still long in coming for Wynter. She therefore formulates the project of the future as "A QUESTION OF GOING AFTER 'MAN,' TOWARDS THE HUMAN" (17) with the help of a "trans-genre-of-the-human perspective" (Scott and Wynter 206). Wynter sees the critical work of Black studies "in the forefront" of this "battle" (Thomas and Wynter 23, see also Weheliye).

Apart from the Black feminist scholars discussed in chapter 1.4., the paradigmatic scholarly trajectory of Afro-pessimism has also responded – even though mostly by indirection – to Wynter's call, developing out of North American Black studies in the last fifteen years, most prominently in the work of Wilderson. While Wynter remains a mostly unacknowledged influence,[27] the work of the Black feminists Hortense Spillers ("Mama's Baby, Papa's Maybe") and Saidiya Hartman (esp. *Scenes of Subjection*) have inspired Wilderson's Afro-pessimism noticeably just as re-readings of Frantz Fanon's writing (see Marriott; Sexton, "Unbearable Blackness"). As Jared Sexton notes, "Wilderson's procedure […] is something like the abstraction of a conceptual framework (regarding structural positionality), a methodology (regarding paradigmatic analysis) and a structure of feeling (regarding the politics of antagonism) that, taken together, remain implicit in the work of various luminaries of black studies but whose full implications only become available when they are rendered explicit and raised to anoth-

---

26 Sharpe uses the term "blackened" in *Monstrous Intimacies* to describe people who are positioned close to the category of Blackness because they are considered "not properly white or nonblack" (191). Through their positional "proximity to blackness (specifically a proximity to the shame, violence, etc. that black bodies are made to wear)," she argues, they "are covered by the shadow of blackness, are […] blackened" (190–91).

27 Greg Thomas criticizes the late acknowledgement of Wynter's contribution to an inquiry of 'the Human' in Afro-pessimism as something close to "a *boycott*" ("Afro-Blue Notes" 301). It is true that until 2016 reference to Wynter's work was blatantly absent in Sexton's and Wilderson's writing and remains marginal (Sexton, "The Unclear Word" endnote 7), while some of the Black feminist scholars whose work I discuss in chapter 1.4. have brought Wynter into an indirect conversation with the basic assumptions of Afro-pessimism (cf. McKittrick, "Rebellion/Invention/Groove"; Sharpe, "Black Studies"; Sharpe, *In the Wake*). Thomas also suggests that Sexton and Wilderson "casually conflate this [Wynter's] 'Man' with 'the Human' that they criticize, rhetorically and discursively, as much as Wynter has insisted upon the radical distinction between 'Man' and 'the Human'" ("Afro-Blue Notes" 301–02). Reading Afro-pessimism together with the work on Black feminist fugitivity discussed in chapter 1.4., I suggest that 'the Human' as conceptualized by Wilderson is at least partially the result of Wynter's concept of 'Man' travelling into and through Afro-pessimist and Black feminist discourse.

er level of theorization" ("Ante-Anti-Blackness"). In doing so, Afro-pessimism goes both beyond refuting the eradication of the "color line" in a 'post-racial age' and beyond merely insisting on the enduring relevance of race. Addressing a particular form of racism that has targeted people racialized as Black in the United States since the transatlantic slave trade – through slavery, the Black Codes (see Blackmon), Jim Crow, and the "(neo)slavery" of the prison industrial complex (James, *The New Abolitionists*) –, Afro-pessimism argues that, as Lewis Gordon was one of the first to claim, we live in a fundamentally "antiblack world" (*Bad Faith*). Afro-pessimism assumes that US society is fundamentally built on and structured by anti-blackness, which has made possible to arbitrarily capture, enslave, imprison, harm, and kill people racialized as Black for hundreds of years. Anti-blackness is therefore understood as inherent to US society and modernity at large.

The "antagonism" and the "ruse of analogy"

In his groundbreaking film study *Red, White, and Black* (2010), Wilderson compellingly argues that on an abstracted structural level Blackness "*ab initio*" meant and still means "Slaveness" in the United States (340). As the title of the book indicates, Wilderson focusses on the filmic representation of the structural positions of people racialized as Indigenous, white, and Black in and outside of US civil society. Rather than the experiences and performances of those three groups of people, Wilderson is concerned with the *structures* that have assigned to them different positions with respect to civil society and have constituted US civil society as fundamentally anti-black at least since the transatlantic slave trade and chattel slavery in North America. Wilderson thus proposes an analytic framework with a high level of abstraction in contradistinction from Marxist, white feminist, and other cultural studies and film studies approaches in order to analyze modernity and its ontic and epistemic foundations in slavery (see, e.g., 14–15), or rather, what Sabine Broeck describes as the "white abjectorship" of "enslavism" (*Gender* 13).

Most significantly, Wilderson rejects analogies between the structural positionality of Black people and other oppressed groups. Instead, he discerns an incommensurability of the structural position of 'the Black' as perpetual 'Slave' located *outside* of US civil society's discourses of ethnicity, class, and gender (*Red, White, and Black* 23). While Native Americans "and Whites can be caught in the grip of slavery without transforming and reracializing the institution itself," he argues, "Blackness cannot disentangle itself from slaveness" (52). Therefore, Wilderson argues, first, that "Black" still means "Slave" (7) or "prison-slave-in-waiting" ("Prison Slave" 18). Second, he contends that "white" refers to the "senior [...] partners of civil society" (*Red, White, and Black* 38). Third, Wilderson describes other groups of people subordinate to the "white" but who fall out of the category of "the Black," such as immigrants of color and to some extent Native Americans as "the junior partners of civil society" (28).[28] In this argument, the white "senior partners" are located at the center of civil society, their "junior partners" at its inside margins, and Black people are positioned "outside of Humanity and civil society" (55). In other words, while Wilderson identifies the "structural positionality" of people racialized as Black outside of the bounds of US civil society, he locates the discourses of ethnicity, class, and gender fully within society's limits. For Wilderson, this means that the "Black or Slave position" is not only defined in opposition to that of whites, the "senior [...] partners of civil society" (38). It is also clearly demarcated from the structural position of civil society's "junior partners" (38), which is how he provocatively refers to racialized non-Black groups in the United States who may become with Wynter and Sharpe potentially blackened. Thus, Wilderson locates Blackness not only "outside the terrain of the White (the Master)" but also "outside the terrain of the subaltern" (65–66) both of whom in this framework belong to what Wynter describes as 'Man.'[29]

In order to explain this condemnation of Blackness to "the outside of Humanity and civil society" (55), Wilderson draws on historian Orlando Patterson's description of enslaved people's "social death" as "generally dishonored, perpetually open to gratuitous violence, and void of kinship structure" (11) in *Slavery*

---

28 In *Red, White, and Black*, Wilderson ascribes the structural position "Red" to Indigenous people in the United States as distinct not only from the positions of the 'white' and 'Black,' but also from the 'junior partners' (cf. 29–30, 48–50). Ever since, he has revised this assumption arguing that "[i]n some ways, American Indians are a liminal category, and in other ways they are more profoundly on the side of 'junior partners' and antagonistic to Blacks" (von Gleich, Spatzek, and Wilderson 14). For recent interventions on the connection between anti-blackness and settler colonialism as well as Indigenous and Black forms of refusal and (un)relation in North America, see, e. g., Day; Maynard; T. L. King; and the contributions in *Otherwise Worlds* edited by Tiffany Lethabo King, Jenell Navarro, and Andrea Smith.

29 This does not mean, of course, that anti-blackness does not find expression in specifically ethnic, gendered, classist, or ableist forms. As M. Shadee Malaklou and Tiffany Willoughby-Herard remind us, for example, it is Black women and Black trans women specifically "who bear the weight of world-structuring anti-black violence most acutely – black women, who in spite of exhaustion and defeat agitate for different horizons of possibility for black life (for all of us)" ("Notes from the Kitchen" 4, 11). For a detailed discussion of the complex relation between Black feminism and Afro-pessimism, see Malaklou and Willoughby-Herard, "Notes from the Kitchen" and the other contributions to their special issue "Afro-Pessimism and Black Feminism."

and Social Death: A Comparative Study (1982).³⁰ Anti-black violence, Wilderson contends with Patterson, is "ontological and gratuitous as opposed to merely ideological and contingent" ("Gramsci's Black Marx" 229); it is arbitrary and "metaphysical" (Douglass and Wilderson 119) in that it does not require any prior act of transgression of society's (legal, moral, etc.) rules to be unleashed. It is *a priori* implicated in Blackness "not [as] a Black experience but [as] a condition of Black 'life'" since the transatlantic slave trade (*Red, White, and Black* 75). This particular 'being in relation to metaphysical violence' accounts for Wilderson's positioning of Blackness as "Slaveness" outside of the realm of civil society and renders analogies and comparisons between the structural positions of people racialized as Black and the "junior partners of civil society" more than problematic. All attempts would fall prey to what he calls the "ruse of analogy," mystifying and erasing the "grammar of suffering (accumulation and fungibility or the status of being non-Human)" that Blackness entails in this argument (37).³¹ The "ruse of analogy," he writes, "erroneously locates Blacks in the world – a place where they have not been since the dawning of Blackness" (37). This is also why Wilderson describes the relation of Blackness to the world and 'the Human' (who is defined as not Black) as "antagonistic" (5, 26), while the "junior partners" have a dialectic and agonistic relation to civil society that leaves room for negotiation, no matter how small this room and the chances to have claims admitted might be.³²

The focus on the structural positionality of Blackness that Wilderson pursues requires "a different conceptual framework, predicated not on the subject-effect of cultural performance but on the structure of political ontology" (57). Wilderson's work is thus not only characterized by a high level of abstraction but also a strict differentiation between the abstracted level of structure, which defines the positions of groups of people to civil society, and the concrete level of performance and experience.³³ Wilderson is primarily concerned with the

---

30 For a critique of the increased use of the concept of social death, see V. Brown.
31 Wilderson draws on Saidiya Hartman's conceptualization of fungibility here which she defines as the "the replaceability and interchangeability endemic to the commodity" that makes the enslaved and "captive body an abstract and empty vessel vulnerable to the projection of others' feelings, ideas, desires, and values" (*Scenes of Subjection* 21).
32 Wilderson argues that in the liminal case of Indigenous people the object of negotiation would be land (*Red, White, and Black* 29–30) and in the case of migrants of color "immigrant rights" (3).
33 I follow Wilderson in describing the Afro-pessimist analysis as structural. The extent to which Afro-Pessimism is structuralist, poststructuralist, or even post-poststructuralist remains to be debated. In an interview, Wilderson stressed that Afro-pessimism challenges "the assumptive logic of semiotics, hence poststructuralism" because it refuses any transformative capacity

former whereas he locates the majority of ethnic and cultural studies research on the latter (von Gleich, Spatzek, and Wilderson 5). He remains skeptical about this research when it comes to Blackness because of its tendency to all-to-easily "make a leap of faith and assert a causal link between the performance and the emancipation of the black people who produced or consumed it" ("Grammar and Ghosts" 121).

The Black border and "the end of the world"

Wilderson's argument that the relation between Blackness and the world should not be understood as a resolvable conflict but as an incommensurable antagonism (*Red, White, and Black* 29–30) inextricably linked with the constitution of the white 'Western' subject makes Afro-pessimism one of the most challenging trajectories of US Black studies in recent years. If we consider this complex argument in relation to border conceptualizations, we may conceive of the antagonistic demarcation – between Blackness as social death outside of civil society on the one hand and non-Blackness as civil life inside civil society on the other – as a distinct border concept, not analyzed as such so far. In fact, Wilderson uses the metaphor of a fortress build around civil society against Blackness to make the argument that "[a]nti-Blackness manifests as the monumentalization and fortification of civil society against social death" (90). He also describes civil society as "a gated community known as the Symbolic Order; gated because it keeps the Slave from entering" ("Black Liberation Army" 198). The structural bordering also becomes apparent, when Wilderson explains that gratuitous violence "against Blacks' lives" is necessary "to actually produce the inside-outside [of civil society]" (von Gleich, Spatzek, and Wilderson 15). The border that demarcates the inside from the outside defines what 'humanness' and the subject concept mean by delimiting 'the Human' – or in Wynter's words "the genre of Man" (Thomas and Wynter 24) – from the "non-Human" at the expense of the subjectivity of people racialized as Black, in other words by ostracizing them beyond the realm of 'the Human.' This epistemic demarcation is absolute because it has not allowed any kind of movement across the border and no relation between the two sides other than as structural antagonism with respect to Blackness. Thus, in Wilderson's theorizing a seemingly insurmountable, absolute,

---

for Blackness (Park and Wilderson 38). Gordon on the other hand criticizes Afro-Pessimism together with "critical theory" and "decoloniality theory" as "rebranded" poststructuralism "attuned with neoliberalism and neoconservatism" ("Shifting the Geography" 45–46).

and antagonistic border comes into focus between 'the Human' and the "Black or Slave position" (*Red, White, and Black* 38).

The absoluteness of the Black border is reflected in the ways in which it withstood any attempts to change its position and structure at least since its consolidation as part of the transatlantic slave trade. The historical changes that have taken place in the United States, for example through emancipation and the Civil Rights and Black Power movements do not figure in the "conceptual framework" and on the level of abstraction Wilderson calls for in his work (10, 57). Indeed from an Afro-pessimist perspective, those endeavors have not effectively eliminated, or even fundamentally altered, the structural positionality of Blackness outside of civil society other than as what Jared Sexton has called "permutations'" ("Social Life" par. 4). Since the socially, culturally, and historically important changes have taken place on the level of experience and performance, Wilderson would argue that they have not disconnected Blackness from "Slaveness" on a structural level (*Red, White, and Black* 11). According to this argument, the constitutive nature of the demarcation of Blackness as "Slaveness" from 'humanness' for civil society makes futile any changes that happen without effectively dismantling the epistemic border structure that has enclosed civil society and demarcated it from Blackness as and to its outside.

Yet, Wilderson does not entirely negate the possibility of overcoming the 'Black border' between 'Blackness-as-Slaveness' and 'the Human' at some point in a future yet to come. He seems, however, much more reluctant than Wynter and the Black feminists discussed below to imagine Black freedom in today's "afterlife of slavery" (Hartman, *Lose Your Mother* 6), because destroying what I suggest to understand as an antagonistic border regime of anti-blackness is, according to Wilderson, impossible without destroying the world as such. As Wilderson argues by drawing on Fanon's work, the world that has for so long ostracized Blackness needs to end for Blackness to no longer entail "social death" (*Red, White, and Black* 337–38; cf. Fanon 71, 168). Sexton appears to speak about what might lie behind this 'end of the world' when he writes:

> If "black is a country" (Singh 2004), it is a stateless country, without birthright or territorial purchase; it is a feat of radical political imagination, the freedom dream of a blackened world in which all might become unmoored, forging, in struggle, a new people on a new earth. ("African American Studies" 223)[34]

---

34 Sexton clearly echoes here some of the deliberations by Hartman, Campt, and Sharpe I discuss as Black feminist fugitive thought in chapter 1.4.

Put in the register of the Black border I propose here, in place of all the borders that categories such as ethnicity, class, and gender implicate, Afro-pessimism suggests that the border between Blackness and the 'the Human world' needs to be destroyed first. By demolishing the 'Black border,' the (modern) world order it has held together would also fall apart, and a truly new, potentially all-encompassing "genre" of what it means to be human in Wynter's sense might (or might not) emerge.

The Black Border and travelling border concepts

As I have shown in chapter 1.2., more often than not, ethnic, area, and cultural studies as well as postcolonial and transnational literary studies that intersect in Chicanx and Latinx border concepts have concentrated their attention on productive processes of negotiating Black, Indigenous, Chicanx, Latinx, postcolonial, migrant, and other non-heteronormative identities and experiences in cultural exchanges on an equal footing. From an Afro-pessimist perspective, however, an inconsolable division, what I suggest to consider as the Black border, comes into focus between "the Black or Slave position" (*Red, White, and Black* 38) and 'the human,' or in other words, between civil life and "social death" (10–11). This ontic division is based on the premise that the history of slavery in the Americas has barred Black people from civil society's human subject position at least since the Middle Passage by making them "anti-Human, a positionality against which Humanity establishes, maintains, and renews its coherence, its corporeal integrity" (6). When we compare and contrast the insights gained about the 'Black border' in Afro-pessimism with Chicanx and Latinx border concepts, at first glance, however, several commonalities catch the eye. All use spatial tropes to conceptualize the relation of differently racialized people and their (im)possibilities to dwell and think as well as communicate with each other within a specific epistemological space. The relation between those groups of people has emerged from colonialism and slavery and their legacies still affect it. Yet, while concepts – such as Pratt's contact zone – construct borders as generally contingent, dialectic, and permeable, however, the 'Black border' I propose to consider in Afro-pessimism appears absolute, antagonistic, and impermeable.

"Borderlands," "contact zones," and "border thinking" have more often than not lent themselves to focus on hybrid, possibly violent processes of navigating non-heteronormative identities and experiences across territorial, cultural, inter-personal, and epistemic borders and boundaries on basically equal terms; they have thus often transported potentially optimistic, hopeful, and sometimes even utopian notions of a better present and future. Accordingly, the reception of these border concepts in North American studies has focused very much on the possibility of border crossings and on the potential hybrid transformations these crossings may involve. No matter whether we consider these concepts as they have been received during their travels or as explicitly conceptualized by Anzaldúa, Pratt, and Mignolo, their underlying logic presupposes extensive comparability, relationality, and the possibility of transforming the "local histories" and "global designs" of this world into a "pluri-versality."

In contrast, the epistemological border concept I read into Afro-pessimist interrogations of the category of 'the Human' very much complicates and questions the possibility of border crossings and the associated structural change with respect to Blackness. Wilderson's work compellingly draws attention to the demarcation between Blackness and non-Blackness that translates into "social death" and civil life, premised on the history of slavery in North America and its continuing legacy. Wilderson's analysis of the role of Blackness for the prevalent conception of 'the Human,' in particular, relies on a strategic focus not on experience and performance but on structural positionalities according to which people racialized as Black are positioned as 'non-human' and all others occupy the position of 'the Human.' If we use the register of the border, this demarcation between 'Blackness-as-non-humanness' and 'non-Blackness-as-humanness' appears to have manifested itself predominantly as an *a priori* absolute and impermeable border that has resisted any attacks against it and has hardly changed its position other than as '(per)mutations.' Considering the demarcation between Blackness and non-Blackness or social death and civil life in Wilderson's work as a border concept makes palpable the seemingly solid and impermeable structure that, as Afro-pessimism suggests, forces Blackness into the position of "Slaveness." It also enables us to understand better the differentiation of the levels of structure and experience with which Wilderson operates.

While Anzaldúa's, Pratt's, and Mignolo's border concepts imply interrelation and comparability – no matter how complex and asymmetrical the power relations may be – the Black border allows for no relation between structural positions on the two sides of its border, none, that is, than what Wilderson theorizes as an antagonism between Blackness and the world. The term *contact* in contact zone, for instance, already implies a relation that the Black border seems to forbid. By foregrounding the possibility of negotiation in a highly asymmetrical space, Pratt's concept assumes that even though different groups of people do not possess the same position of or to power, they can still enter, live in, communicate across, and occupy the socio-symbolic space of the contact zone. Thus, it seems not too far-fetched to compare the position of 'the subordinate' in the contact zone with the position of Wilderson's non-Black "junior partners" located at the inside margins of US civil society. From this point of view, contact zones could be found within civil society as spaces where Wilderson's "junior" and "senior" partners negotiate across asymmetrical power relations, whereas Blackness positioned as 'non-humanness' would provide the basis for these negotiation processes by enclosing civil society with the Black border. From this perspective, Anzaldúa's, Pratt's, and Mignolo's concepts can only account for the "junior partners" who are positioned at or can move to the inside margins of civil society, set off from Blackness through the Black border. By extension, the transformative promise of the relational border concepts exists only for non-Black positionalities. This is why Afro-pessimism implicitly rejects the utopian and celebratory undertones of relational border concepts and points to the limits of the otherwise far-reaching ability of such concepts to travel with respect to Blackness.

Afro-pessimism does not accept an assumed "common ground" (Bal, *Travelling Concepts* 8) of different oppressed people, including Black people, based on the "experience to have to come to terms with modernity/coloniality" (Mignolo, "Delinking" 497). In fact, Wilderson's work suggests that Blackness as social death forms the 'common ground' on which all other power relations play out. As opposed to conceiving border concepts as "the privileged locus of hope for a better world" (Michaelsen and Johnson 2), a notion that assumingly includes Blackness, Afro-pessimism holds that the world would have to end (and its Black border would have to be completely dismantled) in order for a new, all-encompassing "genre of the human" to emerge (Thomas and Wynter 25). The conceptual conflict between Afro-pessimism and ethnic studies disciplines, such as Chicanx and Latinx studies, does therefore not arise from the logic of different degrees of ostracism, oppression, or violence, what some critics would dismiss as "oppression Olympics" (Sexton, "People-of-Color-Blindness" 47). The ultimate point of conflict between "border thinking," "contact zones," and "borderlands" on the one hand and the Black border on the other lies in their different underlying assumptions about what role Blackness plays in the formation (and the potential destruction) of this world and its prevalent conception of 'the Human.'

Returning to the outset of the discussion of border concepts as travelling concepts, it not only proves well founded that to describe something as a border shapes the knowledge discussed in many important ways. It also proves true that an analysis of the travels and immobilities of border concepts reveals their "miniature theories" and underlying logics (Bal, *Travelling Concepts* 22). The confrontation of travelling border concepts with the 'Black border' shows that knowledge produced within civil society tends to assume a far-reaching 'travelability' of different theoretical concepts as well as an almost limitless comparability of people's relations to oppressive powers. Yet, the discussion of the

Afro-pessimist arguments, as contrasted with the logics of travelling border concepts above, confronts us with at least two types of questions that the remainder of this chapter addresses in more detail. First, Afro-pessimism poses important questions to and about knowledge production that interrogate the applicability, not least, of common literary studies terminology while emphasizing the importance of acknowledging the positionalities from which research is conducted. From an Afro-pessimist perspective, much of academic knowledge in and about North America has emerged from within civil society. This side of the Black border still dominates the majority of academic scholarship throughout the world. In fact, this study also takes Black feminist and Afro-pessimist interrogations of the category of 'the Human' and literary representations of captivity and flight from this side of the border into account. After all, I propose to read a border into Afro-pessimist thought that I can see only from my position as senior partner of civil society, while Afro-pessimism seems to suggest that profound knowledge not only about Blackness but about this world built on anti-blackness has always been emerging from the outside-position of civil society. But how can knowledge that is produced on the 'other' side of the 'Black border' be engaged on the inside? I return to this question in chapter 1.5. where I discuss not only the corpus selection but also the possible implications an Afro-pessimist and Black feminist fugitive framework has on common literary studies approaches and the important role positionalities should play in them.

Second, Afro-pessimism's focus on the level of structure has been criticized for potentially pathologizing, conflating, and reductionist tendencies.[35] After all, Afro-pessimism seems to place unconfined freedom[36] out of reach for those posi-

---

[35] For a brief look at the criticism directed at Afro-pessimism, see Scacchi 11–13. She mentions the mostly indirect criticism of premises Afro-pessimism is based on as articulated by Stephen Best ("On Failing"), Vincent Brown, and Kenneth W. Warren (*What Was*), even though the latter two do not engage Wilderson's and Sexton's work. For a more detailed and direct critical engagement of Wilderson's and Sexton's work specifically (to my knowledge a full-fledged critique has not been delivered so far), see, e.g., Gordon et al.; K. Warren, "'Blackness'"; G. Thomas, "Afro-Blue Notes."

[36] For a brief discussion of the concept of freedom in the dominant Western philosophical tradition, see Hames-García (xxxv–xli). His study *Fugitive Thought: Prison Movements, Race, and the Meaning of Justice*, which I will draw on in chapter 2.2., contrasts Western philosophical conceptions of freedom and justice with those proposed by imprisoned writers of color in the United States. With Hames-García, I define 'legal freedom' as the "lack of [immediate] restraint, such as not being a slave or not being in jail or not being bound by an unjust contract" (xxxvi). In my understanding, this term also accounts for the fact that legal freedom for Black people is always already at the verge of becoming legal unfreedom through forms of surveillance, criminalization, poverty, incarceration, and enslavement that make Black people structurally speaking in Wilder-

tioned as socially dead because captivity seems to find no end other than as a (potentially violent) "end of the world" (Wilderson, *Red, White, and Black* 337–38). Yet, instead of refusing to engage with Afro-pessimism and rejecting this trajectory all together, an increasing number of scholars have grappled with the question of Black sociability that happens against all odds on the side of the Black border, where social death seems to deny Blackness any leeway for negotiation in or with civil society. In other words, they have pursued creative methodologies and developed new vocabulary to see, acknowledge, appreciate, and cultivate the social life of the socially dead. I focus on a small selection of this work that seems to find unsettled and unsettling but nonetheless potentially 'common ground' with the travelling concept of fugitivity in the following.

1.4 "Black Feminist Fugitivity" against the Border

To recapitulate, I assume that Afro-pessimism – in theorizing a structurally incommensurable demarcation between non-Blackness and Blackness, civil life and social death, and between "the inside[and]outside of civil society" (von Gleich, Spatzek, and Wilderson 15) – tacitly implies an epistemic border concept that has had very real, i.e., fatal, consequences for Black people in North America and beyond at least since the transatlantic slave trade. Based on this understanding of Afro-pessimism as theorizing a structurally *a priori* incommensurable, absolute, and antagonistic demarcation, the border concept I suggest to consider in Afro-pessimist thought appears decidedly different from well-known conceptualizations of permeable borders as epistemological zones of dialectic cultural contact and conflict developed in American cultural and literary studies over the last thirty years. Yet, while providing an unflinching analysis of the structural delimitation of Blackness from the dominant 'genre of Man,' there seems to be no space in Afro-pessimism to account for the everyday lives of Black people and their battles and negotiations in the United States other than as "permutations" (Sexton, "Social Life" par. 4), as critics of Afro-pessimism have pointed out (see, e. g., Gordon et al.). This is not least because what we may call Black social life figures on the level of experience with which Afro-pessimism as defined by Wilderson is hardly concerned. Concepts that I group around the notion of fugitivity, however, have been reappearing

---

son's words "prison-slaves-in-waiting" ("Prison Slave" 18). It also takes into account, as we will see in part 2, that legally free Black people not only suffer from the constant threat of losing their legal freedom but also experience their legally free, non-enslaved, non-incarcerated lives as imprisoning and confined so that they strive for what I describe as 'unconfined freedom.'

in work that engages Wilderson's insights while trying to think beyond its limits. In the following, I therefore argue that the concept of fugitivity is more suitable – than concepts of borders as epistemological zones – to conceptualize enduring Black social life in the face of anti-blackness as a constant struggle against social death. It is my contention that the 'Black border' I propose to contemplate in Afro-pessimism and the concept of fugitivity together might not only help to better convey the very abstract and theoretically elaborate Afro-pessimist arguments. As concepts, they also make apparent the potential relations and tensions between the Afro-pessimist structural analysis of Blackness and fugitivity's focus on the level of experience and performance, shedding more light on the paradox capacity of social life in social death – not least addressed in Black American literature through the ages.

Whereas only few scholars openly subscribe to Afro-pessimism (i.e., especially Sexton and Wilderson), a growing number of scholars have adopted what I propose to call with Alexis Pauline Gumbs 'Black feminist fugitive thought.' Black feminist fugitive thought draws to different degrees on the insights of Afro-pessimism on structural positionalities, the antagonism, and the 'ruse of analogy,' while developing an analytic language that also accounts for the life in spite of social death and in order to think alternative futures from distinctly Black feminist perspectives. Since the emergence of Afro-pessimism in the first decade of the twenty-first century, Black feminist scholars, such as Hartman, Campt, and Sharpe, have been at the forefront of this endeavor. Their work has attempted to address both Black sociability and the structural position of Blackness in the "afterlife of slavery" (Hartman, *Lose Your Mother* 6) from different disciplinary perspectives and with various analytic tools, corpora, and terminology. Interestingly, their work, which appears closely related to but arguably different from Afro-pessimism, draws to different extents on the long intertwined history of African American fugitivity and Black feminism which Gumbs has aptly coined "Black feminist fugitivity" in her 2016 publication *Spill: Scenes of Black Feminist Fugitivity*. Fugitivity as a concept has travelled, so to speak, through a discourse that engages Afro-pessimism while rethinking its boundaries. I argue that these scholars try to avoid on the one hand an all too optimistic, active notion of resistance, change, and unconfined freedom, and on the other, an all too pessimistic notion of never-changing "Slaveness," striving for a conceptual language that instead accounts for the mundane, less obvious, hidden performances of flight and refusal.[37] They think through what happens when

---

[37] Campt, Hartman, and Sharpe have been collaborating in the "Practicing Refusal Collective" at Columbia University in the City of New York since 2015. The collective addresses anti-black-

we take Wilderson's structural antagonism and the assumption that the 'New World' is built on an anti-black foundation seriously and at the same time try to refuse understanding Blackness only as "Slaveness" by also addressing the performative-experiential level and zeroing in on the perspectives of Black women and Black queer people.[38] As we will see, especially in the juxtaposition of theory developed in close conversation with art, such as music, film, photography, and narrative, an ethical and political aesthetic analytics of the paradox of the social life of the socially dead or fugitive life emerges.[39]

After having looked at the ways in which social death literally captures Blackness from an Afro-pessimist perspective in chapter 1.3., the following chapter thus pursues the concept of fugitivity in Black feminist thought to account for the social life of the socially dead. Both will help me trace literary conceptualizations of captivity and fugitivity in Black American literature in the main part of this study, showing that those narratives provide "fugitive notes" (Cole, *Open City* 138) to the 'Wildersonian' "grammar of suffering" (*Red, White, and Black* 11, 37). Of course, the following discussion of Black feminist conceptualizations of flight is not exhaustive. Instead, it represents a selection of concepts, vocabulary, and notions joined in their consideration of Afro-pessimism's insights about anti-blackness that will prove particularly useful for the literary corpus analyzed in this study.[40]

---

ness and the "fungibility" of Blackness and "black bodies" by focusing on "refusal as a generative and capacious rubric for understanding everyday practices of struggle often obscured by an emphasis on collective or individual acts of resistance" (Columbia Global Centers I Paris). Here the concept of resistance is clearly put aside for the concept of refusal, indicating a reservation to assume the possibility of resistance of people who are structurally positioned as not (or less than) 'Human.'

38 For the long and complex history of Black feminism and Black feminist criticism in the United States, see, e.g., Adrienne Davis; Angela Davis, "Reflections," *Women, Race, Class*; Gore, Theoharis, and Woodard; Hartman, "The Belly," *Wayward Lives*; Hull, Scott, and Smith; Larsen, *Passing, Quicksand*; H. Jacobs; Petry; Spillers, "Mama's Baby, Papa's Maybe"; Shakur, "Women in Prison"; Truth; A. Walker.

39 Of course, oscillating between theory and narrative has a long tradition in African American writing, especially in African American autobiographical texts (see ch. 2.1. and 2.2.) and in Black Feminist criticism (see, e.g., Christian; Hull, Scott, and Smith; A. Walker).

40 For further works that conceptualize notions of fugitivity and refusal with respect to Blackness and Black feminism, not least by experimenting with scholarship, art, the personal, and the vernacular, see, e.g., Bey, *Them Goon Rules*; Gumbs.

The social life of the socially dead

Much of the critique of Afro-pessimism has gathered around an opposition between Afro-pessimism as propagated by Wilderson and the notion of Black Optimism associated with Fred Moten's work (see below and, e.g., Gordon et al., esp. 112–18). The performance studies scholar and poet Moten has become synonymous with an elusive notion of fugitivity and Black Optimism as a trajectory of thought that has entered into a critical conversation with Afro-pessimism about the question of social life and social death (see, e.g., Moten, "The Case of Blackness," "Blackness and Nothingness," "The Subprime"). In his re-reading of Fanon's work in "The Case of Blackness," for example, Moten argues that social life exceeds and escapes social death, actively drawing on the notion of flight, when he asks,

> Can resistance come from such a [pathological] location? Or perhaps more precisely and more to the point, can there be an escape from that location; can personhood that defines that location also escape that location? What survives the kind of escape that ought never leave the survivor intact? (208)

In doing so, Moten's conceptualization of fugitivity remains literally fugitive as it is scattered within and across his theoretical and poetic oeuvre, while posing central questions. The questions pertain not only to the structural position of Blackness in relation to "Fanon's pathological insistence on the pathological" (208) but also to a core critique of Afro-pessimism, i.e., the assumed pathologizing of Blackness as exceptional, which – according to critics – issues from a lack of acknowledgement of persevering social life and the relation of anti-blackness to other forms of oppression (Scacchi 11–12; Gordon et al. 119, 123, 127).[41]

I take a brief look at the assumed opposition between Afro-pessimism and Black Optimism as perceived by Tavia Nyong'o and Jared Sexton as an entrance point into Black feminist fugitive thought that has taken its cues from this debate and has walked a conceptual tightrope by refusing any such opposition. In "Habeas Ficta: Fictive Ethnicity, Affecting Representations, and Slaves on Screen," Nyong'o describes the relation between Moten's and Wilderson's scholarship

---

[41] Whether Wilderson's disregard of the level of performance and experience in his structural analysis of the position of Blackness devaluates the relevance of racism and discrimination besides anti-blackness and people's persistent survival or refusal of and opposition to anti-blackness, racism, and domination remains to be fully debated, just as the question whether Afro-pessimism strategically essentializes, let alone pathologizes, Blackness. See, e.g., Gordon et al.; Sexton, "Ante-Anti-Blackness," "Social Life," "The Unclear Word."

as revolving around the question of "the im/possibility of escape from the slave ship's hold" (292). While Wilderson focusses on "the negation of blackness [as] the basis out of which civil society and its ethno-national cinematic life is animated," Moten – Nyong'o argues – understands Blackness as "the negation of civil society, on the basis of which social life can flourish" (292–93). Nyong'o locates his own research closer to Moten's notion of "negativity," not least because he is "concerned with where the rhetoric of constitutive lack and its *aporetics of loss* [which he appears to ascribe to Afro-pessimism] may be leading black criticism" when lack is "overdrawn and oversimplified" (294). In other words, Nyong'o acknowledges the Wildersonian "constitutive antagonism in the US social order" with which "Wilderson has innovated a distinctive brand of criticism," while rejecting "the slave [...] position of structural lack, [as] a 'nothingness' from which no affirmative or resistant representation can emerge" (287–88). Instead, Nyong'o analyzes "fabrication[s]" of Blackness as ethnicity or nationality in filmic representations of enslaved people from the 1970s to the early twenty-first century and the ways in which what he calls "afrofabulation" may draw on a "zone of indistinction between lack and excess, [...] negation and affirmation" (292, 294).[42] Using the vocabulary that I am developing here, we could say that Nyong'o pursues an analysis that looks for strategies that press at the Black border from the outside, examining 'Motenesque' notions of fugitivity that strive to cross the Black border without necessarily disavowing the Wildersonian structural impossibility of such an endeavor. Yet, Nyong'o's suggestion that in an Afro-pessimist framework "no affirmative or resistant representation can emerge" from the position of Blackness (287–88) needs more unpacking.

In "Ante-Anti-Blackness: Afterthoughts," Sexton argues that Black Optimism and Afro-pessimism actually analyze two sides of the same coin, working in "an intimacy [...] that arrayed them less as opposites and more as conditions of an impossible possibility." Sexton maintains that Wilderson's positing "a political ontology dividing the Slave from the world of the Human in a constitutive way [...] has been misconstrued as a negation of the agency of black performance, or even a denial of black social life" ("Ante-Anti-Blackness," see also "Social Life" par. 19):

> Nothing in afro-pessimism suggests that there is no black (social) life, only that black life is not social life in the universe formed by the codes of state and civil society, of citizen and subject, of nation and culture, of people and place, of history and heritage, of all the things

---

[42] Borrowing from Saidiya Hartman's notion of "critical fabulation" ("Venus in Two Acts" 11–12, see also below), Nyong'o first proposed the concept of "afrofabulation" in his article "Unburdening Representation" (72).

that colonial society has in common with the colonized, of all that capital has in common with labor – the modern world system. ("Ante-Anti-Blackness," see also "Social Life" par. 24).

Thus, 'affirmative or resistant representation' *can* emerge from the position of social death; it just does not figure in the world of the social life of 'the Human' from an Afro-pessimist perspective. While Afro-pessimism focusses on the structural production and preservation of social death as an ontology, Sexton understands Moten's work in (and since) *In the Break* (2003) to revolve "around [both] the impossibility and the inevitability of the 'resistance of the object'" on those different levels ("African American Studies" 222). In Sexton's analysis, Moten poses the question "how black social life steals away or escapes from the law, how it frustrates the police power and, in doing so, calls that very policing into being in the first place" ("Ante-Anti-Blackness," also "Social Life" par. 32).[43] Sexton therefore proposes not to approach "(the theorization of) social death and (the theorization of) social life as an 'either/or' proposition" but suggests to "attempt to think them as a matter of 'both/and'" ("Ante-Anti-Blackness") – even though, as we have seen in chapter 1.3., Wilderson's coining of Afro-pessimism has primarily focused on the former.[44]

Sexton further argues that Moten's affirmation of Blackness, i.e., of "living a black social life under the shadow of social death," should not be misunderstood as "an accommodation to the dictates of the anti-black world" but as a "*refusal* to distance oneself from blackness in a valorization of minor differences that bring one closer to health, life, or sociality" ("Ante-Anti-Blackness," see also "Social Life" par. 23). Black feminists, such as Campt and Hartman have pursued this understanding of social life as 'refusal' in social death and flight towards a position "closer to health, life, or sociality" in refined ways and

---

43 On the relation between the fugitive from slavery and the law of property in the nineteenth century, see Best, *The Fugitive's Properties*.

44 For a discussion of differences in Moten's and Wilderson's work that counter Sexton's argument, see Menzel's contribution in Gordon et al. Menzel argues that while Moten and Wilderson both draw on Spillers's seminal "Mama's Baby, Papa's Maybe," they reference very different aspects of Spillers's complex argument which reflect a "persistent dissension" between Black Optimism and Afro-pessimism around "the maternal." Menzel contends that while Wilderson "powerfully elaborates the essay's account of the violent abjection of Black maternity, [...] he effaces her accompanying gesture toward its insurgent possibilities" which Moten in contrast "amplifies" (Gordon et al. 112). Greg Thomas also accuses Afro-pessimists, especially Wilderson, of misreading oeuvres through selective citation politics, for instance, with respect to the work of the Ghanaian writer Ayi Kwei Armah (1938–), Frantz Fanon, Orlando Patterson, or Assata Shakur ("Afro-Blue Notes" 283–88, 296–97, 303, 307–09).

have indeed attempted to think social death and social life as a matter of both/ and. As Rinaldo Walcott suggests, however, this Black feminist fugitive theorizing has received less attention and should be "more vigorously embrace[d] vis-á-vis the 'big' narrative of freedom's moves," such as Moten's or Neil Roberts's "freedom as marronage" ("Freedom Now Suite" 153).

Walcott argues that Black feminists, such as Simone Browne, Christina Sharpe, and Katherine McKittrick have refused "the return to marronage and fugitivity" that Moten and Roberts have undertaken as an extension of "a kind of freedom" (153).[45] Instead, "black women scholars in their contributions seem to be cautious about the ways the 'big' narratives of freedom might thwart our understanding of what is at stake" (153). In fact, the same may be true for the Black feminists' strategic circuiting of the somewhat heated scholarly debate between Afro-pessimism and Black Optimism. I agree with Walcott that Browne, McKittrick, and Sharpe, but also Campt and Hartman, "have [...] asked us to read differently," "allow[ing] a putting together of the historical and the contemporary in a discontinuous fashion that sheds light on the now of black experience and life" (153). Yet, I do not see a rejection of the concept of fugitivity in their works as such, but a rejection of fugitivity as a term for resistance or freedom specifically. Sidestepping the concepts of agency, resistance, and unconfined freedom, they instead make use of the notion of flight and refusal to conceptualize social life in social death and to participate in past and present strivings for a future different from the one created by slavery (Hartman, *Lose Your Mother* 133).

In "Rebellion/Invention/Groove," for instance, McKittrick discusses Wynter's unpublished monograph manuscript "Black Metamorphosis: New Natives in a New World" about the slave plantation economy and Black cultural production,

---

[45] See ch. 2.1. for a discussion of Neil Roberts's philosophical work on "freedom as marronage." As Roberts explains in the introduction to *Freedom as Marronage*, the term "Marronage (marronnage, maroonage, maronage) conventionally refers to a group of persons isolating themselves from a surrounding society in order to create a fully autonomous community, and for centuries it has been integral for interpreting the idea of freedom in Haiti as well as other Caribbean islands and Latin American countries including the Dominican Republic, Jamaica, Suriname, Venezuela, Brazil, Cuba, Colombia, and Mexico. These communities of freedom – known variously as 'maroon societies,' *quilombos*, *palenques*, *mocambos*, *cumbes*, *mambises*, *rancherias*, *ladeiras*, *magotes*, and *manieles* – geographically situate themselves from areas slightly outside the borders of a plantation to the highest mountains of a region located as far away from plantation life as possible" (4). For a brief summary of the etymology of the term, see N. Roberts 4–5. Roberts broadens and diversifies the definition of marronage/maroonage trans-historically and trans-locally in order to also account for the post-slavery era and for places where marronage/maroonage as defined above did not occur as a form of freedom.

especially Black music, in the Caribbean. McKittrick's mostly indirect connection to Moten's and Wilderson's work becomes clear when she reads Wynter's monograph as "unveil[ing] how the plantation slavery system and its postslave expressions produced black nonpersons and nonbeings (through brutal acts of racist violence designed to actualize psychic and embodied alienation) just as this system generated black plantation activities that rebelled against the tenets of white supremacy" (81). McKittrick's reading of Wynter's work, thus, pursues in a first step an argument not unlike Afro-pessimism that considers slavery and white supremacy (rather than anti-blackness) positioning Black people as "nonbeings" (81). In contrast to Wilderson, however, McKittrick focusses her discussion of Wynter's work on the cultural production as "rebelling" against the Black border (86). While McKittrick does not explicitly make use of the concept of flight, she references "marronages, mutinies," and "revolts" as well as other cultural practices that "invent" Black life in an anti-black world, such as "funerals, carnivals, dramas, visual arts, fictions, poems, fights, dances, music making and listening" (81). Focusing on Black music in particular, McKittrick's argument thus suggests – while not stating this explicitly – that Wilderson's antagonism between Blackness and 'the Human' can be and has been 'rebelled' against, and by doing so spaces are 'invented' that allow Black people to "groove." In line with Moten's notion of fugitivity and Nyong'o's concept of afrofabulation, McKittrick argues that "loving and sharing and hearing and listening and grooving to black music" (81) affirms Black humanity "across and in excess of the positivist workings of antiblack logics" (80).

The conception of fugitive performances that try to escape anti-blackness while moving within and across anti-black geographies, such as the plantation society, is crucial for this study as it supports my engagement with the travelling of the concepts of captivity and fugitivity that, of course, are also related to conceptualizations of space, place, and time. In *Demonic Grounds: Black Women and the Cartographies of Struggle*, McKittrick addresses the ways in which Black female art and testimony in the United States, Canada, and the Caribbean negotiates and, as she would put it later, 'rebels' against traditional, i.e., "white, patriarchal, Eurocentric, heterosexual, classed" geographies (xiii). McKittrick argues that engaging Black women's "geograph[ies] as space, place, and location in their physical materiality and imaginative configurations" helps "make visible [their] social lives which are often displaced, rendered ungeographic" (x). Her analyses show how "locations of captivity initiate a different sense of place through which black women can manipulate the categories and sites that constrain them" (xvi–xvii). As McKittrick notes, "[e]nforcing black placelessness/captivity was central to processes of enslavement and the physical geographies of the slave system" (8). In other words,

> while black people certainly occupied, experienced, and constructed place, black geographies were (and sometimes still are) rendered unintelligible: racial captivity assumes geographic confinement; geographic confinement assumes a despatialized sense of place; a despatialized sense of place assumes geographic inferiority; geographic inferiority warrants racial captivity. (8)

Combining McKittrick's insights about the relation of geography, captivity, and the "struggle toward some kind of sociospatial liberation" (xx) with non-linear, circular notions of time, as discussed below, enables me to consider the spatio-temporal components of confinement in this study,[46] while also opening up the view – in the tradition of Moten and Wynter and the works discussed below – for that which might escape.

Fugitive refusal and futurity

In the lecture "The Sounds of Stillness," Tina Campt reads Moten's concept of fugitivity in "The Case of Blackness" as "a practice that is not defined by opposition or resistance but refusal" which she further qualifies in a later lecture as the "refusal to be a subject to a law that refuses to recognize you" ("Black Feminist Futures"). Trained as a historian of the Black diaspora, Campt's *Image Matters: Archive, Photography, and the African Diaspora in Europe* (2012) examines the ways in which Black diasporic photography functions not only as "*an enactment* of the past," but also as an "*articulation* and *aspiration*" (7) and therefore participates in community and identity formation in hostile environments that negate Blackness. In her examination of vernacular photography of Black German families (1900–1945) and portrait photography of "African Caribbean migrants to postwar Britain" (1948–1960), Campt addresses the broad question of "how do black families and communities in diaspora use family photography to carve out a place for themselves in the European contexts they come to call home?" (14). She puts the concept of fugitivity to direct use in her analysis of "snapshot" photographs of the lives of Afro-German families in Nazi-Germany. Her image analyses of "*the social life of the photo*[s]" (6) reveal the ways in which the "fugitivity of these photos lies in their ability to visualize a recalcitrant normalcy in places and settings where it should not be" (91). The images practice a form of "domestic fugitivity of dwelling and homemaking" (111) by displaying and thereby (re)creating spaces of private refuge for Black German subjects against the odds of and within the Nazi regime.

---

[46] See also McKittrick, "On Plantations"; "Plantation Futures"; "Worn Out."

In her attempt to account for "the possibility if not of agency, resistance, or opposition then, most important, of fugitivity" of Black diasporic subjects (79), Campt describes fugitivity as "articulated through modes of contestation and refusal that reside in the snapshot, in the found image, in the clinical or ethnographic portrait, as well as in the eugenic lecture slide" (80). Consequently in her discussion of definitions of *fugitives*, Campt explicitly includes those who "cannot or do not remain in the proper place, or the places to which they have been confined or assigned" just as those "indistinguishable from the norm through a capacity to undermine its clarity and legitimacy" (87).[47]

Campt's conceptualization of refusal and fugitivity also reappears throughout her 2017 monograph *Listening to Images*. Continuing the work of *Image Matters*, *Listening to Images* further develops fugitivity as a theoretical and methodological concept through Campt's close and wide readings of various photographic corpora from Africa, Great Britain, and the United States. This theorization in progress, i.e., her theorizing while applying, leads to very flexible work-in-progress concepts that remain – like Moten's – mostly out of focus and difficult to pin down. They seem to make use of what Bal describes as the advantages of a concept-driven cultural analysis as opposed to a full-fledged theory as discussed in chapter 1.1. In *Listening to Images*, Campt connects the concept of fugitivity more explicitly with her notion of "the quotidian practice of refusal" to the point where they can hardly be told apart as can be seen in the following two quotes:

> The quotidian practice of refusal I am describing is defined less by opposition or 'resistance,' and more by a refusal of the very premises that have reduced the lived experience of blackness to pathology and irreconcilability in the logic of white supremacy. Like the concept of fugitivity, practicing refusal highlights the tense relations between acts of flight and escape, and creative practices of refusal – nimble and strategic practices that undermine the categories of the dominant. (32)
>
> [...] that practicing refusal means embracing a state of black fugitivity, albeit not as a 'fugitive' on the run or seeking escape. It is not a simple act of opposition or resistance. It is

---

47 The full quote reads: "Fugitives: those who leave, run away, are forced out, or seek refuge elsewhere. Those who by compulsion or choice cannot conform; cannot or will not submit to the law; cannot or do not remain in the proper place, or the places to which they have been confined or assigned. Those who venture into sites unknown or unwelcoming are interlopers and strangers who unsettle our sense of the norm. Yet the fugitive's impact registers not only through difference or through her or his status as an outsider. It registers equally or perhaps even more profoundly in those moments when she or he is indistinguishable from the norm through a capacity to undermine its clarity and legitimacy. Often an elusive presence, the fugitive has an ability to pass that camouflages difference while highlighting the very distinctions on which identity and community are based." (87)

neither a relinquishing of possibility nor a capitulation to negation. It is a fundamental renunciation of the terms imposed upon black subjects that reduce black life to always already suspect by refusing to accept or deny these terms as their truth. It is a quotidian practice of refusing the terms of impossibility that define the black subject in the twenty-first century logic of racial subordination. (109, 113)

Campt explains her concept of refusal with fugitivity and the concept of fugitivity with refusal in a circular manner: "the concept of fugitivity I am invoking is not an act of flight or escape or a strategy of resistance. It is defined first and foremost as a practice of refusing the terms of negation and dispossession" (96). Clearly, this conceptual elusiveness is deliberate and shows the ways in which not only the content and meaning of concepts are relevant but also how their form, application, and 'differing processes' are already an important part of the conceptualization.[48]

Similar to McKittrick's "Rebellion/Invention/Groove," Campt's work engages less directly with Wilderson's Afro-pessimism (note, for example, that both use the concept of white supremacy rather than anti-blackness) and connects her notion of fugitive refusal more explicitly and elaboratively than Hartman and Sharpe towards a "grammar of black feminist futurity" (*Listening to Images* 17). This futurity is expressed in Campt's perspective in "the future real conditional or *that which will have **had to** happen*" and in performances "of a future that hasn't yet happened but must" (17, italicized and bold print emphases in original). Campt differentiates here between hope and aspiration (the latter of which Sharpe also picks up in *In the Wake*),[49] aligning Black feminist futurity with the latter as a "power to imagine beyond current fact and to envision that which is not, but must be" (17). This notion of futurity then allows Campt to read, among other things, missionary and ethnographic portrait photography of Africans at the turn of the nineteenth to the twentieth century as "neither wholly liberatory vehicles of agency, transcendence, or performativity nor unilateral instruments of objectification and abjection" (59–60). Instead, she walks the aforementioned conceptual tightrope of "uncoupling the notion of self-fashioning from the concept of agency" in order to understand self-fashioning

as a tense response that is not always intentional or liberatory, but often constituted by minuscule or even futile attempts to exploit extremely limited possibilities for self-expression and futurity in/as an effort to shift the grammar of black futurity to a temporality that both

---

48 On refusal as a political and ethical alternative concept to recognition in Indigenous communities working against ongoing settler colonial violence, see Simpson.
49 On the role of breath and thinking "otherwise possibilities" in Black studies, see also Crawley.

embraces and exceeds their present circumstances – a practice of *living the future they want to see, now.* (59)

This uncoupling of agency from self-fashioning performances is central to my understanding of fugitivity in social death as it attempts to resist, with Campt's words, "easy categorization and refuse[s] binary notions of agency versus subjection" (59), or rather in this case, social life versus social death, as well as linear notions of time. Campt's notion of fugitive refusal is committed to "accountability" and "responsibility" for a "black feminist praxis of futurity as an existential grammatical practice of grappling with precarity" and anti-blackness today (114) and therefore provides a present- and future-oriented perspective in addition to the unflinching analytic lens that Afro-pessimism provides with which I want to approach the literary corpus.

The "fugitive legacy" of slavery

While Campt focusses on Black feminist futurity based on fugitive practices of refusal in the present and the past, Hartman's work from her first monograph *Scenes of Subjection* (1997), through *Lose Your Mother* (2007), to her most recent book-length publication *Wayward Lives* (2019) looks back in order to tell "a history of the present" ("Venus in Two Acts" 4, see also Wilderson and Hartman 190). *Lose Your Mother: A Journey along the Atlantic Slave Route* is a genre mix of travelogue, autoethnography, autobiography, fictional history, and essay that documents Hartman's visit of Ghana and her repeated re-visiting of the historical archive of slavery and its voids. As such, scholars discuss *Lose Your Mother* as both a literary text for close reading and as a seminal contribution to critical theory for understanding Blackness in the Atlantic world from the historical time of slavery until today. Markus Nehl, for instance, includes *Lose Your Mother* in his study of twenty-first century "neo-slave narratives" that push the boundaries of genres and often take a more transnational outlook. He observes in *Lose Your Mother*, among other things, a "powerful re-negotiation of Paul Gilroy's concept of the black Atlantic and the discourse of roots tourism in Ghana" (81). He also connects the book to a larger discourse about Black America's varied relations to post-independence Ghana by contrasting Hartman's narrative with Alex Haley's *The Roots: The Saga of an American Family* (1976) and Maja Angelou's *All God's Children Need Travelling Shoes* (1986) (82–84).[50] Yet, as Sabine Broeck ar-

---

[50] See also Michelle D. Commander's discussion of *Lose Your Mother* together with fictional

gues, "[b]eyond its autobiographical context, and beyond its intramural address, *Lose Your Mother* needs to be read as a major contribution to theorizing transatlantic modernity as driven by the technological machinery, the economy, and epistemology of enslavement" ("Enslavement as Regime" par. 8). So while it certainly uses typical generic and topical elements of neo-slave narratives (see ch. 2.3.), I approach *Lose Your Mother*, particularly its last chapter, not primarily for its literary and cultural merit but for its theoretical contributions to Black feminist conceptualizations of fugitivity.

In the last chapter entitled "Fugitive Dreams," Hartman discusses the experiences of the descendants of the transatlantic slave trade in Ghana and the United States with respect to captivity and fugitivity. She begins the chapter by telling a story about African "survivors" fleeing from "raiders" during the "global trade in black cargo" (*Lose Your Mother* 226). She observes how those who escaped the raiders had "learned that the settlement in an outlying territory was not the guarantee of sovereignty and that flight was as near to freedom as they would come. And that the gap between what they had dreamed of and what they could have would never be bridged" (227). Subsequently, Hartman notes that during her travels she did not hear stories of those who could not flee and were captured and enslaved. She therefore suggests that the trade poses different legacies for the descendants of the African 'survivors,' that is those who were *not* captured, deported to the other side of the Atlantic, and enslaved, than for the descendants of the captives: "[...] those who stayed behind told different stories than the children of the captives dragged across the sea. Theirs wasn't a memory of loss or of captivity, but of survival and good fortune. After all, they had eluded the barracoon, unlike my ancestors" (232). As part of what Broeck describes as a "demolition project" (*Gender* 13), Hartman thus complicates both African and African American (grand) narratives of captivity and fugitivity and claims "the fugitive's legacy" in today's "afterlife of slavery" (*Lose Your Mother* 234, 6).[51] Hartman foregrounds the related but dissonant legacies of West African and Black American captivity and fugitivity and questions whether unconfined freedom could ever be possible for Black people in an anti-black world, in which "[t]he bloodletting of the modern world allowed for no havens or safe places. The state of emergency was not the exception but the rule. The refuge became hunt-

---

neo-slave narratives, such as Octavia Butler's *Kindred* (1979), and fictional return narratives, such as Reginald McKnight's *I Get on the Bus* (1990) in chapter 1 of her interdisciplinary analysis of speculative *Afro-Atlantic Flight* in literature and tourist cultures.

51 On the notion of 'being left behind' and West African survival of and loss due to the transatlantic slave trade, see chapter 6 of Laura Murphy's *Metaphor and the Slave Trade in West African Literature* that responds and adds to Hartman's perspective here.

ing grounds for the soldiers of fortune whose prizes were people" (227). Importantly, Hartman extends this question into the post-Civil Rights United States:

> If slavery persists as an issue in the political life of black America, it is not because of an antiquarian obsession with bygone days or the burden of a too-long memory, but because black lives are still imperiled and devalued by a racial calculus and a political arithmetic that were entrenched centuries ago. This is the afterlife of slavery – skewed life chances, limited access to health and education, premature death, incarceration, and impoverishment. (6)

Reiterating this insight in her article "Venus in Two Acts," Hartman emphasizes how the "afterlife of slavery" exceeds a mere legacy and the idea of remnants of the past, while also rejecting the notion of "melancholia" (13). Instead, Hartman describes the "afterlife of property" as "the detritus of lives with which we have yet to attend, a past that has yet to be done, and the ongoing state of emergency in which black life remains in peril" (13). Thus, with the notion of the 'afterlife of slavery' Hartman points to the ways in which the transatlantic slave trade, slavery, and their "enslavist" epistemology and ontology have shaped the modern world order and civil societies in North America and beyond (Broeck, *Gender* ch. 2).

Lose Your Mother and "Venus in Two Acts" clearly build on Hartman's earlier historical study of the "terror of the mundane and quotidian" forms of violence against the (formerly) enslaved in nineteenth-century North America in *Scenes of Subjection* (3, see also ch. 2.1.) which Wilderson cites throughout *Red, White, and Black*. But *Lose Your Mother* and *Scenes of Subjection* have been oft-cited not only in Afro-pessimist and Black feminist thought, for their trenchant analysis of the relation between today and the historical time of slavery that "does not allow the reader to think that there was a radical enough break to reposition the black body after Jubilee" (Wilderson and Hartman 183).[52] As Stephen Dillon notes, when Hartman "returns to the slave dungeons on the Gold Coast hoping to find ancestors and a history" in *Lose Your Mother*, she finds "the dust and waste of those who entered the door of no return. She finds the emptiness left by slavery's regimes of unimaginable violence and terror, the nothingness left by the deaths of 60 million or more" (98). *Lose Your Mother*, thus, tries to find a language that addresses how "[s]lavery's mark on the now manifests as the prison, as poverty, as policing technologies, in insurance ledgers, and in the or-

---

[52] See, e.g., Best, "On Failing", *None Like Us*; Broeck, *Gender*; Crawford, "Turn to Melancholy"; Dillon; Goyal, *Runaway Genres*; Levy-Hussen, *How to Read*; Sexton, "African American Studies"; Sharpe, *In the Wake*.

ganization of space" (Dillon 98). Sexton pointedly adds, "we do not simply inherit the aftermath of slavery; we inhabit its *afterlife*" ("African American Studies" 211).

In chapter 2 of *The Physics of Blackness*, Michelle Wright examines the intersecting "spacetimes" of "the past-in-the-present and the present-in-the-past" in Hartman's *Lose Your Mother* as "applying Epiphenomenal time to her consideration of return through the Middle Passage Epistemology" (93). Focusing particularly on the beginning and end of the text (93–94, 96), as I do too, Wright notes how Hartman shows that the two spacetimes, while intersecting, "cannot be subsumed by one spacetime alone," i.e., a "great summing up, a grand conclusion, is simply not possible" (94).[53] Accordingly, Calvin Warren theorizes the time of slavery as not a past "object-event" to "get over with" in the "time of Man" but an ongoing "event-horizon" that he calls "Black time" ("Black Time" 61, 66). Just as geography plays a central role in conceptualizations of captivity and flight, as McKittrick shows, Hartman's work on the afterlife of slavery emphasizes the role of temporal conceptions. McKittrick's work on Black geographies in combination with Hartman's and Warren's concepts of time in the post-slavery era support my working definition of fugitivity to account for the imaginative spaces and material places in and through which captivity and flight

---

[53] Michelle Wright's *Physics of Blackness* criticizes dominant conceptualizations of time with respect to Blackness by focusing on the multidimensionality, transnationalism, and queerness of the spacetimes of Blackness. She argues against a linear casual, progressive narrative of male- and US-centered Blackness from slavery to freedom, a critique that also animates Warren's Black time and Hartman's afterlife of slavery as Wright's reading of Hartman's *Lose Your Mother* shows. However, Wright also criticizes "a reverse linear narrative," which she associates with "Afropessimism" and blames for suggesting that "no *Black* progress has been made because of the continual oppression by white Western hegemonies that began with slavery, moved through colonialism, and now deploy an array of cultural, political, economic, and military power through social and governmental technologies to keep Blacks not only as subalterns – those who are subordinated by power – but also as the (white) Western Other" (8). While Wright's emphasis on the multidimensional, transnational, and queer make-up of Black spacetimes is important to this study, it remains unclear whether her concept of "Afropessimism" refers to the negative conception of Africa or to the work of Afro-pessimists, such as Wilderson and Sexton. For a queer theory perspective on racialized concepts of time in the long nineteenth century in the United States, see also Freeman, esp. ch. 2.

Just as the term 'postcolonial,' 'post-slavery era' does not imply a clear temporal 'after' that would assume slavery as a closed event of a linear past. Instead, the 'post' refers to the ways in which the history of slavery and the epistemology and ontology of enslavism that issues from that history continues to structure today's societies, interpersonal relations, and knowledge productions. I take the term from Sharpe (*Monstrous Intimacies*) and rethink it with Broeck's concept of 'enslavism' (*Gender* 6).

are experienced and performed across circular, non-linear, and expanding time and place in Black American literature.

As the title indicates, Hartman's chapter "Fugitive Dream" in *Lose Your Mother* is, however, not only concerned with theoretically analyzing US anti-blackness in and as the afterlife of slavery. Hartman also engages in a search for a conceptual language with which a past, present, and future reality of Black social life and unconfined freedom become imaginable in the wake of Black social death. In "Fugitive Dream," Hartman opts for the vocabulary of dreaming and flight to describe the unbridgeable "gap" (227) between unconfined freedom and reality. By writing about this 'fugitive' dream both in West Africa and North America, she also evokes Martin Luther King Jr.'s famous public speech "I have a Dream" at the March on Washington in 1963, at the same time criticizing – almost as an aside – the post-racial myth related to Obama's presidential candidacy at the time:

> The legacy that I chose to claim was articulated in the ongoing struggle to escape, stand down, and defeat slavery in all of its myriad forms. It was the fugitive's legacy [...] *It wasn't the dream of a White House*, even if it was in Harlem, but of a free territory. *It was a dream* of autonomy rather than nationhood. *It was a dream* of *an elsewhere*, with all its promises and dangers, where the stateless might, at last, thrive. (*Lose Your Mother* 234, emphasis mine)

In "Venus in Two Acts," Hartman revisits the archive of slavery and her reckoning with it in *Lose Your Mother*, i.e., her attempts to "attend" to "the detritus of lives" of the enslaved (13). "Venus in Two Acts" represents, if you will, a self-commentary on Hartman's attempt in *Lose Your Mother* at reckoning across spacetimes, but also a second attempt at such reckoning, *and* a questioning of the possibility of reckoning through narrative all at the same time. Hartman draws on the historical archive as the primary source that documents enslaved peoples' lives, especially women's and girls' lives, through stories that "are not about them, but rather about the violence, excess, mendacity, and reason that seized hold of their lives, transformed them into commodities and corpses" (2). Wishing to know enslaved people's lives more fully, Hartman tests the possibilities and limits of "critical fabulation" (11) that goes beyond the archive without wanting to "commit[...] further violence in [the] act of narration" (2). Due to "the incommensurability between the experience of the enslaved and the fictions of history" (9), she describes this attempt as "straining against the limits of the archive to write a cultural history of the captive, and, at the same time, enacting the impossibility of representing the lives of the captives precisely through the process of narration" (11). This "critical fabulation" strives for "narrative restraint, the refusal to fill in the gaps and provide closure" or "to *give voice*" (11, 12), because in 'Black time' "[i]t is much too late for the accounts of death to prevent other deaths; and it is much too early for such scenes of death to halt other crimes" (13). Thus, "in the meantime, in the space of the interval, between too late and too early, between the no longer and the not yet," i.e., in today's "afterlife of slavery," Hartman is committed to keep pondering how "a narrative of defeat [can] enable a place for the living or envision an alternative future" (13).[54]

While Hartman's work in *Scenes of Subjection* clearly inspired Wilderson's *Red, White, and Black*,[55] her work on the desire and impossibility to save the captured, enslaved, and murdered from the historical archive of slavery with historical, autoethnographic, and 'critical fabulation' as well as her consistent focus on the genderedness of the afterlife of slavery in *Lose Your Mother* and "Venus in Two Acts" (see also "Belly of the World," *Wayward Lives*) has not taken center stage in Afro-pessimism as coined by Wilderson.[56] Christina Sharpe's "wake work," however, seems to do exactly that, as it brings Afro-pessimism in even closer conceptual conversation with Black feminist fugitive thought.

"Wake work" and/as care in "the hold"

In the first chapter of *In the Wake: On Blackness and Being* (2016), Christina Sharpe continues the strategic blurring of lines between scholarship and other forms of writing, such as memoir, that we already know from Hartman and Campt and in fact also Wilderson (*Incognegro*, *Afropessimism*). Sharpe introduces her study of Black people's proximity to death in the diaspora and the forms of care Black people perform in death's "immanence" and "imminence"

---

[54] Hartman further refines 'critical fabulation' as a critical approach to the historical archive of Black lives in the United States with her most recent monograph. *Wayward Lives* focusses on young Black women's city lives at the turn of the twentieth century with the help of "close narration" (1, cf. Polk, Wooden).

[55] Wilderson thanks Hartman in his acknowledgements to *Red, White, and Black* together with Adrian Bankhead for their "unflinching[...] look] at the void of our subjectivity, [... that] help[ed] the manuscript to stay in the hold of the ship, despite my fantasies of flight" (xi). See also their exchange about *Scenes of Subjection* in Wilderson and Hartman. Sharpe draws on the notion of "the hold" and elaborates it as a concept (*In the Wake*). See below.

[56] I understand gender in relation to the afterlife of slavery and anti-blackness with Patrice D. Douglass – who draws on Hartman and Sylvia Wynter – as that which "for the Human can be performed as possession," while "gender for the Black exposes how Black bodies are possessed, not by their individual or collective (gender) identifications, but by the investments or valuations placed upon gender as a genre for designating Human distinction or 'kind'" (91).

with a personal narrative of several successive deaths of members of her family in a short period of time (*In the Wake* 13). Moreover, like Hartman, Campt, and Moten, Sharpe also theorizes 'in progress' and in close conversation with art. For example, she draws on the ending of Toni Morrison's novel *Beloved* to develop her notion of anti-blackness as "weather," i.e., "the totality of our environments" (*In the Wake* 104; cf. Terrefe and Sharpe, par. 152). In doing so, she emphasizes the continuities of anti-blackness ever since the historical era of slavery as per Hartman and, following Wilderson's main arguments, points to the comprehensiveness with which anti-blackness defines and affects Black being. In an interview with Selamawit Terrefe, Sharpe rearticulates Wilderson's basic assumptions when she explains that "once one accepts that violence precedes and exceeds the Black, that it's not situational violence or a conflict in civil society – that that violence is the grammar that articulates 'the carceral continuum of black life' – then one has to take up the question of what it means to suffer" (Terrefe and Sharpe, par. 26). Thus, Sharpe's work addresses "the ongoingness of the conditions of capture" for Black people across the world as manifested in the twenty-first century (*In the Wake* 20). Walcott notes accordingly that "Sharpe is often read [...] among the Afro-pessimists, and it is clear she is not in opposition to such a reading. However, in some ways she parts with that important strain of radical black political thought because the idea of possibility remains central to her thinking" ("Freedom Now Suite" 152–53).

Like Moten, McKittrick, Campt, and Hartman, Sharpe indeed also looks for another perspective on Black life and its proximity to death and suggests the concept of "wake work" to account for the "intramural" care for Black people by Black people. Even though anti-blackness has created and continues to keep Black life "in the hold," i.e., "containment, regulation, punishment, captivity, [and] capture," Sharpe argues that Black life also "exceeds that hold" (Terrefe and Sharpe, par. 144, 118). "[E]ven as we recognize Blackness as ontology, as structural position," she argues, "something is in excess of that that does not mean that that something exceeds deathliness, but that one might imagine otherwise even as one sees and recognizes captivity" (Terrefe and Sharpe, par. 144, 118; see also *In the Wake* 21). Like Hartman, Sharpe emphasizes that she wants "to distinguish [...] Black being in the wake and wake work from the work of melancholia and mourning" (*In the Wake* 19). Yet, melancholia and mourning are clearly related to Sharpe's concept of the wake since it explicitly includes "*a watch or vigil beside the body of someone who has died*," but importantly also "the air currents behind a body in flight" (10, 21).

In order to capture this complex notion of "Black being in the wake," Sharpe analyses a breadth of narratives in journalistic, filmic, artistic, and photographic material of the Black diaspora that protocol and address "current quotidian disasters" (14). She positions those against a "set of work by Black artists, poets, writers, and thinkers [...] that together comprise [...] the orthography of the wake" (20–21). In close readings of those juxtapositions, Sharpe develops concepts, such as the eponymous "wake work," but also "Black redaction," the "Trans*Atlantic," "Black annotation," the "anagrammatical," and "Black aspiration," that help grasp how "literature, performance, and visual culture observe and mediate this un/survival" (30, 75, 113, 130, 14). "[A]nagrammatical blackness" (75), for instance, describes "how the meanings of words decompose when they are applied to black bodies" (Teshome and Yang 163). In doing so, Sharpe produces what Walcott describes as "a kind of black global painful wail, a voice annunciating not only its victimhood but also its creative responses for living a life with death as an intimate partner" ("Freedom Now Suite" 152), yet in very articulate, specific, and nuanced ways.

Sharpe therefore uses the terms 'fugitive,' 'fugitivity,' and 'refusal' less prominently than Hartman, Moten, and Campt, which may arise not least from the fact that Sharpe indeed follows the Afro-pessimist insights more closely as she thinks rigorously and unflinchingly through the implications of the Wildersonian ruse of analogy and the capture of 'the hold.'[57] As a recent "Editor's Forum" on "Protest and/as Care" edited by Michelle D. Commander with contributions inspired by *In the Wake* and a book discussion in the *ASAP/Journal* show (see, esp., Teshome and Yang), Sharpe's 2016 monograph proves particularly influential in how it addresses the absence or impossibility of care of and for Black people in an anti-black environment as well as in how it develops concepts for forms of care that exist in spite of anti-blackness. Sharpe tackles "'care' as a problem

---

[57] In "Black Life, Annotated," a critique of Alice Goffman's ethnography *On the Run: Fugitive Life in an American City*, Sharpe defines fugitivity as "a powerful way to imagine black life that persists in and in spite of" anti-blackness. She particularly points to "a long history of fugitivity and scholarship on fugitivity as ways of imagining black resistant life lived in captivity that seem to be unavailable to Goffman and to the majority of her readers." *On the Run* is a publically acclaimed, yet highly controversial ethnography built on participant fieldwork Goffman conducted as an undergraduate student about the impact of racialized mass incarceration and surveillance on young Black men in a mixed-income African American community in Philadelphia. Sharpe criticizes the study not only for failing to reference and acknowledge longstanding traditions of critical inquiry into incarceration and surveillance by Black scholars and intellectuals from Frederick Douglass to Angela Davis and in Black studies and African American studies more broadly. She also finds fault with the ways in which the ethnography is written from an insufficiently acknowledged white middle to upper class perspective and for a white middle to upper class readership to whom the insights may appear new and groundbreaking while they are well-known and criticized by people from the communities she writes about.

for thought" and asks "[h]ow can we think (and rethink and rethink) care laterally, in the register of the intramural, in a different relation than that of the violence of the state? In what ways do we remember the dead, those lost in the Middle Passage, those who arrived reluctantly, and those still arriving?" (*In the Wake* 5, 20). She also poses the question, "[w]hat does it look like, entail, and mean to attend to, care for, comfort, and defend, those already dead, those dying, and those living lives consigned to the possibility of always-imminent death, life lived in the presence of death[...]?" (38). Sharpe, thus, defines care "as a way to feel and to feel for and with, a way to tend to the living and the dying" that is always already a "shared risk" (139, 180). As we will see in chapter 2.3., Sharpe draws on the chain gang chapter in Morrison's *Beloved* to develop the notion of "sounding a note of ordinary care" to redirect anti-black violence, such as the rape of Black prisoners through white guards, to the violence of forced prison labor and imprisonment (Sharpe, "'And to Survive'" 173; *In the Wake* 132–33). Commander contends that "[p]aired with the principle of care, which relies not on the production of spectacular gestures, but is instead very much inclusive of the quotidian, wake work is an intentional approach to reckoning with the afterlife of slavery" ("Poetics and Care" 312). Sharpe's 'notes of care' thus indirectly connect Hartman's work on the afterlife of slavery with Campt's notion of quotidian practices of refusal in the Wildersonian anti-black environment of proliferating 'holds' that literally capture Black being.[58]

To recapitulate, conscious of the role which concepts play in analytic work and of their travels, Sharpe, Campt, and Hartman avoid not only the term 'resistance' (particularly in their more recent publications) but also the terms 'Slave' and 'Slaveness' for the status of the non-being of Blackness that Wilderson uses in *Red, White, and Black*. Instead, they describe social death as captivity in the hold of the afterlife of slavery, while social life is fugitive in and of that hold. They allow the ambiguities of fugitivity and their related concepts, such as refusal and the wake, to resonate in their work rather than establish fixed concepts that stop their analyses where social death as captivity comes to show. Ultimately, they also raise questions as to whether and how these concepts could call the clear-cut Afro-pessimist differentiation between structure and experience or performance into question. I suggest that their focus on captivity *and* fugitivity allows for a more nuanced perspective on Black literature in North America and

---

[58] Calvin Warren also draws on Sharpe's *In the Wake*, Spillers's "Mama's Baby, Papa's Maybe," and the 1998 movie based on Morrison's *Beloved* for his definition of Black care as the communal "circulation and sharing" with "endurance" as a "lateral affirmation of injury" as its "objective" ("Black Care" 44–46).

better reflects the literature's cultural, theoretical, and conceptual work as a contribution to a large discourse on Black captivity and fugitivity across genres and centuries.[59]

Admittedly, to relate Afro-pessimism concerned predominantly with the structural positionality of Blackness in the United States to the diverse work of Campt, Hartman, and Sharpe in this way may seem quite a stretch – not only across different levels of abstraction but also across varied geographies, histories, and genres. Nevertheless, when we juxtapose the 'Black border' I propose to contemplate in Afro-pessimism with the work of Hartman, Campt, and Sharpe, we may imagine fugitivity as conceptualizing social life as constant practices of refusal to accept and to remain within the structurally ostracized position of social death. Fugitivity could then be understood as a constant running up against the status of the 'non-human' that – instead of successfully crossing or overcoming the 'Black border' – still remains on the outside of civil society where social death is located. In fugitivity, Black freedom as the supposed end of social death may be expressed and experienced, for example, through photography, music, or narrative, but only as "Fugitive Dreams" as Hartman suggests (*Lose Your Mother* 211), without reaching a position from where to lay claims to civil society that has defined freedom as 'not Black/not Slave' for hundreds of years. In this way, fugitivity points to the historical legacy of physical, mental, and cultural practices of individual and communal forms of refusal of, survival despite, resistance against, and flight from anti-black violence by Black diasporic communities for centuries. Hartman describes these practices as "the ongoing struggle to escape, stand down, and defeat slavery in all of its myriad forms" in search not of integration into civil society but a 'stateless elsewhere' (234). Moreover, fugitivity as a concept enables us to both accept the structural antagonism of Afro-pessimism and reflect on the strategies and expressions of Black survival, perseverance, and sociability in an anti-black world, the latter being unaccounted for in Wilderson's Afro-pessimism and exemplarily discussed in Hartman's, Campt's, and Sharpe's work.

The capacity to flee in this notion of fugitivity appears thus reasonably different from the constrained agency of, for instance, "the subordinate" that the concept of the contact zone adopts. While Pratt's concept would deem negotiating with and self-representing against the white "senior" partners towards change possible, the fugitive practices of refusal and the 'stealing away' of the socially dead assume a more indeterminate form of Black feminist fugitivity towards futurity. Since an Afro-pessimist analysis of the structures that position

---

[59] I burrow the notion that literature does "cultural work" from Jane Tompkins.

Blackness as social death outside of civil society implies an utter lack of symbolic agency in relation to that society, fugitivity comprises the capacity to flee and struggle against the border between social death and civil life. Understood in this way, Black sociability entails the capacity to survive, live, and struggle, in Campt's words, in places "where it should not be" and by extension seems almost congruent with fugitivity in social death. This 'capacity,' however, does not necessarily entail choice or trigger structural change but warrants no more and no less than the enduring social life of the socially dead.

In an Afro-pessimist framework, true Black agency would presumably mean to bring about the end of the world, or in Sexton's words cited above the realization of "the freedom dream of a blackened world in which all might become unmoored, forging in struggle, a new people on a new earth" ("African American Studies" 223). Fugitivity thus may conceptually account for fugitive experiences and performances as Black social life only as long as the 'Black border' remains intact and still positions Blackness outside of civil society. Ultimately, both Afro-pessimism and Black feminist fugitive thought return us to a question that has been posed in different ways by Black studies[60] for hundreds of years: What does it take to dismantle the border erected between people defined as 'Human' and people condemned to 'non-humanness' and to forge a new and truly all-encompassing concept of 'the Human' beyond 'Man' in Wynter's terms? Combining Afro-pessimist with Black feminist fugitive thought enables reflection on this question, on both what has been done (cf. Gordon et al. 126) and what still needs to be done as "a practice of *living the future they want to see, now*" (Campt, *Listening to Images* 59). Thus, to pay heed to the potential realization of the "freedom dream" in the form of the end of the world while focusing on fugitive acts of refusal against social death within this world presents an important challenge of thinking fugitivity and Afro-pessimism together. Therefore, I offer this study's dual focus on captivity and fugitivity as a modest attempt to both embark on Afro-pessimist assumptions and to fathom the possibilities and gains as well as challenges of thinking through and with fugitivity and the rigid and seemingly un-crossable 'Black border.' Instead of overriding the structural antagonism that locates Blackness outside of civil society and condemns it to social death, I argue that fugitivity as a concept can protocol fugitive practices of refusal, but only if we also bear in mind the momentous balancing

---

[60] I refer to Black studies here as not only the discipline that emerged in 1960s in US universities but also the writing and thought traditions of Black Americans in North America inside and outside of academia since the early American period. See, e.g., the slave narrative tradition and Black Power autobiographies as discussed in chapters 2.1. and 2.2. (cf. Harney and Moten; Kelley).

act this fugitive thought experiment necessarily performs. I argue that the concept of fugitivity, as exemplarily discussed in Hartman, Campt, and Sharpe, together with Afro-pessimism bears the potential of regarding both social death and the enduring sociability of Blackness in the literature under examination.

1.5 A Second Note on Method: The Texts, the Tools, and Positionalities

After having developed the theoretical framework, I close this chapter on theory and method with a second note on methodology in order to return to the challenges Afro-pessimism poses with respect to knowledge production, the importance of positionalities in research, and the applicability of common literary studies terminology that will refine my concept-driven cultural analysis for the following literary analyses. In chapters 1.3. and 1.4., I developed a critical theoretical reading lens by studying Afro-pessimism and Black feminist fugitive thought and delineating them from other currently dominant literary and cultural studies approaches. Operating with the theoretical and methodological pluralism of cultural analysis, the following literary analyses will draw on these theoretical deliberations in combination with common narratological terminology, originally established by Gérard Genette and Franz Karl Stanzel, to describe and interpret the narratives in close and wide readings as well as their intertextual connections (see Fludernik and Pirlet; Nünning and Nünning, ch. 5).[61] Narratology thus offers this study tools and terms to describe, for example, the narrative transmission of consciousness, narrative perspectives, representation of time and place, as well as characterization, rather than a wholesome theoretical approach or clear-cut methodology. Together with the concept-driven theoretical approaches of Afro-pessimism and Black feminist fugitive thought discussed so far, such a cultural analysis ensures a theoretically and methodologically well-underpinned analysis of the texts' literary and conceptual contributions to the broader conceptualization of captivity and fugitivity, while also allowing for the necessary flexibility with respect to each writing tradition and individual text.

Of course, any selection of narratives that speak of captivity and fugitivity seems arbitrary due to the long and diverse history of Black American literature.

---

[61] For discussions of the promises and challenges of combining narratological tools with context- and content-based approaches, such as postcolonial, critical race, and ethnic studies, this study pursues, see, e.g., Birk and Neumann; Donahue, Ho, and Morgan; and Prince.

1.5 A Second Note on Method: The Texts, the Tools, and Positionalities — 55

So instead of aiming for comprehensiveness, this study focusses only on a small number of narratives from the 1840s–60s, the 1970s–80s, and the 2010s when Black liberation movements, such as abolitionism, the Black Power movement, and Black Lives Matter, were accompanied by autobiographical and fictional narratives of Black captivity and fugitivity. As Wilderson notes, particularly between the period from the 1800s to the Civil War, on the one hand, and the late twentieth-century movement era, on the other, a "parallelism" or "structural mimesis" can be observed (*Red, White, and Black* 302–04). According to Wilderson, the late 1960s and 70s saw a "brief Nat Turner moment,"[62] during which the Black Power movement, especially the clandestine Black Liberation Army, fought a pronounced battle against the anti-black US civil society in "the spirit of the Slave revolt," a spirit that is also reflected in some of the movies he examines (144, 124).[63] Frank Obenland, Nele Sawallisch, and Elizabeth J. West observe a similarly "precarious continuity between the Civil Rights Movement, the Black Power era," and the twenty-first century (post-)Obama and Black Lives Matter era (226, 224–29). These 'parallelisms' make the selected narratives from the three time periods particularly interesting as they repeat and revise concepts of captivity, fugitivity, and their related freedom dreams of the time.

Certainly, the late nineteenth and early twentieth century, which this study does not cover, also saw important eras of African American literature that address issues of movement and immobility, confinement and escape. Especially the Harlem Renaissance, the 1940s, and the early postwar period produced a wealth of diverse African American literature that has drawn significant attention. Ralph Ellison's "Invisible Man," for instance, narrates his story in the eponymous novel published in 1952 from a hole as a form of self-chosen confinement not unlike Harriet Jacobs (see ch. 2.1.). Richard Wright's protagonist "Bigger Thomas" moves in *Native Son* (1940) from the confining spaces of urban poverty onto death row, while Nella Larsen's *Passing* (1929) addresses racial passing as an ambiguous form of flight from Black communities into white spaces. Yet, the writing from the late nineteenth century to the Harlem Renaissance, as De-

---

[62] Turner (1800–1831) was an enslaved man who "led the only effective, sustained slave rebellion" in US history in 1831. During the rebellion about 60 white people were killed, while in its aftermath many more enslaved people were punished or killed than had been involved in the rebellion. Turner was captured and hanged (Encyclopedia Britannica editors, "Nat Turner").

[63] Arguably, some of the disenchantment and pessimism that finds scholarly expression in Afro-pessimism may issue from the fact that the Civil Rights and Black Power movements did not end in a revolution that dismantled the anti-black structures of civil society, nor did they lead to proper racial equity. For a more detailed discussion of the Black Power movement as represented in autobiographies of (formerly) imprisoned activists, see ch. 2.2.

borah E. McDowell observes, tends to be more explicitly future-oriented with recurring motifs of shame and silence about the history of slavery ("Telling Slavery" 151–52, 165). Tim Armstrong speaks of a "displaced awareness of the legacy of slavery" (and by implication of forms of captivity and fugitivity) in "a wider range of fiction written in the period between 1890 and 1950, in which black subjects find that the 'social death' [...] of slavery applies to their own narrative: they are already dismissed; already guilty and condemned; often already executed or dead" (207). Armstrong further argues that while in novels, such as Richard Wright's *Native Son*, Chester Himes's *If He Hollers Let Him Go* (1945), and Ann Petry's *The Street* (1946), "[f]reedom turns to flight; resistance to punishment; unconsciousness to the consciousness of the oppressed" (207), only the fictional neo-slave narrative in the 1970s "[inaugurated] a new era" (215). Literature in the post-Civil Rights era "self-conscious[ly]" investigates slavery "in historiographically complex and often quite specific terms" that make the texts of the first half of the twentieth century "seem historically muted" with respect to the fugitivity and captivity of slavery and its afterlives (215).[64]

---

64 Apart from the protest novel *Native Son*, Richard Wright's (1908–1960) memoir/fictional autobiography *Black Boy* (1945) also lends itself to a broader analysis with a focus on forms of confinement and the (im)possibility of flight (cf. Sid. Smith, ch. 3). The same seems true of Ellison's (1914–1994) novel *Invisible Man*. Dana Williams analyzes Wright's protagonist Bigger Thomas as "the prototype for the contemporary confined character-in-process" (34), while she describes the protagonist of *Invisible Man* as being in voluntary self-confinement "underground" that enables him "to write his story" (40). The invisible man's self-confinement in a basement clearly recalls Harriet Jacobs's self-confinement in the attic space in *Incidents in the Life of a Slave Girl* (1865). Following Hartman's problematization of the concept of choice with respect to the (formerly) enslaved, however, I question the notion of self-confinement as a choice in my discussion of Jacobs's slave narrative in ch. 2.1. *The Street* about a working-class Black woman in New York City by Petry (1908–1997) became "one of the first novels by an African-American woman to receive widespread acclaim" (Encyclopedia Britannica editors, "Ann Petry"). Larsen (1891–1964, of white Danish and Afro-Caribbean descent) published her two novels *Quicksand* and *Passing* on mixed-race female protagonists that raise interesting questions about the refusal to identify with a racialized community and racial passing as forms of flight from anti-blackness in 1928 and 1929. For an analysis of racial passing in twentieth-century American literature, see Wald. African American writer Himes (1909–1984) began writing when he spent eight years in prison for armed robbery. Apart from his first, autobiographical novel *If He Hollers*, his two autobiographies *The Quality of Hurt* (1971) and *My Life of Absurdity* (1972) also deserve mentioning here. For analyses of Himes's crime fiction, see Drake and H. B. Franklin, *Victim as Criminal* 206–32. An analysis of conceptualizations of confinement and flight in the fictional and autobiographical writing of the African American writers Wright, Himes, and Ellison seems particularly promising in conjunction with the communist autobiography of the formerly imprisoned African American Angelo Herndon entitled *Let Me Live* (1938). For a discussion of Dennis

1.5 A Second Note on Method: The Texts, the Tools, and Positionalities — 57

In his article "'Neo-slave Narratives' and Literacies of Maroonage," Greg Thomas agrees with the significance of the neo-slave novel, but proposes "to examine what happens when slavery or neoslavery is freed from the past and narrative is liberated from the narrow literary confines of [...] the historically bourgeois form of the commercial novel in the modern West" (202). He argues against "separat[ing] fictional novels oriented more exclusively toward the past from fictional or nonfictional texts of resistance oriented more completely toward the past, present, and future of black liberation struggles" (230). Criticizing the predominant definition of neo-slave narratives (201), Thomas discusses Toni Morrison's novel *Beloved* (1987) in relation to the autobiographical writing of the (formerly) incarcerated Black Power activists George Jackson and Assata Shakur (see chapters 2.3. and 2.4. of this study). As Thomas argues,

> for critics to quarantine this militant literature away from Morrison's fiction for reasons of genre or politics is to enforce a selective amnesia or revisionism, not to mention a provincial class-political notion of "literature" – endorsing a normative "evolution" of black literature under the white West from 'antebellum slave narratives' to prescribed "novel-writing." (212)

This study follows Thomas's call by including in its corpus, first, the autobiographical writing of fugitive enslaved people, such as Harriet Jacobs, Josiah Henson, and Frederick Douglass, from the nineteenth century (ch. 2.1. and 2.3.); second, the autobiographical writing of (formerly) imprisoned and/or fugitive Black Power activists, namely Angela Davis, Assata Shakur, and George Jackson of the 1970s and 80s (ch. 2.2.); and third, fictional writing about slavery and its afterlives from the mid-1970s into the new century, such as Ishmael Reed's *Flight to Canada*, Toni Morrison's *Beloved*, Lawrence Hill's *Someone Knows My Name*, and Teju Cole's *Open City* (ch. 2.3.). My corpus selection further follows Thomas's appeal for an expansion of the definition of marronage/maroonage from isolated communities of fugitive enslaved people in rough terrain of the Caribbean and South America to less obvious forms of communal and individual flight in the Americas during and after the historical time of slavery (215–16).[65] Therefore,

---

Childs's analysis of Herndon's autobiography as a predecessor of the Black Power autobiographies, see ch. 2.2.
65 For a definition of marronage/maroonage, see ch. 1.4. Greg Thomas, like Neil Roberts, suggests subcategories to the already existing notions of *petit* and *grand marronage* (G. Thomas, "'Neo-Slave Narratives'" 218; cf. N. Roberts 10, 98–99). Thomas lists active and passive maroonage; urban, marine, aerial, and transnational maroonage; erotic, religio-spiritual, and supernatural maroonage and revolutionary maroonage (219–23), whereas Roberts coins the notions of "*sovereign marronage* and *sociogenic marronage* [in order] to supply a resource for describing

the analyses address not only physical or literal captivity and flight in the narratives but also more elusive forms of confinement and escape that may accompany, precede, or follow from the literal experience. By juxtaposing a range of fictional and non-fictional texts about explicit and less obvious forms of captivity and fugitivity, this study opposes the dominant dualism of fight or flight (cf. 205) and accounts for the gray area of fight in/as flight. In order to gain a broad "literacy of maroonage" (212), the following chapters read "black revolutionary literature" of the 1970s and 80s which is "routinely eliminated from literary history, or the literary record, literary criticism and the very conception of 'literature' in a fashion that [...] short-circuit 'memory' and 'imagination' and severely limit the interpretation" alongside more established literary writing traditions (213).[66] This juxtaposition allows for the tracing in in-depth analyses of an intertextual web of motifs and tropes of captivity and fugitivity that travel (Bal) through time and different writing traditions of Black American literature, "signifyin(g)" on each other (Gates, "The Blackness of Blackness"; *The Signifying Monkey*).

By focusing on slave narratives, Black Power autobiographies, and fictional neo-slave narratives, the study methodologically combines the study of fiction with that of autobiography. Novels and autobiographical texts share many characteristics stemming from their common narrativity, such as plot, dialogue, characterization, and setting, but differ with respect to their "claims about a referential world" (Smith and Watson 10).[67] The novels discussed in this study are mostly committed to a realist and truthful account of slavery and its afterlives and explicitly draw on historical records or autobiographical writing for their narration.[68] The autobiographical texts, however, claim not only a truthful but

---

the activity of flight carried out by lawgivers, or sovereign political leaders, and agents of mass revolution" (10). For a brief discussion of the pitfalls of using fugitivity (or marronage/maroonage) as a metaphor for people without the actual experience of flight and the dangers of romanticizing flight as a form of freedom, see Gordon, "Some Thoughts."

66 This study focusses on autobiographies and novels, but short fiction represents, of course, another field of literature that would lend itself to this analytic approach. For a discussion of "enclosure" and "fugitivity" in African American poetry, such as Harryette Mullen's, see, e.g., Tremblay-McGaw. For analyses of fugitivity in African American drama, see McCormick, *Staging Black Fugitivity*; "Fugitivity and Neo-Slave Performance."

67 I define *narrativity* with Ansgar and Vera Nünning as "those characteristics that distinguish the narrative text from other genres," i.e. poetry and drama (101). These characteristics are usually located on the "level of the content (the story) or the level of the narration (discourse)," such as plot, narrative transmission, or experientiality (Nünning and Nünning 103).

68 Toni Morrison used a newspaper clipping about the fugitive Margret Garner's infanticide as inspiration for *Beloved*. Ishmael Reed's *Flight to Canada* uses, among others, Josiah Henson's slave narrative as foil. Lawrence Hill consulted slave narratives and other autobiographical writing of the eighteenth and nineteenth century for *Someone Knows My Name*, and Teju Cole's *Open*

1.5 A Second Note on Method: The Texts, the Tools, and Positionalities — 59

factual relation between "*the* world" and the narrator (10) as they are defined by the assumed joined identity of the historical author, narrator, and protagonist (what Philippe Lejeune has famously described as "The Autobiographical Pact"), while still relying on processes of narration and fictionalization to transform the author's memories of their lives into a plotted, stylistically embellished narrative (Smith and Watson 11; Schwalm, par. 18). To account for the thematic, stylistic, formal, and historical specificities of each writing tradition adequately, the three analyses in part 2 are dedicated to only one group of texts at a time. Their literary histories, definitions, and the state of the research on those traditions will be discussed in each chapter introduction in more detail. Yet, the analyses in part 2 are less interested in the ways in which the texts under scrutiny fit into and abide by clearly distinguished genre traditions. Rather, they focus on the ways in which the tropes of captivity and fugitivity as well as specific narrative forms and styles connected to these tropes stay the same or change, while travelling in Bal's sense from text to text, between writing traditions, and across time periods. In fact, as we will see, the texts always already destabilize any clear divisions between the writing traditions and fundamentally question concepts of progressive movement across linear time and through clear-cut geographies of confinement and freedom.[69]

Thus, in the following chapters, I approach both autobiographical writing traditions, such as slave narratives and political autobiographies of the 1970s and 80s, and neo-slave novels of the late twentieth and early twenty-first centuries primarily as narrative texts based on a common narrativity and therefore with the same methodological toolbox of literary analysis mentioned above. Having said this, however, autobiographical narratives present specific narrative situations that emerge more directly than fiction from the historical author's experience and a particular "site of storytelling" (Smith and Watson 70), such as the site of the prison or places of exile, that need to be acknowledged. After all, when formerly enslaved writers, such as Jacobs and Douglass, still fear re-capture while writing their narratives after having fled from enslavement into the so-called free North, this influences the narrative production and meaning mak-

---

City exhibits a very broad referential frame made of, among others, Nigerian, US American, and European history, literature, art, culture, and theory. All the texts have a realist foundation, but some include fantastic or speculative elements. See ch. 2.3.

69 In addition to Wynter's concept of the "genre of Man" (see ch. 1.3.), I use the term *genre* (or *writing tradition*) also as an "organizational category for literary texts" that "sort[s] texts into groups based on common sets of characteristics" (Nünning and Nünning 189). Genre descriptions frequently used in the study are, for example, the novel, autobiography, slave narrative, and neo-slave narrative.

ing process in significant ways (see ch. 2.1.). The same seems true for fugitive Black Power autobiographers, such as Assata Shakur, who is still in hiding in Cuba today, a fact that necessarily impinges on any interpretation of her autobiography's concluding claims to freedoms gained (see ch. 2.2.). To better differentiate between author, narrator, and protagonist in autobiographical texts I use the last name to refer to the historical author of the text (or to the historical person who told their story to an amanuensis) and the first name to refer to the autobiographical subject who narrates a story (the narrating I) about themselves (the narrated I; Smith and Watson 71–76).[70]

Moreover, the Afro-pessimist and Black feminist theoretical framework established in chapters 1.3. and 1.4. also poses challenges for the use of terminology of narrative theory and the study of autobiographies. As we have seen in chapter 1.3., Afro-pessimism exposes the assumptive logics of predominant cultural and literary studies, such as Border studies, as not being able to account for Blackness positioned outside of the realm of 'the Human.' Together with Black feminist fugitive thought, it questions the applicability and meaningfulness of much seemingly universal terminology when used for the (formerly) enslaved and their descendants, such as 'choice,' 'resistance,' 'agency,' 'woman,' 'mother,' 'rape,' 'care,' and 'African diaspora' (see, e.g., Hartman, *Scenes of Subjection* 51–65, 79–101; Sharpe, *In the Wake* 78–81; Wilderson, "Grammar and Ghosts" 124).[71] Thus, Afro-pessimism's basic insights also question the extent to which well-established narratological concepts enable literary scholars to discuss the presence or absence, role, and influence of Blackness in/on narratives. Most prominently, Wilderson has shown how the typical plot progression of narratives from equilibrium via disequilibrium to equilibrium restored does not account for Blackness since it has neither a prior "equilibrium" to start with nor can equilibrium be "restored" in the end (Wilderson, "Prison Slave" 19; *Red, White, and Black* 26; "Black Liberation Army" 178). Instead, Blackness exists in a constant "state of emergency" (*Red, White, and Black* 6–7; see also Hart-

---

[70] Of course, the terms as suggested by Smith and Watson are only approximating the complex narrative situations of autobiographical texts and the lines between the terms often remain more blurry than the terminology suggests (Hames-García 102). See Smith and Watson for an in-depth introduction to the study of autobiography.

[71] Various Afro-pessimist and Black feminist scholars have also shown the limits of white feminist, Marxist, and psychoanalytic approaches when they address Blackness or fail to differentiate between addressing people positioned as non-Black and Black (see, e.g., Broeck, *Gender*; Douglass and Wilderson; Wilderson, "Gramsci's Black Marx"; Terrefe).

man, "Venus in Two Acts" 13; Sharpe, *In the Wake* 100).[72] From an Afro-pessimist perspective, thus, Blackness proves 'unplottable' because, as Hartman notes with respect to her historical study *Scenes of Subjection* that Wilderson then builds on in *Red, White, and Black*, "every attempt to emplot the slave in a narrative ultimately resulted in his or her obliteration" (Wilderson and Hartman 184; Wilderson, *Red, White, and Black* 26).

Since an elaborate analytic narratological toolkit that acknowledges the basic Afro-pessimist insights has not been developed yet,[73] this study relies on the well-established literary studies tools while also drawing on the concepts of Black feminist fugitive thought in order to focus specifically on instances where the 'unplottability' or impossible narratability of Blackness comes to show.[74] In combining Bal's cultural analysis with a narratological toolbox as a common language that ensures comprehensibility and a theoretical lens indebted to Afro-pessimism and Black feminist fugitive thought, I thus strive for an analytic practice that looks for the assumptive logic of anti-blackness and flight in texts by way of their stylistic makeup on a micro- and macro level. I pursue a reading against the grain of the narratives that accounts for absences, contradictions, and hints both on the thematic and stylistic level of the narratives that may counter the typical narrative drive for transformation, development, and closure. The analytic practices of scholars discussed above will serve as blueprints. In his analysis of the movie *Bush Mama* (1976), Jared Sexton, for instance, argues that the film "indexes" the "ontological condition of gratuitous violence exterior to the interlarded rationales of the colonial enterprise (including its systems of patriarchy and class warfare)" even though the "diegesis cannot sustain" that awareness ("People-of-Color-Blindness" 46).[75] Thus, Sexton identifies the

---

[72] Nevertheless, narratives by and about Black people frequently use this plot structure. See, e.g., Hill, *Someone*.

[73] Such a toolkit could provide alternative narratological terminology, for example, for the 'Black (non-)being' in narrative beyond the terms 'character'/ 'narrator' or for the narrative construction of Black time and place on the basis of the work discussed in chapter 1.3. and 1.4. (see, e.g., Sharpe, *In the Wake*; C. Warren, "Black Time"; McKittrick, "Commentary").

[74] *Merriam-Webster* defines the term *narratable* as "capable of being narrated" ("Narratable"). For the purpose of this study, the term *narratability* thus refers to the ability or possibility to create a narrative. It differs from the concepts of *tellability* or *reportability* which are sometimes interchangeably used in narrative theory together with narratability to refer "to features that make a story worth telling, its 'noteworthiness'" (Baroni par. 1).

[75] Sexton specifically shows how "the atrocities" the main character Dorothy suffers are not fully "comprehensible by way of the analogical gestures of anticolonialism that animate the freedom dreams of the prison letters between Dorothy and her imprisoned lover, Ben" ("People-of-Color-Blindness" 46).

ontology of Blackness in his primary material while acknowledging that the text labors against that awareness. The filmic text is not capable to reveal the antagonism completely; instead, it only hints at it. The same proves true for the texts under scrutiny in this study as they account for the role of anti-blackness and the Black border in the making of 'the Human' as well as the (im)possibilities of escape in indirect and latent ways.

Of course, analyzing Black American narratives of captivity and fugitivity with the help of an Afro-pessimist and Black feminist theoretical framework from the position of whiteness in German American studies bears further challenges. My academic white anti-racist engagement with the Black border and fugitivity that speaks to African American studies and Black (Diaspora) studies discourses unavoidably falls prey to the pitfalls of re-centering whiteness as life and reinforcing the anti-black border structure it wants to critique. As Terrefe explains, Afro-pessimism appears particularly appealing to young white (male) scholars as "an invigorating theory because it's a purely intellectual enterprise for them," while Black scholars like herself "cannot *be* Afropessimists since the idea and reality of being is foreclosed" to them (Terrefe and Sharpe, par. 81, emphasis mine). Moreover, this endeavor is further complicated by

> a longstanding ethnographic gaze in German [American studies] scholarship on race where the structure and history of white supremacy [and anti-blackness] is addressed primarily through indirection, i.e.[,] in analyses of cultural forms of expression that are deemed 'ethnographically' representative of non-white life forms. An underlying assumption of such studies often seems to be that literature, film, art, and theory of and about African Americans, Native Americans, Asian Americans, Latinx Americans, and other minoritized communities in the United States can only speak for and of themselves, while white cultural production is understood as not racialized and therefore capable of speaking universal truths about the human condition. (Essi et al. 11)

As I have argued in chapter 1.3., the premise of this study is that Black knowledges produced from the outside of civil society is essential not only to understanding anti-blackness, but the modern world built on anti-blackness, and this understanding is the prerequisite for possibly moving towards its dismantling. As Sara Ahmed argues,

> the task for white subjects would be to stay implicated in what they critique, but in turning towards their role and responsibility in these histories of racism, as histories of this present, to turn away from themselves, and towards others. This 'double turn' is not sufficient, but it clears some ground, upon which the work of exposing racism might provide the conditions for another kind of work. We don't know, as yet, what such conditions might be, or whether we are even up to the task of recognizing them. ("Declaration of Whiteness," par. 59)

1.5 A Second Note on Method: The Texts, the Tools, and Positionalities — 63

This study is committed to this challenge with "critical vigilance" (Applebaum 3–4, 18; cf. Yancy), offering analyses of autobiographical and fictional narratives that center on the different roles Blackness and anti-blackness play in the formation and persistence of the prevalent conception of the non-Black 'Human' as well as the texts' contributions to 'dreaming' alternatives. What follows, thus, is an investigation of how Black American narratives of flight and captivity have tried – similar to the theoretical contributions discussed so far – to critique anti-blackness and move toward an all-encompassing notion of 'the Human' beyond the dominant genre of the white, male 'Man' in different forms, styles, and genres for centuries.

Part 2 Practices of Flight: Captivity and Fugitivity in Black American Literature

This part makes up the main body of this study with three chapters on captivity and fugitivity in slave narratives (ch. 2.1.), Black Power prison and flight autobiographies (ch. 2.2.), and neo-slave narratives (ch. 2.3.). An overview of the state of research on the writing tradition introduces each chapter after which close readings of two, three, and four narratives follow respectively. Chapter 2.1. discusses slave narratives as founding texts of Black American captivity and fugitivity narratives that function as central intertexts for the following two writing traditions and therefore creates an important basis for the in-depth discussions in chapters 2.2. and 2.3. by identifying ambiguous forms of confinement and refuge, concepts of time and place, the role of community, and narrative strategies of evasion and reiteration as central focal points for the following analyses.

2.1 Fugitive Forbears

> What do reciprocity, mutuality, and the recognition of the captive's humanity mean in the context of slavery? In other words, who is protected by such notions – the master or the slave?
>
> Saidiya Hartman, *Scenes of Subjection* 53.

This chapter presents a discussion of two of the most well-known slave narratives that are taught widely in schools in North America and elsewhere (Fisch, "Introduction" 1) and have been well-known and well-studied since the 1980s and 90s as major reference points to understanding chattel slavery and its afterlives. Harriet Jacobs's *Incidents in the Life of a Slave Girl* (1861) is probably the most renowned female-authored slave narrative today and as such a primal text of this narrative tradition from the perspective of enslaved women and mothers. Jacobs's slave narrative recounts her life as "a slave girl" and fugitive woman who escaped slavery by hiding in an attic space on her legal owner's plantation for seven years before fleeing to the northern states. Frederick Douglass's *Narrative of the Life of Frederick Douglass, an American Slave, Written by Himself* was first published in 1845 and is today probably the most recognized slave narrative, even though by the late 1970s it was still considered "a virtual nonentity" outside of African American studies (Franklin, *Victim as Criminal* 6). It shares Douglass's experiences as a boy who inherited legal enslavement from his enslaved mother in Maryland and who escaped slavery as a young man in 1838 to become a famous orator, editor, and writer of three autobiographies and a novella. The autodiegetic narrator of the *Narrative* goes through major transformations in its eleven chapters as he gains increasing awareness of his enslavement and develops a desire for freedom, first by witnessing gendered anti-black violence (the beating of his aunt Hester), second by mentally escaping slavery through literacy, third through a physical fight with the plantation owner Edward Covey, and ultimately through physical flight. The two following autobiographies, *My Bondage and My Freedom* (1855) and *Life and Times of Frederick Douglass* (1881, revised 1892), not only provide revised versions of the original *Narrative* which became "more sentimental, more tear-inducing than its predecessor" (St. Smith 192), but also expand Douglass's story to cover his life after his escape.

In "Beyond Douglass and Jacobs," John Ernest criticizes slave narrative research and teaching that understands Jacobs and Douglass as "exceptional-representative" female and male slave narrative writers (219) whose narratives are read primarily for their 'authentic' representation of slavery at the cost of an in-depth study of their form. He also bemoans that the sole study of Douglass's and Jacobs's writing often stands in for a large and much more diverse body of texts.[76] The following analysis will combine a symptomatic thematic inquiry with a formal analysis of Jacobs's and Douglass's writing which are chosen specifically because of their assumed "exceptional-representative" status. In doing so, this

---

76 Apart from the two prominent nineteenth-century slave narratives, set in what is today the United States discussed here, the slave narrative by Josiah Henson (probably the most well-known fugitive in Canada) will be discussed in chapter 2.3. Other famous slave narratives of the nineteenth century in North America and the Caribbean are *History of Mary Prince, A West Indian Slave* (1831), *Narrative of William Wells Brown, an American Slave* (1849), *Narrative of the Life and Adventures of Henry Bibb, An American Slave, Written by Himself* (1849–50), and *Twelve Years a Slave: Narrative of Solomon Northup, a Citizen of New-York, Kidnapped in Washington City in 1841...* (1853) (see M. Prince, Osofsky). Of course, autobiographical narratives about experiences of enslavement were also published elsewhere and during other times. The *Interesting Narrative of the Life of Olaudah Equiano* (1798) represents an early contribution to the genre. For an overview of the diversity of the genre and its research, see Ernest "Introduction." For discussions of slave narratives and slave testimony from the Caribbean, South America, and Canada, see Aljoe; and Siemerling, "Slave Narratives."

Even today, displaced, trafficked, or captured persons forced into unpaid labor use the tradition of slave narratives in their (co-written) memoirs (V. Smith, "Neo-Slave Narratives" 183–84), such as Francis Bok and Edward Tivnan's *Escape from Slavery: The True Story of My Ten Years in Captivity – and My Journey to Freedom in America* and *Slave* by Mende Nazer and Damien Lewis. See also Murphy, *Survivors of Slavery*. For discussions of the slave narrative as a formal "template" and framework for contemporary narratives about human rights abuses around the world and as "the site for the reinvention of form" in Black diasporic world literature, see Goyal, *Runaway Genres*; Li 30–55; Murphy, *New Slave Narrative*.

chapter lays the foundation for the analyses of the Black Power autobiographies and neo-slave narratives in the following chapters. As Sidonie Smith has observed, "the earliest black American autobiographies, the slave narratives, established certain prototypical patterns, both thematic and structural, that recur again and again in subsequent black autobiographies" (ix) and, as I would add, other Black American narratives. These patterns are, among others, plots that emerge from "public-political impulses" to "develop an analysis of America's racism," involve "masking" as a risky strategy for survival and a drive "to *break away* from" an enslavist, confining environment and to "break into" a free, unconfined community (10, 15, ix). As we will see in chapters 2.2. and 2.3., slave narratives not only form important intertexts of the autobiographical writing of African American political prisoners and fugitives of the 1970s and 80s. They are also the autobiographical forerunners of the fictional neo-slave narratives of the late twentieth and early twenty-first centuries. In-depth analyses of these later narratives' literarization of fugitivity and captivity clearly require a prior understanding of the central narrative patterns of the most widely read slave narratives in order to identify continuities and discontinuities in their literary processing and fashioning of flight.

After a brief overview of the slave narrative as a literary form and the state of slave narrative studies, the chapter will explore the conceptualization of confinement, flight, and freedom in the two narratives and their critical scholarly reception. As we will see, both Douglass's and Jacobs's narratives strive in different ways for an unconfined form of freedom that promises more than a legally free status, which, however, remains out of reach even in the so-called free northern states. As I have elaborated in chapter 1.4., I refrain from directly connecting flight with freedom and instead question the relation between the liminal performance of flight to freedom and unfreedom. If a liminal space exists on the Black border between life inside civil society and social death outside, I propose to understand this space as performances or practices of fugitivity that cannot completely rid themselves from social death. Fugitivity, as developed in chapter 1.4., addresses the ways in which flight in these narratives functions as a refusal of slavery and anti-blackness that does not necessarily involve agency and neither result in unconfined freedom but that remains undergirded by anti-blackness and social death. Thus, I argue that these slave narratives show through formal characteristics that flight and emancipation in nineteenth-century North America do not equal freedom for Black people. While their escape plots remove the protagonists geographically from enslavement, the narratives show on the levels of form and content that the formerly enslaved still exist in the historical, ontic, and epistemic time of slavery.

As the oldest African American genre constitutive to African American and in fact North American literary history (Gates and Davis xxii; Olney 65), slave narratives[77] tell of personal experiences of slavery and flight from the perspective of the (formerly) enslaved. They are an early autobiographical form of "protest literature" (Sid. Smith 8) by formerly enslaved Africans and their descendants who fled slavery before becoming writers and orators for the movement to abolish the transatlantic slave trade and slavery, many of which were also active participants in the Underground Railroad.[78] Most narratives are set in the Americas, especially in the United States, the Caribbean, and Canada as well as the British Isles, in the eighteenth and nineteenth centuries.[79] The first-person accounts were either written by the fugitives themselves or told to abolitionist amanuenses and editors who published the narratives mostly in the United States and Great Britain to gather support for the abolitionist movement (cf. Gould). "Flight is an inevitable element of slave narrative," Kerry Sinanan reminds us, "since the very existence of the form depends on the successful escape of the narrator" (71). It ensures a suspenseful plot, with disguise and concealment as central strategies for survival and 'the North Star' frequently acting as a symbol for the desire for

---

[77] I acknowledge the debate over the use of the term 'slave narrative' and particularly the use of the term 'slave' which I avoid as a description of people and characters whenever possible as it levels out any other personal defining characteristics and reinforces the enslavist anti-humanization of Black people. Alternative terms for slave narratives and subcategories are, for instance, "freedom narratives" (Lovejoy, "Freedom Narratives") or "emancipatory narratives" (Mitchell). Yet, since I follow Afro-pessimism and Black feminist fugitive theory in questioning whether what formerly enslaved and imprisoned Black people gain by refusing, enduring, and fleeing from physical enslavement can be described as freedom, I refrain from using the alternative terms.

[78] The Underground Railroad was a network of formerly enslaved and legally free Black people as well as white people in enslaving southern and non-enslaving northern states and Canada that helped enslaved people to flee. "Both a material and psychic map, the Underground Railroad contained and signified secret knowledge and secret knowledge sharing. [...] The Underground Railroad was an emancipatory lifeline if untold/unwritten, and site of violence/death if told/written. A covert operation, which was developed through human networks rather than scientific/cartographic writings, the Underground Railroad illustrates how historical black geographies are developed alongside clandestine geographic-knowledge practices. These practices signaled that spaces of Black liberation were invisibly mapped across the United States and Canada and that this invisibility is, in fact, a real and meaningful geography. The life and death of black subjects was dependent on the unmapped knowledges, while the routes gave fugitives, Frederick Douglass wrote [in the *Narrative*], 'invisible agency'" (McKittrick, *Demonic Grounds* 18).

[79] For a discussion of the challenges and the interpretative relevance of locating slave narratives in diverse localities and literary histories (e.g., with respect to their place of production, the narrator's/writer's origin and/or racialized and gendered identity, the places where they experienced slavery, travelled through, and settled in), see R. R. Thomas.

freedom in the northern states or Canada.⁸⁰ At the same time, the slave narrative arguably always already has the colonial captivity narrative in the background (Tawil, "Introduction" 3; cf. Pierce).⁸¹

The production context and the purpose (i.e., the support of the abolitionist movement through illustrations of slavery) created a complex web of communication and mediation between fugitive, amanuensis, editor, sponsor, and target audience (V. Smith, "Form and Ideology" 222–23). Instead of focusing on the literary representation of individual experiences, slave narratives were to communicate relatable, authentic first-hand accounts of slavery that propagated for abolition (Sid. Smith 8). In order to appeal to the target audience, the content and style of slave narratives catered to the reading habits, aesthetic expectations, and cultural norms of the white liberal population of the northern states and the United Kingdom at the time. Prefaces and introductory letters by white abolitionists, such as those preceding Douglass's *Narrative* by William Lloyd Garrison (1805–1879) and Wendell Phillips (1811–1884), were to attest to the trustworthiness and authenticity of the narratives and integrated them into white abolitionist discourse (cf. Gould). Gaps and silences in slave narratives illustrate the immediacy of the writing and publication and the ways in which their authors and/or narrators were quite literally writing for their lives and those of their communities at the time. Information about the practicalities that led to successful es-

---

80 Of course, Black flight from the United States into other national, legally free territory was more complex through the ages. For a certain period of time "enslaved Black people in Upper Canada fled to free regions in the United States, including the former Northwest Territory (which included parts of what is now Michigan and Ohio), Vermont, and New York – states that banned slavery in 1777 and 1799, respectively" (Henry, "Black Enslavement"). Moreover, Mexico as a destination for enslaved people fleeing slavery around that time is entirely absent from slave narratives, even though scarce historical documents prove that fugitives who resided close to the southern border fled in large numbers to Mexico as the nearest place of refuge. The absence of a substantial historical archive of this fugitive legacy issues from a lack of literacy and an interested publishing industry and readership in Mexico (Adams 61–62; 73). The African American writer Gayl Jones (1949–) addresses this mostly unknown history of Mexican-Black American hemispheric relations and migrations in her novel *Mosquito* (1999). For an analysis of the novel's inter-American perspective on the legacy of Black people in Mexico and Central America, see Adams (91–96). For a discussion of borders and their dissolving in the novel, see MacDonald.

81 The colonial captivity narrative was a popular first-person account of white colonialists who were held captive in Native American communities in seventeenth-, eighteenth-, and nineteenth-century North America. The narratives (also known as Puritan or American captivity narratives) served to legitimize the colonization of Native land (Pierce 84–85). For a discussion of the relation between the captivity narrative and the slave narrative tradition as well as the spiritual autobiography, see Pierce.

cape as well as names were often kept secret or changed in order to protect the writers and their family and friends from recapture and to allow further fugitives to take similar means or routes of escape. They also formed their narratives in particular ways in order to comply with the Victorian standards of decency, respect, and the dominant conceptualizations of masculinity and femininity. Fittingly, Sinanan describes Jacobs relating "the sexual realities of her life as a female slave subject to her master's desires" as "walk[ing] a rhetorical tightrope" (77).

Research on slave narratives first gathered momentum in the 1970s when publishing houses made the texts newly available and historians started to (re-)discover the autobiographical writing by formerly enslaved people as historical sources in response to the Black Power movement's reconceptualization of Blackness as empowering and defiant (Rushdy, *Neo-Slave Narratives* 26, 39–40). According to Elisabeth A. Beaulieu, historical studies, such as John Blassingame's *The Slave Community: Plantation Life in the Antebellum South* (1972), "revolutionized the writing of history by reclaiming slave narratives, both written and oral, as primary source material" (5, see also St. Smith 189–90). These sources counteracted dominant racist historical narratives about slavery and emancipation that focused on the contributions of white people to the fight against slavery as benevolent acts and characterized enslaved people as docile and comfortably enslaved as well as passive receivers of emancipation (Rushdy, *Neo-Slave Narratives* 42–43). African American studies scholars, Black feminists, and Black Power activists were at the forefront of re-discovering slave narratives as founding narratives of a distinctly African American history, culture, and literature (cf. 62–63). Today slave narratives are appreciated not only as central historical documents testifying to the history of slavery and its abolition from the often-ignored perspectives of the formerly enslaved (Bruce). The study of slave narratives as a literary genre and an essential part of literary history, especially of the United States, also represents a central line of inquiry today. Slave narratives' contributions to political and social theory and philosophy were increasingly taken note of since the 1990s (N. Roberts 56–84; Hames-García, ch. 3). Scholars across the humanities continue to draw on slave narratives till this day, not least because as the beginning of African American writing about slavery, captivity, and flight, slave narratives serve as central intertexts for African American and Black diasporic cultural production and thought through the ages.

Due to the production context discussed above, many slave narratives share similar plot lines and formulaic elements. They typically start by stating or approximating the time and place of the narrator's birth and by identifying in so far as was possible the familial and kinship relations they were born into in order to enforce an understanding of themselves as humans with narrative agency and then go on to describe the enslaved person's path to freedom.[82] Of course, this was no easy undertaking since most enslaved people in the Americas were deprived of knowledge about their date of birth and their familial relations as families were systematically ripped apart soon after the children had legally inherited the status of enslavement from their mothers with birth (*partus sequitur ventrem*). Usually, the information provided at the outset of slave narratives is incomplete. Douglass's *Narrative*, for example, starts with the famous 'I was born'-formula (Olney) and with an exposition of his family situation at the large plantation complex in Maryland. Yet, he can only estimate his birthday and provide little information about his enslaved mother Harriet Bailey and his white father, who was probably his legal owner Captain Antony (*Narrative* 12–13).

Apart from the 'I was born'-formula that a majority of slave narratives uses, James Olney points to the centrality of gaining literacy in many slave narratives as the "literacy-identity-freedom" paradigm, with Douglass's *Narrative* as the prime example (Olney 65). For enslaved people to learn to read and write was largely forbidden. It does not surprise then, that – as Olney observes –, most slave narratives consider literacy as a prerequisite to acquire a consciousness about their enslavement, an identity, and ultimately freedom (65). Frederick, for instance, manages to learn to read and write first with the help of Sophia Auld, the wife of his legal owner Hugh Auld, and then by studying alone and with the help from poor white kids on the streets of Baltimore (*Narrative* 29, 31–31). His legal owner's aggravation about his wife teaching Frederick the alphabet shows him the role of "consciousness as the means to freedom, [and] the written language as a means to increase consciousness" (Franklin, *Victim as Criminal* 14). However, as Frederick still lacks an understanding of how to become free, this consciousness also becomes "his greatest torment" (15). "To

---

[82] While the growing body of slave narrative literary studies research has provided more nuanced analyses of the slave narrative as a literary form than Olney's formula, John C. Havard summarizes a typical slave narrative plot in similar terms: "The narrator often begins with sketchy reminiscences of childhood, then describing the initial exposure to whippings, rapes, and slavery's other injustices, an exposure that spurs the desire to attain freedom. The narrator then relates one or more failed escape attempts, building narrative tension by revealing the challenges the quest for freedom involved. The tale finally culminates with the long-desired successful escape and the beginning of a new life in freedom. In addition to exhibiting similar structures, slave narratives tend to focus on the same injustices, with the breakup of families and the inconsistencies of religious slaveholders taking a central role alongside the whippings and rapes. The narratives that were presented as being written rather than dictated by the slave narrator also usually contain elaborate accounts of how the author learned to read and write" (86).

'steal' learning" clearly threatens the system of slavery (Beaulieu 8). Gaining literacy, however, proves to be only a partial form of escape as long as Frederick remains physically and legally enslaved. Pursuing literacy in slave narratives thus represents intellectual, mental attempts at escaping the enslavist system that remain ultimately limited.

But as we will see in the following discussion of Jacobs's narrative, not all slave narratives emphasize literacy and a solitary fight for freedom as central to escaping slavery. The majority of known slave narrative writers or narrators were formerly enslaved men who tended to deemphasize familial relations (Beaulieu 8) and represent "life in slavery and the escape as essentially solitary journeys" towards freedom as full "manhood" (V. Smith, "Form and Ideology" 229). However, some slave narratives, especially those that narrate flight from a couple's or a woman's perspective, focus on a more familial or communal approach to gaining freedom (e.g., William and Ellen Craft, Harriet Jacobs, Mary Prince). Yet, until Black feminist scholars, such as Angela Davis and Deborah E. McDowell, called for a female-centered study of slavery and the integration of their slave narratives into the canon in the 1980s and 90s (Davis, *Women, Race, and Class* ch. 1; McDowell, "In the First Place"), scholars of slave narratives primarily focused on Douglass's oeuvre and to a lesser degree other male-authored slave narratives as exemplary. Following in Davis's and McDowell's steps, Elisabeth Ann Beaulieu genders Olney's well-known formula by drawing on Jacobs's narrative as introducing a "family-identity-freedom" paradigm (9). Recent companions to the genre of slave narratives consequently discuss both male- and female-authored/narrated slave narratives and include analyses that go beyond Douglass's and Jacobs's well-known oeuvre by also adopting a hemispheric approach and by discussing lesser known narratives (see, e.g., Ernest, *The Oxford Handbook*; Fisch, *The Cambridge Companion*).

Both female- and male-authored/narrated slave narratives depict the violence with which slavery was inflicted on people of African descent by its main agents, i.e., legal owners, plantation overseers, 'slave catchers,' traders, and their families in order to explain their reasons for escape and push for abolition. They reflect how during chattel slavery sexual and physical abuse as well as murder of enslaved people were common, legally "licensed" practices (Hartman, *Scenes of Subjection* 82). The systematic ripping apart of families and communities, the forced coupling of enslaved people, rape of enslaved women for economic reproduction purposes and for the personal pleasure of the perpetrators, harsh physical work and living conditions, brutal punishment, and constant surveillance were common practices across the Americas (Browne; Adrienne Davis; Angela Davis, "Reflections") that found their ways into the narratives in different forms and conspicuous silences.

Relying on the notion of "violence beyond limit" by slavery historian David Eltis, Wilderson argues that

> a) in the libidinal economy [of slavery and its afterlives] there are no forms of violence so excessive that they would be considered too cruel to inflict upon Blacks; and (b) in [the] political economy [of slavery and its afterlives] there are no rational explanations for this limitless threatre of cruelty, no explanations which would make political or economic sense of the violence. ("Black Liberation Army" 204)

Slave narratives (and slave narrative studies) grapple with this gratuitous violence against enslaved bodies. By sharing acts of anti-black violence in writing, both reproduce the spectacle of the painful abjection of Black people to the status of the non-human as a point of reader identification (cf. Hartman, *Scenes of Subjection* 20). Many slave narratives, such as Douglass's and Josiah Henson's, also reveal the anti-humanization by repeatedly comparing enslaved people to farm animals (Douglass, *Narrative* 12, 26, 35; Henson 23–24). As Howard Bruce Franklin notes, Douglass's *Narrative* is a "book created by a being who was once considered an animal, even by himself, to an audience not quite convinced that he is in fact a fellow human being. So it should come as no surprise that animal imagery embodies Douglass's deepest meanings" (*Victim as Criminal* 7).[83] Of course, the reproduction of abjecting anti-black violence runs parallel to the narratives' attempts to emphasize the humanity of the autobiographical self and their community and create an agential subject narrator (who is literate, knowledgeable, independent, authentic, and trustworthy) in an environment that systematically denied them that status.

Reiterations of life and "comparative freedom" against social death in Frederick Douglass's *Narrative* and *My Bondage*

In slave narratives, gratuitous violence often appears inflicted on enslaved female bodies, such as the malnourished and physically abused bodies of Henriette, Henny, and Mary in Douglass's *Narrative*. In fact, Frederick becomes first aware of his status as enslaved as a young boy not through violence inflicted on his own body but through witnessing, hidden in the enclosed space of a cupboard, the torturous physical punishment of his aunt Hester. Douglass (1817/18–1895) shares this traumatic memory in the first chapter after the *Narrative*'s for-

---

[83] Slave narratives also reflect the ways in which slavery dehumanizes the perpetrators by comparing them to "predatory beasts" (H. B. Franklin, *Victim as Criminal* 17).

mulaic inception. The narrator describes how he witnesses his legal owner Anthony Auld lashing his half-naked aunt and how he experiences the witnessing as passing through the "blood-stained gate, the entrance to the hell of slavery" (*Narrative* 15). The graphic scene that has clear sexual-predatory undertones has provoked an ongoing discussion about the ethics and aesthetics of representing the spectacle and banality of anti-black violence and its gendered forms in narratives about slavery and its scholarly discussion. Hartman begins her study of "the terror of the mundane and quotidian" forms of violence against the (formerly) enslaved in the nineteenth century in *Scenes of Subjection* by acknowledging the scene as "one of the most well-known scenes of torture in the literature of slavery" that "establishes the centrality of violence to the making of the slave and identifies it as an original generative act equivalent to the statement 'I was born'" (3). Yet, she refuses to recite the scene in order to call attention to the ways in which its reiteration "reinforce[s] the spectacular character of black suffering" (3) and obscures "the more mundane displays of power and the border where it is difficult to discern domination from recreation" (42). In referencing Hartman, Sharpe, too, acknowledges the "possible deadening effects into the present of the repetition of the spectacular brutality of Douglass's account" (*Monstrous Intimacies* 2). In contrast to Hartman, however, Sharpe quotes extensively from the scene in the *Narrative* and in *My Bondage* as part of her introduction to her first monograph *Monstrous Intimacies* from 2010 (5–8). Like Hartman, Sharpe considers the beating of Hester as a "primal scene" (6) that when looked at in detail can help "explore slavery's inherited and reproduced spaces of shame, confinement, intimacy, desire, violence, and terror" that reoccur in narratives of the twentieth century (4). The beating of Hester, who is called Esther in Douglass's second autobiography, "approximates his mother's rape and his own monstrous conception and birth, [as well as] his rebirth into a new subjectivity in which subjectification equals objectification" (6).

Importantly, Sharpe describes the transitions from the Hester scene in the *Narrative* to the Esther scene in *My Bondage* as a sign of "a shift in the possibilities for black freedom" in Douglass's register (11). Apart from the change of the name,[84] in *My Bondage*, Douglass leaves out the "blood-stained gate" and provides more detail about the closet from which the narrated I observes the beating. Esther also receives a voice as she pleads for mercy (cf. Sharpe, *Monstrous Intimacies* 7). Sharpe ascribes these changes to the introduction of the Fugitive Slave Act, which allowed slaveholders to track down and capture fugitives

---

[84] For a more detailed interpretation of the name change, see Sharpe, *Monstrous Intimacies* 9–11.

even in what was considered 'free' territory and came into effect five years after the publication of the *Narrative* and five years before the publication of *My Bondage*. While he was writing *My Bondage*, Douglass had to confront "new political and representational developments and realities, he had to resignify black, slave, and free" (12). Sharpe also draws on Douglass's revision of his aunt's name to illustrate her notion of the anagrammatical as it "is literally the anagrammatical (the same letters are rearranged) because the violence will not hold" (Terrefe and Sharpe, par. 120).

Douglass's re-signification of what it means to be Black, enslaved, and free from his *Narrative* to *My Bondage* also finds expression in the revised representation of another form of violence that issues from resistance and self-defense. Frederick's physical struggle with Covey appears as "another primal scene" in both texts (*Monstrous Intimacies* 9). After being introduced to his status of enslavement as a child by witnessing gendered anti-black violence and having gained a freedom dream through literacy, exercising self-defense accelerates Frederick's path to non-enslavement. Thomas Auld sends Frederick, about 17 years old, to work in the fields of Covey's plantation in order to have his rising freedom dream shattered through everyday forms of anti-black violence, such as hard fieldwork, constant surveillance, and regular physical punishment. After six months at Covey's, Frederick is left brutalized and demoralized until he regains a sense of self-worth through a physical fight with the villain and antagonist Covey. As Sharpe notes, the fight "symbolically reverse[s] his brutalizing passage through the blood-stained gate" (9) as he moves from witnessing violence to violent resistance. In *My Bondage*, the fight becomes an even more "staged and ritualized battle" (Stauffer 211, cf. *My Bondage* 187–89), after which he gains "manly independence" (*My Bondage* 190). Because he is no longer "*afraid to die*," he declares himself "a freeman in *fact*" and a slave only "in *form*" (190, 191), pushing for a redefinition of freedom that he already begun as "degree[s] of freedom" at the end of the *Narrative* (*Narrative* 75, cf. Stauffer 211, Weinstein 130).

After claiming the status of a free Black man "in *fact*," the escape from his enslaved "*form*" is mostly left out of the two autobiographies for the author's own protection (N. Roberts 59). The ellipsis stretches between the time Frederick earns money when his legal owner rents him out to work in Baltimore and his arrival in New York City as a fugitive vulnerable to recapture. Douglass reflects on the ellipsis between part 1 "Life as a Slave" and part 2 "Life as a Freeman" at the end of the first part's last chapter "My Escape from Slavery" and the beginning of the second part's first chapter "Liberty Attained" where the narrating I explains the whereabouts of the narrated I metaphorically during the escape ("in a flying cloud or balloon" and geographically after the elided escape ("the great city of New York"):

> Disappearing from the kind reader, in a flying cloud or balloon (pardon the figure), driven by the wind, and knowing not where I should land – whether in slavery or in freedom – it is proper that I should remove, at once, all anxiety, by frankly making known where I alighted. The flight was a bold and perilous one; but here I am, in the great city of New York. (261)

As Neil Roberts summarizes, the actual flight involved Douglass "impersonating a sailor, boarding a section of a North-bound train reserved for blacks, exiting the train in Delaware, transferring to a steamboat headed to Philadelphia, ultimately making his way to New York City, then New Bedford, Massachusetts, and later Rochester, New York" (59).[85] Even though he had been bought by abolitionist friends, and therefore had gained legal freedom after the publication of the *Narrative* in 1846, Douglass only shared these details surrounding his flight in his third autobiography published more than forty years after his escape (Douglass, "Douglass on His Escape" 104–11).

In his philosophical study of "freedom as marronage," Roberts draws on Douglass's middle autobiography for its "structural analysis of slavery and strategies of overcoming the enslaved condition" (56). Accounting for the role of experiential knowledge of enslavement and flight from transatlantic slavery, he uses Douglass's insights into "the process by which people emerge from slavery to freedom" to define what he calls "freedom as marronage" (4). According to Roberts, "freedom as marronage" focusses on the "liminal space" between what is otherwise perceived as stable dichotomous categories of either freedom or unfreedom. In his discussion of Douglass as a "Slave Theorist" (51), Roberts observes that Douglass "adds the words *comparative* and *comparatively* before nouns in sentences throughout [*My*] *Bondage*, mirroring identical passages in [the] *Narrative* as well as novel scenario descriptions that the first autobiography does not mention" (71–72). Roberts argues that by adding "the adjectival political term" Douglass relativizes his notion of freedom, family relations in slavery,

---

85 Douglass frames his route to freedom in his autobiographies primarily as an individual effort and only circumstantially mentions the role Anne Murray played in his escape (cf. *Narrative* 70–71, *My Bondage* 265; cf. McDowell "In the First Place" 175). Research on Douglass and his writing frequently reproduce this neglect. Murray (1813–1882) was a legally free-born Black woman, domestic worker, and member of the East Baltimore Mental Improvement Society when she met Douglass in Baltimore (*Narrative* 70, editorial footnote 5). She helped him facilitate his escape to New York City financially and by providing sailor clothes for disguise. After they married in New York, she supported her husband's work as an abolitionist orator by working as a laundress and shoemaker (Keita and Jones 124).

and improved treatment of enslaved people as stated in the *Narrative* from the retrospective of the Black ex-slave and perpetual fugitive (71). The revisions and additions demonstrate to Roberts "the disjuncture between normative philosophical ideals and human social practice, [according to which] Douglass formulates a relativistic notion of freedom that contains fundamental baseline requirements while being attentive to the comparative experience of freedom in different settings" (73).

In "Frederick Douglass's Self-fashioning and the Making of a Representative American Man," John Stauffer looks more closely at the transitions between the *Narrative* and *My Bondage* that Sharpe and Roberts also take note of. Stauffer describes the shorter *Narrative* as formally "more lyrical," mediated by white editorship, and thematically focused on "bondage and alienation" from the perspective of the still legally enslaved (208, 214). In accordance with Roberts, Stauffer considers the four times longer *My Bondage* as the "more complex [...] and politically and intellectually more compelling" book in which "Douglass explores the psychology of slavery and freedom" as well as the issues of "race and the [...] transition from bondage to freedom" from the independent perspective of a free Black intellectual (208, 210–11, 214). Like Roberts, Stauffer traces the changed perspective back to "changes in Douglass's life," especially his legally free status and a larger distance to the experiences of slavery (208). Stauffer then identifies "a crisis of language and aesthetics" in the representation of Douglass's life after the successful flight which he sees induced by the unprecedented challenge to narrate "Life as a Freeman" (211–12). Tackling this crisis, Stauffer argues, Douglass "created a new genre that follows the form previously reserved for white men: a narrative that describes a life in freedom, rather than ending at the moment of freedom, as all previous slave narratives had done" (212). He describes Douglass's new writing strategy in the second part of *My Bondage* first as "emphasizing his continued subaltern status and struggle for freedom in the face of Northern racism" (212) and second as illustrating his career as a "successful black orator" through the addition of unmediated transcripts of his speeches in the appendix (213). Importantly, Stauffer discusses this self-fashioning process through the lens of performativity when he describes not only Frederick Douglass's fight with Covey and his work as an orator as performances, but also claims that he "performs his freedom" throughout the text (211, 213, 211). This notion of performing freedom, I would argue with Sharpe, issues from the fact that with the Fugitive Slave Act of 1850 "slavery was legally extended into so-called free spaces, and the already constricted possibilities for black freedom in the

North were made more insecure" (*Monstrous Intimacies* 11).⁸⁶ Thus, the "crisis of language and aesthetics" seems to stem less from the challenge to narrate freedoms gained without precedent (as Stauffer argues) and more from the fact that the freedom gained was precarious and at risk of being invaded by enslavist forces at all times (as Sharpe and to a lesser degree Roberts suggest).⁸⁷ It was a confined form of freedom in the 'hold' of slavery upheld by the Black border.

Roberts's and Stauffer's discussions of *My Bondage* both make persuasive arguments about how Douglass's conceptualization of freedom and his self-fashioning evolved on the level of performance. Their analyses, however, disregard the underlying level of positional structure that Sharpe addresses. This structure, as I argue with Afro-pessimist theory discussed in chapter 1.3., remains unchanged and supported by the Black border despite the narrator's and author's performative transition from legally enslaved to legally free status. After all, as Joy James points out, "[a]nti-black violence and terror also exist within maroonage[,] complicating the enterprise of freedom" ("Afrarealism" 125). Roberts's notion of freedom as marronage does not fully account for this complication, while Stauffer's focus on performance indirectly reflects on it. Yet, the question to what degree the term freedom applies to what the fugitive gains through flight from places of enslavement in the historical time of slavery remains unaddressed in both Stauffer's discussion of Douglass's self-fashioning and Roberts's study of concepts of freedom in white European and Black diasporic thought.⁸⁸ Roberts also fails to address the question of what a differentiation between freedom as marronage/maroonage and freedom in white Western philosophy means for the former's efficacy.

Stauffer and Roberts grant that Douglass addresses anti-black racism in the North (Stauffer 212) and "prophetically foreshadows post-Civil War debates on the comparative freedom of ex-slaves who find themselves excluded from voting, working as underpaid wage laborers, or relegated to the proto-prison convict lease system" (N. Roberts 80) that we will look at in more detail in chapter

---

86 Sharpe grants, however, that "the Act and its enforcers did not staunch the flow of enslaved people seeking freedom, nor did its passage institute a wholly new practice; rather it formalized the mechanisms of the capture and return (and spurred new forms of resistance) of enslaved people that were already in place" (*Monstrous Intimacies* 11).

87 I take the notion of enslavist forces invading any form of freedom, community, or kinship relations available to the (formerly) enslaved from Broeck's concept of enslavism (*Gender*) and Spillers's inquiry into the processes of 'ungendering' of (formerly) enslaved people in "Mama's Baby, Papa's Maybe."

88 Neil Roberts draws on Hannah Arendt and Phillip Pettit as well as Frederick Douglass, W. E. B. Du Bois, Frantz Fanon, and Édouard Glissant, among others.

2.2. and 2.3. Yet, they do not sufficiently address the influence this anticipated post-emancipation sense of unfreedom and imprisonment has on Douglass's re-iterations and re-conceptualizations of what it means to be free or unfree pre-emancipation. Confusing Blackness with race (in particular in Roberts's conflation between Fanon's *damnés* and Aimé Césaire's Caliban as 'Natives' on the one hand and chattel slaves on the other [20–26]) further complicates the matter. If at all, it leads to a focus on race rather than the specifics of Blackness and anti-blackness as products of the Middle Passage and obscures the role anti-black violence has played in enforcing perpetual social death onto African descended people.

Trying to make claims to acceptance, reform, and integration, Douglass's revisions in and additions to the *Narrative* in his second autobiography reflect that the vocabulary of freedom from white Western enlightened modernity[89] does not adequately describe the status of perpetual social death of Black people in mid- to late nineteenth-century North America. That Douglass "would spend his life revising and rewriting the autobiographical events that thrust him into the spotlight," Autumn Womack argues, "can be understood as a refusal to reduce narratives of black life and enslavement to a singular event." His constant re-narration of "comparative freedom," I argue, also questions a clear definition of freedom after physical escape. Repetition, as Hartman notes in *Scenes of Subjection*, "points to that which can never be fully recollected and to the impossibility of restoring that which has been breached. The constancy of repetition is catalyzed by the inadequacy of redress and the regularity of domination and terror" (*Scenes of Subjection* 76). This seems also true for other slave narrative writers, such as Josiah Henson or William Wells Brown, who – like Douglass – published several versions of their life stories. Ernest agrees that "differences between the various versions of Josiah Henson's life (largely influenced by white writers, and in which Henson is increasingly identified with Harriet Beecher Stowe's character Uncle Tom), and the various versions of Douglass's life (as he increasingly redefined himself over time)," "can raise serious questions about the possibilities and limitations of black public identity in the nineteenth century" ("Beyond

---

[89] I refer here again to Hames-García's overview of concepts of freedom in dominant Western philosophical thought from Aristotle to Hannah Arendt (xxxv–xli) that he defines as primarily concerned with "individual autonomy, mastery over objects/others, and lack of external restraint" (xxxvii), i.e., "freedom from others" (xli). Hames-García juxtaposes these notions of freedom with those of African American thinkers and activists from Douglass and Jacobs to W. E. B. Du Bois, Martin Luther King Jr., Angela Davis, and Assata Shakur who define freedom as primarily relational and communal, as the "freedom to have connections to others" (xli, xlii; see also Hames-García, ch. 1 and 3).

Douglass and Jacobs" 229). Ernest, however, reduces the differences to Henson's and Douglass's changing relationships to the abolitionist editors and levels of literacy. But, if we follow Wilderson, it is a narrative impossibility for writers/narrators of slave narratives to narrate 'the lived experience of Blackness' and its structural position in narrative form, which usually assumes equilibrium at the beginning, disequilibrium as the narrative crisis, and equilibrium restored as its end (see ch. 1.5.). Wilderson describes the fact that the "three-point progression of a drama for the living cannot be applied to a being that is socially dead (natally alienated, open to gratuitous violence, and generally dishonored)" as "the quintessential problem of the oxymoron *slave narrative*" (*Red, White, and Black* 27). Deviating reiterations of autobiographical (or fictional) accounts of slavery and flight search for freedom in what Hartman describes as "reiterative acts that undermine and discursively reelaborate the conditions of subjection and repression" (*Scenes of Subjection* 56). They are as close as writers and narrators approximate narrating Black flight towards freedom.[90] While Douglass rewrote his fugitive life at least three times, Jacobs published only a single version of her slave narrative but she nevertheless found other stylistic forms to address spectacular and mundane forms of gendered anti-black violence and to question the concept of freedom just as well.

"Loophole[s] of retreat" from the garret to the North: Harriet Jacobs's *Incidents in the Life of a Slave Girl*

Originally published in 1861, the narrative was first reissued in 1969 under the dominant assumption that it was a fictional narrative, a sentimental novel of slavery written by a white female writer, such as its declared editor Lydia Maria Child (1802–1880). It gained wider scholarly acknowledgement in the 1980s when Black feminist scholars increasingly (re)discovered writing by and

---

90 The various reiterations of Henson's and Douglass's written lives also recall the ways in which about one hundred years later Malcolm X's autobiography would "journey through his various personas – Malcolm Little, Detroit Red, Satan, Minister, Malcolm X, and finally, El-Hajj Malik El-Shabazz – represent[ing], in many ways, his valiant attempts to grapple with the reality of a black manhood birthed in the crucible of anguish and detestation" (C. Henderson 23). The need to reiterate autobiographical narratives of forms of captivity and flight can also be found in Chester Himes's complicated publication and editing history of his crime fiction series and autobiographical writing about his experience in prison in the late 1920s and early 1930s (cf. Drake; H. B. Franklin, *Victim as Criminal* 206–32).

about Black women. In 1981, the historian Jean Fagan Yellin was able to verify that *Incidents* was actually "Written by Herself," i.e., a first-hand autobiographical account by the formerly enslaved Harriet Jacobs (1813–1897) who had written the narrative around 1853 and published her story under the pseudonym 'Linda Brent' at the beginning of the Civil War (Yellin, "Written By Herself"; St. Smith 189, 191, 193).[91] *Incidents* finally secured its spot in the American literary and historical canon alongside Douglass's *Narrative* in the 1990s when it was increasingly included in research and teaching (Zafar 2, 4–5). As Chinosole shows, particularly in its representation of an intelligent Black fugitive woman as writer, narrator, and caring mother, *Incidents* "provided the counter-iconography to abolitionist and sentimental narratives" like Stowe's well-known sentimental reform novel *Uncle Tom's Cabin* (1852), which Chinosole describes as "a mutation of the slave narrative" (*The African Diaspora* 133, 106).[92] As one of very few female-authored slave narratives and by focusing on questions of femininity, family, kinship, and motherhood in slavery and flight, *Incidents* acts as a fundamental "Ur-text" for neo-slave narratives by Black female authors (Mitchell 9, 18) and autobiographical writing by Black female prisoners and fugitives (Hames-García, ch. 3) (see chapters 2.2. and 2.3.).[93]

Jacobs's narrative consists of 41 chapters, edited and introduced by Child. Valerie Smith describes Child as "the voice of form and convention in the narrative [...] who revised, condensed, and ordered the manuscript" ("Form and Ideology" 233, cf. *Incidents* 6). Child's editing ensured that Jacobs's manuscript sufficiently catered to the expectations of Northern white women, the target audience, with respect to form, style, and content. *Incidents* uses the sentimental novel as a foil, frequently adopting romantic plot lines (V. Smith, "Form and Ideology" 230–35) and elements of "the gothic romance" (Tawil, "Introduction" 3).[94] At key points, however, the narrative departs from the language of domesticity and dominant courtship and seduction plot lines in order to report on en-

---

[91] In order to differentiate between the author of the text and its fictionalized autobiographical narrative voice, 'Harriet Jacobs' or 'Jacobs' refers to the author of the autobiography, while the pseudonym 'Linda Brent' or 'Linda' marks the narrator and narrated I in the text.
[92] Incidentally, Jacobs had asked Stowe to edit her narrative. "But Stowe responded to the request by demanding the right to borrow her life story instead. [...] She intended to use Jacobs to prove that, *Uncle Tom's Cabin* was truer to life than the flesh and blood stories of living enslaved Africans" (Chinosole, *The African Diaspora* 106).
[93] For a brief discussion of other female-authored or female-narrated slave narratives, see Santamarina.
[94] On the relation of slave narrative and sentimental literature, see Weinstein. For a discussion of how *Incidents* navigates sentimentalism and the cult of true womanhood, see Patton, *Women in Chains* 53–74.

slaved people's experiences of slavery (5) and "agitate for social justice" (Gunning 333). Jacobs uses the "literary conventions associated with [sentimental] courtship and marriage stories to emphasize the extent to which her condition removed her from the world in which many of her white female readers lived" (Ernest, "Beyond Douglass and Jacobs" 223). Apart from conveying Jacobs's personal story, sides and digressions, reader addresses, and her own preface criticize "slavery, slave holders, and northerners who do not actively oppose and resist enslavement of blacks in the South," making *Incidents* "primarily a political work" to serve the abolitionist movement in the disguise of a sentimental novel (Hames-García 116; cf., e.g., *Incidents* 61–62). Michelle Burnham adequately describes Jacobs's use and indirect critique of sentimentalism as a narrative strategy of "camouflage" (290). As I will show, Jacobs's strategy of physical escape through self-confinement and flight and her narrative strategy of using and manipulating dominant narrative styles for her own purposes "are finally quite alike in their advertent but fortuitous use of camouflage" (290). In other words, *Incidents* not only addresses flight on the level of content by first seeking refuge in confined spaces on the plantation and then by physically escaping to the North but uses disguise, a central element of fugitivity, as a narrative form to narrate her story of escape. Yet, while the textual 'loopholes' prove very effective in concealing and conveying Jacobs's political message, the hiding places available to Linda on the plantation and the northern states turn out to be only places of refuge at best and places of continued confinement at worst.[95]

Even though slave narratives such as Douglass's *Narrative* and Jacobs's *Incidents* seem to ultimately reassure, in Joy James words, "that America works to fulfill on some level its democratic promise – despite its racial rages" ("Black Revolutionary Icons" 136), *Incidents* – like the *Narrative* and *My Bondage* – also questions the possibility of freedom in the enslavist and anti-black world of nineteenth-century North America. The primary concern of Linda, the autodiegetic narrator and Jacobs's alter ego, is the safety and freedom of herself, her children (see, e.g., *Incidents* 66, 118), and by extension her people.[96] Joanne Braxton understands Linda as an "outraged mother," an archetype of protagonists in Black female autobiography who not only "seek[...] brighter futures for their children; [... but] also critique patriarchy, freedom, and the hypocritical but accepted notions of legitimacy and ideal womanhood" (131). Arranging

---

[95] For an analysis that instead emphasizes choice, agency, and liberation in *Incidents*, see, for example, Mitchell 22–41.
[96] As Chinosole admits, despite its community-centered approach *Incidents* is not free of "[c]ultural, class, and caste bias [which] is apparent throughout the narrative" (*The African Diaspora* 106).

the narrative not solely around the Black female narrator but other, often maternal characters and their attempts to save family members from harm is a central tradition of Black female autobiography (Braxton 131–32). As Michael Hames-García points out, Linda does not consider herself completely free as long as her family and her community remain captive. While Jacobs and Douglass thus share a future-oriented notion of freedom, Jacobs's freedom concept is in contrast to Douglass's distinctly relational (116).

On the plantation in a southern town, Linda suffers from her legal owner Dr. Flint's attempts to rape and physically abuse her, and from his threats to send her and her children to work in the fields if she does not give in (*Incidents* 26–28). As Rafia Zafar writes, Dr. James Norcom (the real name of the fictionalized character Dr. Flint) was the "antagonist" of Jacobs's life because he did not only want power over Jacobs's physical labor but also "her complicity in her own sexual degradation" (3). According to Linda, her "life in slavery was comparably devoid of [the] hardships" the majority of enslaved people suffered from (such as overworking and whippings [*Incidents* 92]). She therefore includes in her narrative stories of other enslaved people who were physically punished, imprisoned, or killed for disobeying orders, for causing what white people considered disturbances or nuisances, or for flight attempts.[97] These more explicit stories of anti-black violence, in which some of the victims are given a voice in their vernacular (cf. Y. Johnson 23; Mullen 262–63), also insinuate what might await Linda and her children at any time (see, e.g., *Incidents*, ch. IX). The risk of rape is, for example, indirectly addressed when Linda describes the "inevitable destiny" of enslaved girls as "indicated plainly enough" in the vignette of enslaved sisters whose violation of their "purity" became clear to the legal owner's wife through their resemblance to her husband (*Incidents* 45, 44).[98] Thus, Jacobs shows that

---

[97] Incarceration is a constant threat. Among others, Linda's uncle Benjamin spends six months in prison after failing to escape his legal owner (21), while her brother William, her aunt, and children are also held in prison after Linda goes into hiding (82).

[98] If read primarily for its autobiographical and historical content, one could question whether it was likely that Jacobs was really spared from sexualized violence and rape. Jacobs and her editor Child certainly had to make strategic decisions about the narrative's content and style with respect to its expected readership that might have led them to avoid such details. As Gunning notes, "Brent's compelled silence forces her to be the guardian of the white reader's morality, since she keeps to herself all the unpleasant, unseemly details that insult a pure sensibility" (343). Chinosole finds fault with this line of argumentation, however, asking that "[i]f Jacobs/Brent was bold enough to ignore the cult of true womanhood when she told of the illicit sexual liaison with one slaveholder [Mr. Sands], why would she necessarily lie about the rape of another? More pointedly, why would she strain to explain that Dr. Flint did not physically and sexually assault Brent? To claim Jacobs omitted such a gruesome fact as rape in deference to the

"[t]he enslaved black woman in the house, [...] often in a better material position than the black women in the field, is nonetheless positioned in the midst of the everyday intimate brutalities of white domestic domination, positioned within a psychic and material architectonics where there may be no escape from those brutalities but in the mind" (Sharpe, *Monstrous Intimacies* 9). In other words, Jacobs shares "her [own] H/Esther story" in disguise and passes through her "own blood-stained gate" when her narrated I understands in the chapter "A Perilous Passage in the Slave Girl's Life" that "while she has in some measure and for some short time been protected there is no position that she can occupy openly that will stop this violation" (*Monstrous Intimacies* 10). Flint represents the legal, physical, and psychological reach of the enslavist system into any space the enslaved inhabits, may it be architectural, bodily, familial, or mental.

In the chapter "A Perilous Passage" Linda not only indirectly addresses the risk of rape, but also explains her initial strategy to escape when she 'chooses' the white neighbor Mr. Sands as her romantic and sexual partner. Before physically hiding from Dr. Flint in the "garret" for seven long years, Linda 'steals herself' from her legal owner's grip into what appears to be a consensual relationship in which she conceives her two children, Benny and Ellen, and finds "something akin to freedom" (*Incidents* 47). Valerie Smith emphasizes Linda's choice of one type of confinement over another as a gradual empowerment and progression from slavery to freedom ("Form and Ideology" 226, 228). According to Smith, Linda 'chooses' Sands to father her children over being forced to act as so-called mistress to her legal owner; she later 'chooses' concealment in the attic space over her own and her children's continued enslavement; and ultimately she 'chooses' "Jim Crowism and the threat of the Fugitive Slave Law in the North" over enslavement in the South (226). In chapter 3 of *Scenes of Subjection*, Hartman explores the (im)possibilities of the enslaved to choose and give consent in the legal archive of slavery and problematizes the notions of choice and consent that Smith emphasizes through a close reading of Jacobs's "Perilous Passage" (*Scenes of Subjection* 102–12). Hartman shows that choice "is a legal entitlement beyond the scope of the enslaved, who are reduced to chattel, unprotected by law," while at the same time "the narrative endeavours to represent

---

mores of the times is debatable at best, but to ascribe to the theory that she would fabricate Brent's insistence on her victory over Dr. Flint's sexual designs and falsely elaborate on the reasons for his defeat rather than simply omit the event of rape represents the fast-forwarding of contemporary feminist psychological theory with a Eurocentric bias. It confuses the goals of runaway slave narratives with those of a sentimental novel and conflates the anxiety of middle-class European American domestic narrative with the nothing-to-lose reality of Black slave women's lives" (*The African Diaspora* 109).

Linda's choice, precisely in order to make claims for freedom" (110). Linda "reinscribes the status of the self as property in order to undo it" and in order to differentiate her "relation with Sands from her relation with Flint or choice from nonconsent" (110). Yet, as Hartman notes, this effort is "fraught with perils precisely because there is no secure or autonomous exteriority from which the enslaved can operate or to which they can retreat" (102).[99]

Despite numerous interpretations of the "garret" and "the loophole" that suggest otherwise (cf. V. Smith, "Form and Ideology"; see also below), I argue that the impossibility of retreat that Hartman's work suggests here is also true for "the garret," the small attic space in the cabin of Linda's grandmother Martha and the so-called free North in *Incidents*. The ambiguity of captivity, flight, and refuge and their complicated relation to freedom is most pronounced in Linda's retreat from the gendered threats of slavery into the attic space. However, with her physical flight, this ambiguity is expanded to include the North that does not offer unconfined freedom, but merely a larger place of refuge at risk of being invaded by enslavist forces.

The "garret"

The relationship with Sands as an initial strategy of escape proves fleeting as Flint continues to threaten Linda. So to escape rape and because she fears that her children, fathered by Sands but owned by Dr. Flint, will be send to work on the plantation (*Incidents* 78), Linda decides to go into hiding. She first hides in the house of "a friend" (79), then in the attic and under kitchen floorboards of a white enslaving "benefactress" (89). She also spends a night disguised as a sailor on a boat and in the "Snaky Swamp" (89–90), until finally settling in the small attic space in her grandmother's cabin on the premises of the plantation. She stays in the space that is about nine inches long, seven inches wide, and three inches high at the highest point (i.e., 3 x 2 x 1 meters) with "no admission of either light or air" (92), relying on her grandmother's, uncle's, and aunt's support to "save herself from the white sexual predator" (James, "Black Revolutionary Icons" 137). Since Linda's grandmother Martha asked

---

[99] In her analysis of *Incidents* as an "emancipatory narrative," Mitchell acknowledges Hartman's critique of the possibility of choice in the chapter "A Perilous Passage" by granting that choice "in the Black woman's literary tradition" is "never unencumbered" (34). However, Mitchell then discounts Hartman's argument because Linda stays "hopeful" after all (35), missing the momentous impact the questioning of choice has for the interpretation of *Incidents* at large.

Linda to stay for the sake of the family (75), hiding in the attic space can be understood as a concession to this wish and to feelings of motherly responsibility – a compromise between living in slavery with her family and fleeing without them. She also hides on the plantation instead of escaping to the North for fear of being caught and killed during an escape attempt and in hopes that a later escape – when she is no longer searched for – proves less risky and that she can later take her family with her.

The attic space keeps Linda safe from Flint's sexual threats and allows her to read, sew, and observe her children from afar through a small "loophole" that she bores through the attic wall (*Incidents* 92). It also provides time and a small space to make escape plans by writing letters to her legal owner that suggest she already successfully escaped to the North. But due to a lack of shelter from extreme heat in the summers and extreme cold in the winters and because she cannot properly stretch her body in the small space, Linda falls seriously ill during her confinement, with lasting physical and mental effects (96–97, 116). As Katherine McKittrick summarizes, Linda's "limbs are benumbed by inaction; she loses the power of speech; she remains unconscious for sixteen hours; she questions her spiritual beliefs and the seeming indifference of God; she becomes delirious; and dark thoughts fill her mind" (*Demonic Grounds* 40). So, at times, the attic enables creative work, critical thinking and writing, as well as strategizing, in other words a place and space as close as possible to a "room of one's own" (Woolf) for an enslaved Black women in the historical time and place of slavery. At other times, the space, that is also infested by rodents and insects (92–93), seriously endangers not only her mental health but her physical well-being and becomes life-threatening. After all, the "garret" could turn into a death chamber at any given time through illness or if she were found. Depending on her physical and mental state, Linda describes her hiding place varyingly as "my little den" (93), "so good a place of concealment" (94), "my cell," "my prison," "my dungeon" (105), and "a living grave" (116).

As Burnham points out, just like Linda's hiding place is located "in the center of her master Dr. Flint's domain" on the plantation, the chapter entitled "Loophole of Retreat," in which Linda describes her first few months in the attic, is located "in the exact center" of the narrative, acting "as the hinge which balances twenty chapters on either side" (278). While acknowledging the ambiguity of the attic space as a physical place of confinement, many scholars have interpreted the "garret" and its "loophole" as the pivotal point in the narrative and as physical and metaphorical space of transition from slavery or captivity into freedom. With a tendency to downplay or pass over the clearly confining and potentially lethal aspects of the "garret," these scholars focus on the ways in which it opens up epistemological spaces in the narrative and the enslavist system through which Brent/Jacobs can share her story and theorize the meaning of enslavement, womanhood, and the process by which she becomes free.

Burnham, for instance, shows on the one hand how in order to escape "surveillance" by her master, Linda's self-confinement "in many ways enacts the condition of slavery on a hyperbolic scale" (283). "The absence of freedom, the physical hardships, the separation from children and family, and the secrecy that all mark the slave's condition are repeated and exacerbated by Jacobs's confinement 'in her dungeon'" (283). On the other hand, she also points out that the "garret" enables Linda to return the surveilling gaze of the white enslaver from the loophole to the plantation because she is "concealed and contained within the semblance of black enslavement and female powerlessness" (284).

> Harriet Jacobs inhabits a structural site where the practice of power seems so incredibly unlikely that she is able to get away with the resistance to and manipulation of her master. Thus, by inhabiting a loophole in the first, more spatial, sense of the word as a defensive and enclosed space, Harriet Jacobs enacts the second, more textual, definition of loophole as "a means of escape"; she has discovered and retreated into a loophole in the patriarchal institution of slavery. Although Jacobs' loophole of retreat is the most confining space imaginable, it is finally a space of escape. (284–85)

Similarly, Valerie Smith – while emphasizing Linda's repeated self-confinement and further identifying narrative "equivalents of the garret" in Jacobs's strategic "ellipses and ironies" ("Form and Ideology" 225) – also admits that Jacobs "dates her emancipation from the time she entered her loophole, even though she did not cross into non-enslaving territory until seven years later. Smith discerns in the phrase 'loophole of retreat' "an ambiguity of meaning" that signifies "withdrawal," "avenue of escape," and "confinement" which "renders the narrator spiritually independent of her master, and makes possible her ultimate escape to freedom" (226). Focusing on "the cramped, hidden spaces in which black women's self-expression moved toward literary production" (266), Harryette Mullen argues that through "successive recombinations of tropes on the home as woman's shelter and prison" (from the "garret" on the plantation to Linda's "servant's room in the North"), Jacobs "manipulates the ideology of domesticity" in order to "question and revise the figure of the woman whose interiority is derived from her confinement in domestic space" (260).

Clearly, Mullen, Smith, and Burnham try to reckon with the fact that Linda chose self-confinement in a narrow space over slavery and physical flight, or in McKittrick's words, that she considered "[d]isabling, oppressive, dark, and cramped surroundings [...] more liberatory than moving about under the gaze of Dr. Flint" (*Demonic Grounds* 41), while also looking for ways in which the self-imprisonment was enabling for Linda. In doing so, they question whether Linda's "garret" is a place of either resistance or prolonged captivity and their answers – while acknowledging the ambiguities – tend towards an emphasis of the freedoms gained. In their readings, Jacobs's act of hiding in a confined space produces an empowering freedom narrative disguised as sentimentalism and romance. In *Demonic Grounds*, McKittrick also describes Jacobs's "garret" as a "usable and paradoxical space" that positions Linda "across (rather than inside or outside, or inevitably bound to) slavery" (xxviii, 41). Yet, for her, like Hartman, "[t]he question of geographic freedom is wrapped up in the racial, sexual, and bodily constraints before and during her retreat to the attic" (40). Jacobs's "garret" as part of Black women's geographies "creates a way to think about the histories of black women as they are wrapped up in a legacy of unprotected public bodies situated across the logic of traditional spatial arrangements – on slave ships and auction blocks, in garrets, under a white supremacist gaze, in white and nonwhite places" (45). Extending the perspective into the post-slavery era, McKittrick draws on Hortense Spillers's notion of "garreting" (77) in order to argue that "black women's geographies (post-Jacobs) are garretings – they are still unresolved because of the impact the black female body does and does not have upon traditional geographic arrangements" (62).[100]

In 2019, Tina Campt and Saidiya Hartman focused in the opening speech of their conference "Loophole of Retreat" at the Guggenheim Museum on the ways in which Jacobs's concept of the loophole provides a space for reflecting on and working towards future freedom. Campt and Hartman drew on "the legacy of Jacobs in stealing away and anticipating freedom in the confines of enclosure" and described Jacobs's "loophole of retreat"

> as a dark hole and an attic space where she planned her escape and nurtured dreams of freedom. It is a transient space of reprieve from the assault of the world, a zone of refuge from the weather, a perch from which to fashion a runaway plot. We invoke the multiple registers of the loophole of retreat as a preparation ground for cultivating possibility. We seek to explore its potential as a site of isolation and a paradoxical space of refuge. It is a site Jacobs claimed as simultaneously an enclosure and a space for enacting practices

---

[100] McKittrick quotes from Spillers "Mama's Baby, Papa's Maybe" in which Spillers describes Jacobs's narrative as a story of "garreting" that allows for a better understanding of Black women's histories and lives as "not-quite spaces" (77). Spillers attributes the term "garreting" to a conference paper by Valerie Smith that was revised and published as "'Loopholes of Retreat': Architecture and Ideology in Harriet Jacobs's *Incidents in the Life of a Slave Girl*" (1990). As McKittrick notes, however, "Smith does not use the word 'garreting' in her published article (although it is implied)" (*Demonic Grounds* 160).

of freedom – practices of thinking, planning, writing, and imagining new forms of freedom. It is a place we mobilize in an effort to revalue black women's intellectual labor.

Following Hartman's questioning of 'choice' in *Scenes of Subjection* and McKittrick's inquiry into the (post)slavery processes of "garreting," Campt and Hartman describe the potential of Jacobs's "loophole" and the ways in which her confinement and strategies of escape endure and work towards "possibility" with notions of fugitivity and refusal familiar from their works, again refusing to clarify the ambiguity of the concept.[101] Ultimately, Linda's "garret" forms a temporary place of refuge from the gendered anti-black violence of slavery that enables "dreams of freedom" and a form of care for herself and her family, while at the same time remaining a place of confinement and isolation that seriously risks Linda's mental and physical health and could turn into a death chamber at any given time. In fact, the weather that threatens Linda's health in her hideout can be understood, with Sharpe, as the reach of anti-blackness, the "atmospheric condition of time and place," into the place of refuge, where in Sharpe's words "the air of freedom might linger [..., but] does not reach into the hold" (*In the Wake* 104).

The North

The "air of freedom" remains out of reach at the end of the narrative as well. When Linda finally flees to New York City via boat in the company of another female fugitive (123–25), she learns that due to the Fugitive Slave Act she "was, in fact, a slave in New York, as subject to slave laws as I had been in a Slave State. Strange incongruity in a State called free!" (150). Linda acknowledges that living physically removed from places of enslavement represents an "improvement" in her "condition" (159). Yet, Linda's need for refuge does not end after the arrival in the 'Free' North, as she remains in constant danger of re-enslavement as Flint still tries to locate and reclaim 'his property' (cf. 135–36, 150). Even after Flint's death, his daughter Emily and her husband Mr. Drodge continue to search for Linda wishing to sell 'their inheritance' (152). Moreover, Linda's daughter Ellen, who had been taken to the North to Sands's cousin before Lin-

---

[101] Jasmine Syedullah took a similar approach in her conference paper "Loopholes in the Fugitive Heart of Harriet Jacobs," presented at the 2016 annual meeting of the American Studies Association in Denver.

da's arrival, still works under slavery-like conditions (131).[102] Sandra Gunning observes a parallelism here between Linda's experiences as a "slave girl" in the Deep South and her daughter's experience in the supposedly free North (350). Last but not least, even though Linda rejects the practice of buying one's freedom as it follows the logic of rightful ownership, ultimately Linda's employer, Mr. Bruce's second wife, buys Linda from Flint's daughter against Linda's will in order to give her a legally free status (155).

Consequentially, at the narrative's ending Jacobs describes herself and her children as "free" in a legal sense but without a safe home, compromising and upending the conclusive and potentially happy ending:

> Reader, my story ends with freedom; not in the usual way, with marriage. I and my children are now free! We are as free from the power of slaveholders as are the white people of the north; and though that, according to my ideas, is not saying a great deal, it is a vast improvement in *my* condition. The dream of my life is not yet realized. I do not sit with my children in a home of my own. I still long for a hearthstone of my own, however humble. (156)

While Linda clearly acknowledges the relevance of legal freedom as an "improvement" in her "condition," she also shows awareness of this freedom's constraints in the North and the ways in which it remains "continually deferred by new setbacks" (Hames-García 114). Not only will the physical and mental consequences of her seven-year long imprisonment (and the prior life in slavery) impair her physical and mental well-being for the rest of her life (Weinstein 130), Northerners' complicity with and inactivity about slavery in the South limit even the legal freedom granted to her (Hames-García 114). Gunning rightfully points to the frustration of Linda's search for freedom "by Northern conditions that duplicate almost exactly those survived by Brent and her family in North Carolina" despite "the shift in region" as "an active critique of the 'freedom' offered to blacks by the North" (350). Thus, Linda "refuses to validate the freedom offered to her in the North" and instead positions "herself as a survivor and an active critic of both the North's and the South's peculiar forms of social justice" (353).[103]

---

[102] While she is in hiding on the plantation, Linda tricks Flint into selling the two children to their biological father Sands (ch. XIX) and asks him to free them (100). He agrees to do so, but after allowing them to live with their great-grandmother for years, Sands sends Ellen to care for relatives instead (108–09). Benny is sent to join Linda and her brother William in the North later (135).

[103] Originally, Jacobs wanted to end her narrative with a chapter on John Brown (1800–1859) and his armed rebellion against enslavers in 1859. Yet, Child asked Jacobs to exclude this chapter

Ultimately, the fact of slavery in the South captivates her family's existence in the North that remains a confining "garret" with larger 'loopholes of retreat.' The North as a space of refuge is still at risk of being invaded or infiltrated by the reach of enslavist forces at any time. Slavery's epistemologies and ontology still endanger herself, her family, friends, and other Black people. They shape being free in the northern states as little more than an improved condition and actual, unconfined freedom remains a mere freedom dream. Cindy Weinstein confirms that "[w]e see this over and again in slave narratives, whether it is a realization that life in the North has its own set of racist constraints, or the acknowledgment that one's pursuit of freedom must lead one still further from one's place of birth and into Canada or to England, or the sense that one can never fully free oneself of the horrible memories of slavery" (130, see also Sid. Smith 24–25). Thus, *Incidents* clearly shows that freedom for Black people anywhere in the historical era of slavery is not true freedom at all as long as enslavist forces can reach Linda and her children in the extended place of refuge that is the North and while others are still exposed to the system's gratuitous violence in the South. Slavery in one place corrupts freedom in another place in the historical era of slavery.

Before and during their escape, the enslaved narrator conveys a substantial amount of hope for freedom in the northern states "precisely because there was a 'place' to escape to" (Sid. Smith 17). In other words, until their successful arrival in the northern states "[t]he geographical line between Kentucky and Ohio, between Maryland and New York or Pennsylvania, was [considered] the point of the radical break from an enslaved to a free identity" (17). Yet, with their arrival or after living in legally free territory for some time, they acknowledge that fleeing to the North in nineteenth-century America only changes the scenery and improves their condition but does not change their structural position outside of civil society. While the white heroines of white-authored sentimental narratives indulge in the closure of past sufferings, formerly enslaved fugitives stay "in a perpetual state of escape. They may escape, but they also may be recaptured. They may have freed themselves from their masters, but they never escape from the past, which is always with them, haunting them physically and psychically" (Weinstein 131). Jacobs's and Douglass's nuanced conceptualizations of forms of (un)freedom and flight create small loopholes of possibility disguised in sentimental forms while not losing sight of the fact that they remain small

---

because she considered it an inappropriate ending ("Lydia Maria Child" 194) and thus prevented the narrative to end on an even more openly political note (Hames-García 274) than the disguised sentimental ending that Child approved.

holes in the otherwise solid Black border through which they can peek into civil societies' freedoms while remaining on its outside. In Ernest's words, slave narrative authors and narrators

> wrote not to display the extent to which they had escaped slavery; rather, they wrote to *get into the realities of slavery*, and to force their readers to recognize that, in fact, there *was* no escape from slavery, not for African Americans born into it, and not for white Americans in the North who had never experienced it. Authors of slave narratives did not write simply to celebrate their escape; they wrote because so many others remained enslaved, a condition that would not change for many until the nation addressed the economic, political, social, and legal structures that supported slavery and the racial assumptions that extended from slavery. ("Beyond Douglass and Jacobs" 224–25)

Surely, that the United States had not sufficiently "addressed the economic, political, social, and legal structures that supported slavery and the racial assumptions that extended from slavery" into its afterlife by the 1960s gave rise to the Civil Rights and Black Power movements. Unsurprisingly, the understanding of legal freedom as merely a 'loophole of retreat' from perpetual confinement and flight finds also expression in the writing of formerly imprisoned or fugitive Black Power intellectuals as well as African American novelists more than one hundred years after the height of the slave narrative. As we will see in the following two chapters, the reiterations and rewritings of the attempts to gain freedom from captivity through flight continue in these texts.

2.2 Fugitive Black Power

> State, master, and slave in an interminable battle over freedom created the language of the fugitive or incarcerated rebel – the slave, the convict. The language of the illegal or criminalized in turn created the conditions for freedom not rotted in captivity.
> Joy James, "Introduction: Democracy and Captivity" xxv.

In "Black Revolutionary Icons and 'Neoslave' Narratives" from 1999, Joy James criticizes slave narratives such as Jacobs's and Douglass's for ultimately trusting the "promise" of American democracy despite "its racial rages" (136) and therefore turns to the revolutionary writing of activists from the Black Power movement for a more radical critique of the state. As I have shown in the last chapter, while slave narratives indeed appear overtly reformist, they covertly question the possibility of unconfined freedom in an enslavist North America. Slave narrative writers and narrators posed radical questions about their position, time, and place disguised in sentimentalism, working towards expanding their own and all Black people's ability to live less confined lives. In doing so, Black fugitive abolitionists of the nineteenth century were literally writing for their newly gained, utterly unstable legal freedom as well as for their own survival and the survival of their communities. As we will see below, (formerly) imprisoned and fugitive Black Power activists of the early 1970s were also most literally writing for their lives and those of Black and poor people as well as people of color in the United States (and elsewhere). Thus, turning to autobiographical narratives of (formerly) imprisoned and fugitive Black Power activists published in the 1970s and 80s is a logical next step not only in James's investigation of "Black Revolutionary Icons" but also in my examination of fugitive narration in Black American literature.

In this chapter, I analyze the autobiographical writing by three Black Power activists and (former) fugitives and prisoners with respect to their representation of captivity and flight. The analyses focus on the ways in which the life writing of George Jackson (1941–1971), Angela Davis (1944–), and Assata Shakur (1947–) take up and rewrite tropes of Black fugitivity and captivity established in the slave narrative tradition and show how the criminalization, incarceration, and forced prison labor of African-descended people in North America superseded slavery after its abolition. The shared characteristics include (1) editorship and the question of authenticity; (2) the use of camouflage as physical disguise and narrative strategy; (3) forms of confinement inside and outside of the prison, the concept of the prison as plantation and of the society as prison; (4) self-censorship to protect themselves, others, and their political projects; (5) a critical analysis of anti-black racism and classism and an anti-racist and socialist/communist revolutionary pedagogy; (6) a focus on the role of the community, family, and motherhood; and (7) tensions between individuality, exceptionality, and representation.

George Jackson's letters published in two collections, *Soledad Brother* (1970) and *Blood in My Eye* (1971), document not only his life in prison in the 1960s and early 70s. The letters to his family and friends also share his critical analysis of Black people's "neoslavery" in the twentieth-century United States and his propagation of the need for an anti-racist, socialist revolution (*Soledad Brother* 55, 236). In *Soledad Brother*, I argue, Jackson re-conceptualizes flight in the post-slavery era as a maroon search for forms of refusal and self-defense from the position of the *a priori* captive and from within the walls of the physical prison as plantation as well as the American society as prison. Jackson's letter collection is connected to Davis's and Shakur's autobiographies not only because their shared experiences of imprisonment, flight, and revolutionary fight during the movement era intersected (cf. Perkins, ch. 6), but also because Davis and Shakur drew on Jackson's arguments and their literary form while developing their own, specifically Black feminist perspectives. In fact, Angela Davis's activism around the case of the 'Soledad Brothers,' of which Jackson was a part, indirectly led to her going underground and her incarceration.[104] The analysis of *Angela Davis: An Autobiography* (1974), written after Davis was released from jail and acquitted of all charges, examines how Davis draws on the communal history of enslaved people's escape and conjures a community of prisoners to refuse state violence while indirectly acknowledging its inescapability. Davis's "political autobiography" (*Angela Davis* xvi) uses her life story leading up to her activism, her flight from law enforcement, her incarceration, and the trials as lessons about US racism and classism and as a basis from which to rally for support of the Black Power movement. *Assata: An Autobiography* shares the political impetus and revolutionary pedagogy as well as some common themes and formal characteristics with Davis's text. The discussion of forms of flight in *Assata* from teenage runaway and "astro-travel" via prison escape to perpetual fugitivity in Cuban exile draws attention to the ways in which flight – literal, unconscious, or mental – and the search for forms of refusal and revolutionary possibilities perform futurity. While her attempts remain – like Jackson's fugitive self-defense – attacks from the outside of civil society, they nonetheless perform forms of familial and communal care "in the wake" (Sharpe, *In the Wake*). The chapter ends with a brief look at Patrisse Khan-Cullors's *When They Call You a Terrorist: A Black Lives Matter Memoir* (2018), co-authored with asha bandele. The memoir's explicit intertextual references to Davis's and Shakur's work illustrate how their Black feminist fugitive care and futurity continue to provide a vocabulary and revolutionary legacy for today's intersectional movement for Black lives.

Before delving into the close readings, I provide an overview of the different strands of research on Black Power prison autobiographies. Autobiographical writing by formerly imprisoned Black Power activists can be broadly located at the crossroads of two popular intersecting genre traditions and their respective scholarly inquiries, i.e., African American and Black women's autobiography on

---

104 The prisoners George Jackson, Fleeta Drumgo, and John Clutchette became known as the Soledad Brothers, when they were charged with murder of a prison guard in retaliation for the deaths of three Black prisoners during a prison uprising in the yard of Soledad prison, California, in January 1970. Fay Stender formed the Soledad Defense Committee that was rallying for support for their case led by Angela Davis. In August 1970, George Jackson's younger brother Jonathan presumably attempted to free the Soledad Brothers by taking over a courtroom that resulted in the deaths of Jonathan, the prisoners James McClain and William Christmas as well as Judge Harold Haley. Because the guns Jonathan used were registered under Davis's name, Davis was afterwards charged with murder, kidnapping, and conspiracy and went underground (James, *Imprisoned Intellectuals* 63, 85–86). For a summary of the events in the courtroom and a discussion of conflicting news accounts as well as the representation of the incident in Shakur's and Davis's autobiographies, see, Perkins, ch. 4.

the one hand and African American and Black women's prison writing on the other. Slave narratives, such as Jacobs's *Incidents* and Douglass's *Narrative*, of course, represent main precursors to both of these traditions. By addressing questions of confinement, refuge, flight, and freedom as well as community and kinship, *Incidents* and *Narrative* represent central intertexts for autobiographical writing by Black Power prisoners and fugitives, such as Jackson, Davis, and Shakur whose writing observe similar constraints to Black life more than one hundred years later. A third strand of research that plays an important role in understanding Black Power autobiographies are critical prison studies that draw on the autobiographies for their idiosyncratic critique of the carcerality of society and its relation to slavery, settler colonialism, and genocide.

While autobiographical narratives by formerly enslaved fugitive abolitionists have widely been discussed as founding texts of the African American literary tradition and history as well as their contribution to theorizations of freedom and captivity have been widely acknowledged (see ch. 2.1.), autobiographical writing by (formerly) imprisoned and fugitive African Americans from the twentieth century has received less attention. And that is despite the fact that, as Howard Bruce Franklin pointed out already in 1978, African American prison writing is not "some peripheral cultural phenomenon but something close to the center of our historical experience as a nation-state" in which "the main lines of American literature can be traced from the plantation to the penitentiary" (*Victim as Criminal* xxii). As Ashraf H. A. Rushdy notes with respect to the emergence of the fictional neo-slave narrative tradition, historians and activists "increasingly used the discourse of slavery" during the Civil Rights and Black Power movement "to describe social conditions and political processes in the sixties" (*Neo-slave Narratives* 6). The autobiographical writing of Black Power activists discussed below stands proof. In fact, one "romantic appeal" of these (formerly) imprisoned and fugitive writers that is rather tragic "lies in the clear parallels that can be drawn between them and the suffering of their [enslaved] predecessors" (Loeb 154).

Even though American prison literature has drawn increasing attention in recent years, it remains, as Kerstin Knopf notes, "far from being an established literary subgenre," even more so with respect to women's prison literature and, as I may specify, autobiographical prison writing by Black women (212).[105] Reasons for this oversight are manifold. Among other things, as a sub-genre of Afri-

---

[105] Another sub-genre of African American captivity narratives that has received little attention are memoirs by formerly captive African American soldiers from the Vietnam War, for instance (cf. Loeb).

can American literature, African American prison literature has often been relegated to the less appreciated genre of popular culture (Fontenot) and its stylistic form and political purposes have posed challenges to longstanding aesthetic expectations and literary approaches (H. B. Franklin, *Victim as Criminal* xxi; Chernekoff 44). Just as Katharina Gerund attests fugitives a key role in African American literature that is not "adequately reflected in scholarly work" (285), confinement and incarceration play a central role in African American literature that tend to be overlooked not least due to the disparity of the body of literature and the diversity of fugitivity and captivity narratives. Also, public interest in the Civil Rights and Black Power movements has mostly gravitated towards narratives by male activists, such as Martin Luther King Jr. (1929–1968), Malcolm X (1925–1965), Eldridge Cleaver (1935–1998), George Jackson, and Mumia Abu-Jamal (1954–) (Loeb 154).[106] In order to (re)write the histories of the Civil Rights

---

[106] The autobiographical texts of the Black activists mentioned here, of course, would also lend themselves to my analysis of captivity and fugitivity narratives as the following brief outlook on their work proves. Due to the overrepresentation of male voices in the discussion of the Black Power movement, I discuss Davis and Shakur instead. Moreover, Jackson's writing will be discussed as a lesser-known example of male Black Power life writing.

King and X (born Malcolm Little) were the most well-known Black political leaders during the Civil Rights movement until their assassinations in 1965 and 1968 respectively. King repeatedly spent days in jail for nonviolent protest, while X was incarcerated from 1946 to 1954 for burglary before his career as a political Muslim leader advocating Black pride and nationalism (Gillespie 361–62). Cleaver was Minister of Information and Head of the International Section of the Black Panther Party after he had spent eight years in prison on rape and assault charges in the 1960s. Mumia Abu-Jamal (birth name Wesley Cook) was a member of the Black Panther Party as a teenager from 1968 to 1970. He later worked as a journalist and supported MOVE, a Black liberation group in his hometown Philadelphia that suffered from state persecution. Abu-Jamal was sentenced to death in 1982 for killing a police officer. After spending almost twenty years on death row, his sentence was changed to prison for life in 2001. *The Autobiography of Martin Luther King, Jr.* is a collection of King's papers edited and published by Clayborne Carson in 1998. It includes the famous "Letter from Birmingham Jail" which King wrote in jail after protests in Birmingham and which James considers an important early contribution to "the genre of contemporary black protest and prison literature" ("Introduction" 14). *The Autobiography of Malcolm X* was written by Alex Haley based on interviews with X and published posthumously in 1965. James describes the autobiography as "creatively embellished" and points to the fact that Haley had worked for the FBI "which sought to discredit Malcolm X" ("Framing the Panther" 139). Cleaver wrote *Soul on Ice* (1968), a memoir and essay collection, during his time in prison. Abu-Jamal has not published an autobiography but several collections of opinion pieces, poems, and essays that address the history of slavery and the US prison system, life in prison and on death row (see, e.g., *Death Blossom*, *Death Row*) which made him an internationally known voice for (political) prisoners. Abu-Jamal continues to comment on current events such as fatal police shootings of African Americans and the emergence of the Black Lives Matter move-

and Black Power movements, historians have frequently drawn on their autobiographical writing, leaving female activists' perspectives on the sidelines.[107]

Apart from historical studies, a substantial number of studies focus on the (autobiographical) writing of male activists, such as King, X, and Cleaver, as philosophical sources and literary works (see Joseph; Hames-García, ch. 1; C. Henderson; Sid. Smith, ch. 4 and 5). Franklin's *The Victim as Criminal and Artist: Literature from the American Prison* (1978) represents an early contribution to the study of predominantly male prison writing in America as a literary form. It analyzes texts written by men "who have become literary artists through their prison experience" (234), such as white and Black convicts' autobiographical and fictional writing,[108] but also slave narratives, spirituals, and prison songs. While the most well-known prison writing of the 1960s and 70s had just been emerging, Franklin studied, among other things, how African American prison literature of the 1960s and 70s movement era developed as a distinct form out of the communal African American history in which "American society as a whole [always already] constitute[d] the primary prison" (244). As we saw in chapter 2.1., slave narratives in the nineteenth century had already "explor[ed] what it means to wake into consciousness as a Black slave imprisoned within the boundaries of America" (236–37), laying the ground for the tropes of the prison as a plantation and American society as a prison. Like slave narratives and Black prisoners' songs before them,[109] Black Power prison writing argues that "the [imprisoned] author's individual experience is not unique or even extraordinary, but typical and representative" (249–50). They are communal experiences of Black people outside of the prison, where African Americans suffer

---

ment from his prison cell (see *Have Black Lives*). For brief summaries of King's, X's, and Abu-Jamal's lives and political activism as well as excerpts from their writing, see James, *Imprisoned Intellectuals* chapters 1, 2, and 12. For a discussion of Cleaver's memoir, see C. Henderson.

107 For an overview of historical studies of the Black Power era in its breadth that also clearly shows the marginalization of women's perspectives involved in the struggle, see Joseph. For studies that highlight women's historical contributions to the Civil Rights and Black Power movements, see Gore, Theoharis, and Woodard; Collier-Thomas and Franklin.

108 Franklin argues that early white male prisoners' writing mostly framed the prisoner as an individualized criminal in the confessional or picaresque mode (*Victim as Criminal* 127–29). With the establishment of the modern prison system in the late nineteenth century and early twentieth century, the focus shifted to also account for the convict's place in the "social subclass" of white prisoners and "the main exploited and oppressed social class, the proletariat" (244, 137).

109 The role of autobiographical writing by Black political prisoners in the first half of the twentieth century as predecessors to Black Power autobiographies has only very recently received scholarly attention. See Childs's article "Angelo Herndon" discussed below.

from state surveillance and violence similar to that systematically inflicted on the imprisoned (cf. 246).[110]

Observing "certain unifying and predominant formal characteristics, determined not only by the background of the writers but also by their [political] intentions," Franklin described the writing by (formerly) incarcerated Black activists during the time of its emergence in the 1970s already as a "coherent body of literature" (234–35) that deemed further investigation. Yet, only at the turn of the new century, when the rise of the prison industrial complex gained wider scholarly attention (see below),[111] publications, such as Tara Green's edited collection *From the Plantation to the Prison* (2008) drew attention to this narrative form in more detail. Following imprisoned writers' claims about captivity as a communal experience of Black people inside and outside of prisons, Green significantly extends Franklin's definition of prison writing as writing by "prisoner-artist[s]" (*Victim as Criminal* 243) and describes confinement as a central trope in African American literature at large (2). Green defines confinement broadly as the experience of "individuals who are placed within boundaries – either seen or unseen – but always felt," "who are imprisoned and who are unjustly relegated to a social and political status that is hostile, rendering them powerless and subject to the rules of those who have assumed the position of authority" ("Introduction" 3). Confinement narratives in Green's volume are "set either entirely or partly in a place of confinement, including plantations, prisons, and segregated societies, and tell of African Americans' experiences with incarceration" (4). Green's broader definition proves useful for the Black Power life writing discussed below as it accounts for forms of confinement that precede and follow as well as exceed physical incarceration. However, Green's collection focusses exclusively on Black men's fictional and autobiographical writing from the 1940s to 70s, such as Richard Wright's *Native Son*, Ralph Ellison's *Invisible Man* (1952), and Malcolm X's autobiography (1965). It largely neglects Black women's writing – even though, as James notes, "perhaps the most celebrated

---

110 Franklin describes *Look for me in the Whirlwind: The Collective Autobiography of the New York 21* (1971) by Kuwasi Balagoon as a prime example of a community-centered prison autobiography because it is "a collective autobiography of not only [the imprisoned Black Power activists] themselves but their people in the same era" (*Victim as Criminal* 250). Davis's and Shakur's autobiographies discussed below exhibit a communal focus as well.

111 Judy Sudbury defines the prison industrial complex as a "'political and economic chain reaction'" characterized by "a symbiotic and profitable relationship between politicians, corporations, the media and state correctional institutions that generates the racialized use of incarceration as a response to social problems rooted in the globalization of capital" (166).

contemporaries associated with the iconography of revolutionary antiracism are female" ("Black Revolutionary Icons" 139).[112]

As James criticizes in the aforementioned article, the women of the Black Power movement who *have* received public and scholarly attention are more often than not "recognised [primarily] as the lovers or partners of black male revolutionaries, [...] who radicalised and linked them, at least in the public mind, to antiracist armed struggle" (140). This identification promoted "the image of black female militants as sexual and political spectacles" (140) who were not primarily known for their political work, but "were romanticised as icons, noted for a particular form of physical appearance tied to 'fashion,' skin-colour, and youth that led to their commodification" (141). Margo Perkins is one of the first few scholars who have paid attention to female Black Power autobiographical writing and their contributions to the literary traditions they draw on. Perkins's monograph *Autobiography as Activism* from 2000 discusses Davis's and Shakur's Black Power autobiographies as well as Elaine Brown's *A Taste of Power* (1992)[113] "as texts" (xx) that represent what Davis describes in her preface as a "*political* autobiography" (*Angela Davis* xvi).

Perkins shows how Davis's, Shakur's, and Brown's texts share a considerable number of themes as well as formal, genre, and stylistic characteristics with each other and with other autobiographical writing by male Black Power activists as well as slave narratives as their literary forbears (see esp. Perkins, ch. 1, 2, and 6). Perkins clearly reveals the ways in which Shakur, Davis, and Brown use and rewrite the genre of autobiography by drawing on the traditions of the slave narrative and Black autobiography. With clear similarities to the aforementioned tropes and characteristics, which connect Davis's, Shakur's, and Jackson's writing with Douglass's and Jacobs's, Perkins defines the genre of the political autobiography through (1) a common focus on the community over the autobiographical subject; (2) the purpose of the autobiography to fur-

---

[112] Apart from Davis and Shakur, James mentions Harriet Tubman as well as "Native American Anna Mae Aquash, killed during the 1976 American Indian Movement occupation of Wounded Knee, and former political prisoner Lolita Lebron, of the Puerto Rican Independence movement, [as] significant but lesser known [female revolutionary] figures" ("Black Revolutionary Icons" 139). As examples for "black female predecessors and elders in the southern civil rights movement," she lists Fannie Lou Hamer, Ella Baker, Rosa Parks, and Septima Clark (141). She also indicates that the few well-known female activists and revolutionaries eclipsed the work of the majority of Black female organizers who supported "the medical, housing, clothing, Free Breakfast and education programmes" of the Black Panther Party mostly unnoticed by the public and underappreciated by the Party (139).

[113] Elaine Brown (1943–) was chairwoman of the Black Panther Party from 1974 to 1977 after Huey P. Newton fled to Cuba.

ther the struggle and historicize its development; (3) the use of the autobiographer's voice to address the issues of others; (4) the inclusion of strategic silences to protect the movement and individuals involved in the struggle; and (5) the exposure of oppression to intervene and educate (Perkins, ch. 1; cf. Gerund 287).

Following Perkins, Janice Chernekoff also discusses the autobiographical writing by Davis and Brown and juxtaposes them with the writing of the white Civil Rights activists Sally Belfrage and Mary King in "Resistance Literature at Home: Rereading Women's Autobiographies from the Civil Rights and Black Power Movements" (2005).[114] Chernekoff successfully shows that the writing by female Black Power activists is not only closely connected to that of Black men in the movement but also to the writing of white female Civil Rights activists. She shows on the one hand how Brown's and King's narratives – both published with more distance to the events described – can be understood as "retrospective oppositional narratives" that "want to retell the history of the organizations they were involved with, and [...] correct the historical records and popular myths about this period in history" (51). On the other hand, Chernekoff aptly describes Davis and Belfrage as "in medias res authors" who use the "battle rhetoric" of the political discourse from which their narratives directly emerge and which they want to actively contribute to (45, 49). While Chernekoff's pointed analysis provides useful literary studies terms to also discuss formal characteristics of Jackson's, Davis's, and Shakur's writing, her analysis completely disregards the specific role race, in particular Blackness and whiteness, have played in the "US oppositional narratives" (40) she examines.

Apart from the study of the literary traditions of Black (female) political autobiography and prison writing, a growing number of studies on the US prison system as a prison industrial complex has contributed to the study of Black Power prison autobiographies by taking into account the role of racism, antiblackness, and the legacy of slavery. As one of the few Black Power activists who survived the FBI's Counter Intelligence Program and imprisonment, and was acquitted of all charges, Angela Davis has dedicated her activist and research career since her release to the study of incarceration, the role of (anti-

---

114 Sally Belfrage (1936–1994) was a journalist and writer of several memoirs. *Freedom Summer* was published in 1965 and describes Belfrage's experience as a participant in the 1964 Mississippi Summer Project, in which activists from across the United States joined local Mississippians to teach at Freedom Schools, assist in voter registration, and support the Mississippi Freedom Democratic Party. Mary King (1940–) was staff member of the Student Nonviolent Coordinating Committee (SNCC). She also volunteered in the Mississippi Summer Project. Her memoir about her experiences as a Civil Rights activist entitled *Freedom Song: A Personal Story of the 1960s Civil Rights Movement* was published in 1987.

black) racism, and the possibility of prison abolition. Her article "From the Prison of Slavery to the Slavery of Prison" (1998) provides an important overview of how in the late nineteenth and early twentieth century the convict lease system, the Black Codes, sharecropping,[115] and lynch laws systematically policed and criminalized Black people's lives, in effect replacing slavery as a controlling system.[116] It has functioned as a research catalyzer for more recent historical and legal studies that support Davis's claims. Douglas A. Blackmon, for instance, provides a journalistic historical study that illustrates how the post-emancipation convict lease system in the American South was in fact *Slavery by Another Name*, a system that systematically forced Black people into incarceration and debt peonage after they were no longer legally 'enslaveable' (2008). Moreover, legal scholar Michelle Alexander shows in the *New Jim Crow* (2012) how the US criminal justice system even today reproduces a legalized form of segregation, a "racial caste system" that systematically criminalizes, incarcerates, and disenfranchises Black people and people of color who become "permanently barred by law and custom from mainstream society" (13–14).[117] Alexander reveals how beginning in the early 1970s the so-called war on drugs facilitated the transition from Jim Crow segregation to so-called mass incarceration[118] as another controlling system of "legalized discrimination" that continues to put Black people and people of color in 'their place' even today (17, see esp. ch. 2 and 5). The "'law and order' rhetoric and campaigns" of the Reagan and Nixon eras developed "in response to [the] radical and progressive social movements" of the 1960s and "fed the contemporary imprisonment crisis fueled by resistance and backlash to the turbulent decades of protest against the prevail-

---

115 For a brief explanation of sharecropping in America after the Civil War until the mid-twentieth century, see Gillespie 383.

116 In "Prison of Slavery," Davis also shows the extent to which Black intellectuals, such as Frederick Douglass, W. E. B. Du Bois, and Ida B. Wells-Barnett (1862–1931) were aware of and criticized convict leasing alongside lynching. For more of Davis's critical prison studies research, see, e.g., her essays collected in Part I of James's *Angela Davis Reader* and Davis's monograph *Are Prisons Obsolete?*.

117 While Black boys and men have been particularly affected by incarceration, Black girls, women as well as queer and trans people of color also form an increasingly overrepresented group in the US prison population in recent years (cf. Sudbury 164). On the role of gender and sexualized violence in the prison system in the United States, see Davis, *Are Prisons Obsolete?* ch. 4. On the global scope of the gendered rise of the prison industrial complex, see Sudbury.

118 As Rodríguez notes in a recent public statement, "[t]he 'mass' of 'mass incarceration' is not an undifferentiated cross-section of the US demography, but is in fact a targeted, profiled, carcerally segregated population that reflects the nation's racial chattel and racial-colonial foundations *and their present tense continuities*" ("'Nuance' as Carceral Worldmaking").

ing order" in its aftermath (James, "Introduction: Democracy and Captivity" xxxii).[119]

Studying the legal, historical, and sociological dimensions of racialized imprisonment in the twentieth and twenty-first century, scholars have also increasingly drawn on autobiographies and essays by (formerly) imprisoned Americans for their understanding of how "politicians, the courts, and business interests largely engineered the modern US prison system as a replacement to the system of chattel slavery" (Hames-García 98). In fact, "imprisoned writers, poets, and activists have come to occupy a discrete – if marginal – discursive space within academia and the literary 'market' at precisely the time of the prison's transformation into a primary economic, political, and social institution of American life" (Rodríguez, *Forced Passages* 81).[120] Such research no longer reads prisoners' writings primarily for their important 'insider perspectives' that may inform readers about "(human) rights violations in prisons and thus contribute to such discourses on prisons and rights from the non-authoritative perspective of the objects of incarceration policies" (Knopf 211; see, e.g., Davis, *Are Prisons Obsolete?* ch. 4). Black studies, critical ethnic studies, and critical race studies scholars like Joy James, Michael Hames-García, Dylan Rodríguez, and Stephen Dillon instead "argue that the knowledge produced by the prisoner exposes a truth about the United States that cannot be accessed from elsewhere. The prisoner could name what others could not even see" (Dillon 12). In other words, they draw on these texts for their long-standing critique of incarceration in the United States and its political, societal, and historical functions – a critique that exceeds and precedes the insights of sociological, political, and historical research until the turn to the twenty-first century. James, Rodríguez, and Dillon show how at least since the 1960s non-white prisoners' narratives have made visible to non-incarcerated white people the management, regulation, criminalization, and immobilization of Black, poor, and Indigenous people, and other people of color considered "surplus or expendable" (Dillon 116) as a systematic and structural problem that reaches back to the racist and anti-black history of slavery, genocide, and settler colonialism in the United States.[121]

---

119 See Dillon, ch. 1 for a discussion of "how the law-and-order politics of Goldwater and Nixon connected the prison to the free market" that "depended on the immobilization of those rendered surplus, resistant, or fugitive to new racialized economic regimes structured around privatization, deindustrialization, deregulation, and finance" (22–23).
120 For examples of prison writing collections at the turn of the twenty-first century, see, e.g., Chevigny; Chinosole, *Schooling*; Churchill and Ward; H. B. Franklin, *Prison Writing*.
121 For a more detailed overview of the legal and political history of captivity and forced labor of Black people in the United States from slavery through Reconstruction, and Jim Crow to ra-

In *Forced Passages* (2006), for example, Rodríguez shows how the writing of (formerly) imprisoned intellectuals has created a growing body of knowledge about "the chattel logic" of imprisonment in the United States (42) that arose from a genealogy of the slave ship's "effective mass capture, immobilization, and bodily disintegration" during the Middle Passage (241). He considers the "subversive" writing by Davis and Jackson, "two of the most widely recognized imprisoned liberationists of the twentieth century," "paradigmatic" as they "established a vibrant political legacy on which other imprisoned radical intellectuals would thrive" (114). Their writing represents a "political and theoretical point of departure for the genesis of [the post-1970s formation of] radical prison praxis" (113) which Rodríguez defines "as an active current of political-intellectual work shaped by a condition of direct and unmediated confrontation with technologies of state and state-sanctioned (domestic) warfare" (2).

Michael Hames-García studies theorizations of justice, solidarity, and freedom in the writing of African American and Latinx imprisoned intellectuals, arguing that their "experience of unfreedom [...] can give rise to concrete notions of freedom's possibilities that are more enabling and expansive than those that have preoccupied, indeed dominated, the Western philosophical tradition" (xxxvi). He describes this notion of an alternative form of freedom clearly connected in the texts of the (formerly) imprisoned to performativity and relationality that emerged from the history of slavery and may remind readers also of Roberts's concept of "freedom as marronage":

> The prisoners whose writings I examine argue that, as a consequence of the relationship between one's own freedom and that of others, freedom is something not to be possessed but rather to be enacted and practiced through struggle for the freedom of others. This is a position much indebted to a tradition of theorizing freedom-in-struggle on the part of black slaves, former slaves, and their descendants in the United States. (xliv)

Joy James's work on "Black revolutionary icons" and "imprisoned intellectuals" significantly contributed to work, such as Hames-García's and Rodríguez's that reads prisoners' writing for their critical analyses of US imprisonment and freedom, by (re)publishing the writing in the early 2000s and re-interpreting it as "(neo)slave narratives" by "new abolitionists." James's collections of essays, excerpts from longer works, and interviews of imprisoned intellectuals (*Imprisoned Intellectuals, The New Abolitionists*) as well as her edited volume *Warfare in the American Homeland: Policing and Prison in a Penal Democracy* (2007) shed light

---

cialized mass incarceration, see, e.g., the section "Law and Master-State Narration" in James's "Introduction: Democracy and Captivity" (xxv–xxx) and Alexander (ch. 1).

on the United States as a "penal democracy" at war with its minoritized communities.[122] James's editorial work thus not only emphasizes the diversity and political relevance of writing "by Native American, African American, Latino[/a/x], Asian, and European American prisoners incarcerated in the United States during the twentieth and twenty-first centuries" and links the movement for the abolition of slavery to prison abolitionism. It also paved the way for understanding the increasing number of writing by the (formerly) imprisoned – from imprisoned activists of the twentieth century to writers emerging from prison in the twenty-first century as a radical critique of the classist, racist, anti-black world instead of as creative expressions "of individual enlightenment and personal transcendence" (Rodríguez, *Forced Passages* 115).[123]

James's use of the term 'neoslave narrative' departs from the dominant definition as twentieth-century novels about slavery that focus perspectives of the (formerly) enslaved in the tradition of slave narratives (Rushdy, *Neo-Slave Narratives*; see ch. 2.3.). Instead, she uses it as an umbrella term that includes not only (1) the writing by (formerly) imprisoned social justice activists since the 1960s but also (2) historical legal documents ("state-master narratives") that have defined and racialized freedom and imprisonment since slavery as well as (3) "[a]dvocacy abolitionism and its narratives by nonprisoners" ("Introduction: Democracy and Captivity" xxii). With respect to the first category, James further differentiates between narratives that seek emancipation "(parole, clemency)" and those that seek freedom "(liberation from racial, economic, gender repression)," the latter of which she describes as "abolitionists' insurrectionary narratives" (xxii). While a strong connection between slave narratives, novels about slavery

---

[122] *Warfare* also contains articles from Rodríguez ("Forced Passages"), Sexton ("Racial Profiling"), and Wilderson ("Prison Slave") whose later works significantly changed the field of Black studies, critical ethnic studies, and critical race studies with respect to the role of anti-blackness and captivity (see ch. 1.3.).

[123] Rodríguez criticizes non-insurrectionary prison writing that often emerges from prison writing programs "as a domesticating and contained literary genre" stemming from a long history of "prison philanthropy" (104, 86; cf. 87–92) that the state, publishing industries, and academia have used as a reformist and pedagogic tool (*Forced Passage* 81–86). He also draws attention to the risks of considering prison writing as a genre: "The academic and cultural fabrication of 'prison writing' as a literary genre is [...] a discursive gesture toward order and coherence where, for the writer, there is generally neither. Structuring the alleged order and coherence of imprisonment is the constant disintegration of the writer's body, psyche, and subjectivity – the fundamental logic of punitive incarceration is the institutionalized killing of the subject, a process far more complex than even the spectacle of physical extermination emblematized by the death-row execution. This logic is precisely that which is obscured – and endorsed – by the inscription and incorporation of prison writing as genre" (85).

of the twentieth and twenty-first centuries, and the writing by (formerly) imprisoned Black Power activists clearly exists – and in fact builds the basis of the corpus selection of this study (see ch. 1.5.) –, in order to unmistakably identify the three different text types discussed in this and the following two chapters, the term 'neo-slave narratives' will refer only to novels focusing on the life of (formerly) enslaved and their descendants. I describe the three autobiographical texts discussed in this chapter as Black Power autobiographies by (formerly) imprisoned and/or fugitive activists that belong to what James considers "insurrectionary narratives."

Following James's broad definition of neo-slave narratives, in "'An Insinuating Voice': Angelo Herndon and the Invisible Genesis of the Radical Prison Slave's Neo-Slave Narrative," Dennis Childs rediscovers and reevaluates *Let me Live* (1938) by communist, former prisoner, and chain gang member Angelo Herndon as the first autobiography by a Black political prisoner in the United States.[124] Childs shows how Herndon's text functions as "a critically important – if almost completely ignored – progenitor" of political autobiographies by Black Power prisoners, such as Shakur, Jackson, and Davis before the "Movement Era" (31).[125] Childs focusses his analysis on moments "when Herndon's Communist pedagogical narration gives way to a vernacular of Black abolition – and [on] how this expressive shift tends to occur in moments of intensive racial terrorism and prison torture" (49). He shows how "Herndon transmutes the individual predicament of political imprisonment into a platform for radical social analysis – both of the regime of official incarceration as it exists in the prison world, and the system of social incarceration experienced by Black and poor people in the 'free world'" (31). By including stories of other captives, a "chain gang song," and the detailed description of the lynching of Hayes Turner and

---

124 For a brief explanation of chain gangs in America after the Civil War until the mid-twentieth century and its connection to the system of sharecropping, see Gillespie 285.
125 According to a largely accepted definition of the term 'political prisoner,' "political prisoners and prisoners of war are imprisoned as a direct result of their political activities in civil society – community organizing, political education, public speech, armed self-defense, artistic production, guerrilla warfare – that foster or manifest insurrection against socially embedded forms of domination and subjection" (Rodríguez, *Forced Passages* 4). However, this definition does not account for so-called common prisoners who "remain broadly unrecognized by the activist public [...], while generally presumed to be outside staid and elitist conceptions of the 'political'" (5). These prisoners, according to Rodríguez, "compose the vast majority of those who have become activists and political intellectuals *while imprisoned*, many of whom were and are engaged in unprogrammatic (or nonorganizational) varieties of liberationist-directed antisystemic activity prior to, during, and after their incarceration" (5). See also Davis, "Political Prisoners."

his pregnant wife Mary in 1918, Herndon indirectly reveals "irrefutable differences between the capitalist state's sovereign right to exploit and dispossess *the worker*, and the white supremacist state's necropolitical right to kill, dehumanize, and desecrate *the n[*****]*" (45).[126] Of course, the "foreboding three-word title" of the autobiography already gestures towards the risk of "living entombment and premature death within a carceral system" (38) that resonates with both the slave narrative and the political Black Power autobiography tradition. Childs consequently locates Herndon's text at a "critically significant nexus connecting the antebellum slave narratives of those such as Harriet Jacobs and Frederick Douglass with the Black Power era prison narratives of those such as George Jackson and Assata Shakur" (32). Slave narratives, Black Power prison autobiographies, and earlier autobiographies, such as Herndon's, Childs argues, "use the unendurable situation of imprisoned living death as a prism through which to unveil the structurally unfree nature of Black life outside of official zones of incarceration" (40).[127]

Childs pointedly describes "pre-1865 chattel slavery" as "the original prison industrial complex of the Western hemisphere" and uses the term "neo-slavery" for the "post World War II era through our current moment of mass incarceration and legally justified murder of Black people" ("Angelo Herndon" 52, 48). Childs's discussion of Herndon's autobiography as a progenitor of the Black Power prison autobiographies reveals that the persecution and imprisonment of Civil Rights and Black Power activists and the systematic criminalization of Black and non-white poor urban communities since the 1970s does not represent a watershed moment in the history of incarceration. Instead, "Herndon's elaboration of the ubiquitous quality of formations of neo-slavery in the 1930s disturbs today's consensus in respect to the current apparatus of industrialized punishment that sites the late-1970s as the birth of mass racialized imprisonment in the United States" (42). As Childs convincingly argues, "Herndon and other Black social and political prisoners experienced the postbellum version of *America's centuries-old complex of chattelized incarceration* for generations before the arrival of the term 'prison industrial complex'" (42). From the perspectives of Herndon,

---

[126] I refrain from spelling out the n-word in the original.
[127] Another contribution to the study of Black male prison writing that critically inquires racialized mass incarceration in the United States from a literary studies perspective is Kristina Graaff's *Street Literature: Black Popular Fiction in the Era of US Mass Incarceration*. It explores the "street-prison symbiosis" by analyzing not only the literary form but also the socio-political, material contexts of the texts' production and circulation.

Jackson, and most recently, Black Lives Matter activists, such as Khan-Cullors,[128] slavery, convict leasing, Jim Crow, the persecution of Black Power activists, and racialized mass incarceration represent 'permutations,' in Sexton's sense of the word, of the anti-blackness that has held Black and blackened people captive in the Americas, at least, since the transatlantic slave trade.

In *Fugitive Life* (2018), Stephen Dillon discusses how the cultural production of post-Civil Rights fugitive and imprisoned, specifically "feminist, queer, and antiracist" activists, such as the George Jackson Brigade,[129] Davis, and Shakur, anticipate, address, and criticize the late twentieth-century rise of racialized mass incarceration and its relationship not only to slavery but to neoliberalism (4).[130] Dillon argues that in the 1980s a "neoliberal-carceral state" formed in the United States that utterly "engulf[ed]" poor, Black, and Brown lives inside and outside of prisons (10, 12). Dillon draws among other things on the essayistic work of Shakur ("Women in Prison: How We Are" 1978), Safiya Bukhari ("Coming of Age: A Black Revolutionary" 1979),[131] and Davis ("Reflections on the Black

---

[128] See Khan-Cullors's personal account of the ways in which the 'war on drugs' and the 'war on gangs' have affected poor Black communities in urban California in the 1980s and 90s, especially Black queer people and Black people with disabilities.

[129] The George Jackson Brigade was an underground revolutionary group of formerly imprisoned men and women of different races, genders, and sexual orientations, and working-class backgrounds. They were active in the second half of the 1970s in the United States and dedicated their work to "Native sovereignty, domestic national liberation movements, workers' rights, feminism, gay liberation, and, most centrally, prison abolition" (Dillon 1).

[130] By drawing on the work of Hortense Spillers, Ronald Judy, and Cathy Cohen, among others, Dillon describes the conceptualization of Black women as other than human and other than woman since the historical era of the transatlantic slave trade and chattel slavery as a forced 'queerness' (see 13–18). This "gender nonconforming" or "ungendered" position produces "relational forms of [gendered and sexualized] difference that make possible social, civil, and premature death" (24, 15, 18). However, Dillon also acknowledges that which escapes subjection as queer, "the fugitivity of caged human beings" (21). For a brief discussion of the usage of the term "queer" and its role in queer studies, see Somerville.

[131] Safiya Bukhari-Alston (1950–2003, born Bernice Jones) was a member of the Black Panther Party and the Black Liberation Army in the 1970s. She was imprisoned on felony murder charges (among others) and spent eight years in prison. After her release, Bukhari-Alston dedicated her activism to political prisoners and co-founded the Free Mumia Abu-Jamal Coalition in New York City. Her writings and speeches that deem further critical attention as Black Power political life writing were published in 2010 with a foreword from Angela Davis and an afterword from Abu-Jamal under the title *The War Before: The True Life Story of Becoming a Black Panther, Keeping the Faith in Prison and Fighting for Those Left Behind*. For a brief summary of Bukhari-Alston's life and work as well as her aforementioned essay, see James, *Imprisoned Intellectuals* ch. 9. For Shakur's abovementioned essay and an abridged version of Davis's essay, see James, *The*

Woman's Role in the Community of Slaves" 1971) as well as the neo-slave narrative *Dessa Rose* by Sherley Anne Williams.[132] He shows how these texts "connected an emergent neoliberalism to chattel slavery through a queer conception of temporality" (25). Dillon calls this queer temporality "possession" (25) in order to account for the ways in which slavery not only haunts the present prison regime and the market as a legacy, but influences and ruptures the present with its presence (25) reminiscent of Calvin Warren's concept of "Black time." The enslavist past "takes hold of it [the present], determining the contours and possibilities of the now" (25). Dillon argues that Black feminism, having "merged amid the crises of global capitalism, white supremacy, heteropatriarchy, and state power that spanned the 1960s and 1970s" (93), particularly Black feminist fugitives and prisoners as well as writers "can help make connections between past and present that other epistemologies leave unthought and unthinkable" not least through the use of imagination (98).

Dillon's concept of possession in the neoliberal-carceral state adequately describes the complex relation between past and present with respect to slavery, imprisonment, and the lack of unconfined freedom beyond the prison that Black Power activists observed in the 1970s and Rodríguez and James as well as Wilderson, Hartman, and Sharpe so pointedly theorize. As Dillon shows, their critical insights allow us to understand this relation as a spatio-temporal carceral continuum that brings the aforementioned insights of the critical prison studies research discussed above together. This continuum undermines dominant temporalities of past and present and "undoes normative conceptions of space" so that captivity and enslavism exceed "the walls of the prison proper" and the historical era of slavery and form "a symbiosis between the deindustrialized landscape of the late twentieth-century urban United States and the gendered racisms of an emerging prison-industrial complex" (10).

Dillon assumes that the position of the 1970s Black feminist fugitive produced knowledge that used cultural expression to imagine future possibilities and "negated national narratives of progress, equality, and justice" (5). He argues that this "queer figure [...] is the site of a dramatic reimagining of freedom that points the way out even as life is increasingly surrounded" (5). Dillon understands the fugitive's underground as a space that involves "a spatial and temporal shift as well as an epistemological one" because it is positioned "literally and metaphorically beneath the neoliberal-carceral state" (12). Black feminist fugi-

---

New Abolitionists 79–89, 99–112. For a full version of Davis's "Reflections," see James, *Angela Davis Reader* 99–112.
132 Sherley Anne Williams (1944–1999) was an African American writer. She based her historical novel of slavery *Dessa Rose* (1986) on two historically documented incidents.

tives in the underground of the 1970s, Dillon argues, "produced a conception of freedom founded on running away" (22). Dillon further joins the space of the underground and the figure of the fugitive with the space of the prison and the figure of the prisoner because "the prison is constantly at war with the possibility of affective, emotional, and physical escape," making the prisoner "always already a potential fugitive" and the fugitive "a future prisoner on the run" (21). Thus, Dillon insists on the possibilities of flight, actual, epistemological, and affective (25), even in the "impossibilities of the present" (22), because the "fugitive runs even when it seems there is nowhere to go" (22). While indeed "[t]housands of people evaded capture for years, sometimes decades, and many are still running" (21–22), Dillon admits that "[t]here may be no outside to power because it is already there, but that does not mean power is totalizing in its effects" (18).[133] As we will see in the following analyses, especially of Jackson's and Shakur's texts and their lives in prison and in exile, being on the run does not equal freedom in a world where there is no getting inside the safety of civil society for the socially dead bordered out.

This detailed look at the state of research on Black Power autobiographies has allowed us to gain an understanding of which texts have been neglected as well as the different angles from which the texts have been approached as literature, history, and radical knowledge production. The following analyses combine these perspectives, examining not only the idiosyncratic fugitive and captive knowledge Jackson's, Davis's, and Shakur's autobiographical writing produce and their contribution to an understanding of white supremacy and anti-blackness, but also the ways in which their literary styles use and manipulate established (life) writing traditions, especially that of the slave narrative, that shape this knowledge production in important ways. Thus, I follow Perkins in describing autobiographical writing by (formerly) imprisoned Black Power activists as a sub-genre of African American writing that deserves rigorous literary analysis, while also considering this writing with James, Rodríguez, and Dillon as an important "political and theoretical point of departure for the genesis of radical prison praxis" (Rodríguez, *Forced Passages* 113). This writing wants to es-

---

[133] It seems telling that for his further examination of "queer futurity" (25), Dillon abandons the writing of formerly imprisoned and fugitive activists to follow "a less identitarian conception of fugitivity" (26). In the last chapter of his study, Dillon includes an analysis of two poems by the non-imprisoned, non-fugitive Audre Lorde and June Jordan and their conceptualizations of desire. The Black queer writers Lorde and Jordan allow Dillon to "explore fugitivity as an epistemological and affective force, as opposed to a social, political, and legal location" that he deems "foundational to undoing the reign of the carceral and the terror of neoliberalism" (26, 25).

cape domestication as popular entertainment or mere documentation of lived experience and instead theorizes the captive post-slave position of Black people in the United States while textually dreaming of unconfined freedom.

"A slave in a captive society": George Jackson's prison letters

George Jackson, born 1941 in Chicago, was convicted of armed robbery aged 18 for his involvement in stealing seventy dollars at gunpoint from a gas station and sentenced to serve one year to life in prison in 1961. His sentence was continuously extended as Jackson politicized himself and others in prisons in the state of California, becoming involved in prison resistance that cost several lives, including eventually his own in 1971.[134] His political organizing and writing functioned as a central catalyzer for the prison movement in the late 1960s and early 70s in the United States in which activists, such as Angela Davis, became involved. In fact, Franklin describes Jackson as "the leading theoretician of the prison movement" (273).

Jackson's writing was published in the form of two letter collections, *Soledad Brother: The Prison Letters of George Jackson* in 1970 and a year later *Blood in My Eye*, which collects letters Jackson wrote in 1970 and 1971 and which appeared only a few days before he was killed. While the focus of the letters in *Soledad Brother* slowly shifts from personal correspondence to political discourse on the oppression of Black people, imperialism, fascism, and socialism, letters in *Blood in My Eye* mostly take the form of political opinion pieces and instructional essays that argue for a communist revolution in coalition with people of color in Africa and Asia. Greg Thomas describes *Blood in My Eye* as "a practical manual of guerilla warfare and a text which would land its author a place of comparison alongside Ernesto 'Che' Guevara and even Mao Tse-tung in global political theory or intellectual studies of world revolutionary change" ("'Neo-Slave Narratives'" 203). According to Franklin, Jackson's second book attempts "to solve the main theoretical problems of revolutionary strategy in America by applying the lessons learned in prison" (*Victim as Criminal* 236). While *Blood in My Eye* is the lesser known of Jackson's publications and deserves a detailed political analysis, I am interested in Jackson's no less political but more explicitly autobiographical and literary writing in *Soledad Brother* that intersects stylistically and thematically more closely with the other two autobiographies discussed in this chapter.

---

[134] For a brief summary of Jackson's life in prison and the events that eventually led to his death, see James, *Imprisoned Intellectuals* ch. 5.

As a letter collection, *Soledad Brother* is divided into numerous letters of varying length that are entitled with the date of the correspondence, followed by the salutation of the addressee, and are closed by the writer's valediction. The book is further divided into two sections: "Recent Letters and an Autobiography" (1–33) is the shorter, first part; the second part "Letters 1964–1970" (35–330) includes the large majority of letters ordered mostly chronologically and sometimes clustered around one addressee at a time. Especially the second part clearly creates the immediacy of one side of a "dialogical and dialectical process" during which George's imprisonment continuously transforms not only his consciousness and self-conception but his correspondents' as well (Perkins, ch. 3). The letters function as snapshots that "chronologically illustrate[...] (as opposed to narrate[...]) Jackson's move through changing levels of political consciousness" (Perkins, ch. 6), resulting in a narrating I whose conceptions of self, of his correspondents, and of society keep changing (cf. also Smith and Watson 88–89).

The letters in *Soledad Brother* span the period from 1964 to 1970, ending a few days after the death of Jackson's younger brother Jonathan during his armed take-over of a courtroom. The majority of letters are directed to Jackson's family (especially his parents and his brother) as well as to members of the Soledad Defense Committee. Besides reflecting on his daily life and struggle to stay sane and alive in prison, the narrator shares his changing self-conceptions and political convictions with the correspondents, connecting his coming-of-age story, his imprisonment, and the situation of Black people in the United States and across the world with imperialism and the legacy of slavery. Sidonie Smith consequently argues that *Soledad Brother*, together with Cleaver's *Soul on Ice* and Amiri Baraka's *Home: Social Essays* (1966), "are not so much histories of individuals as they are *social analyses* of American racism" (119, emphasis mine).[135]

Focusing primarily on himself and his generation of young African American men, particularly in his early letters George reproduces stereotypes of African American women, making some of the letters appear admittedly "sarcastic, condescending, and dogmatic" (Perkins, ch. 6). In particular, early correspondence with and about his mother Georgia tend to reproduce sexism and the racist and misogynist myth of the 'Black matriarch,' infamously propagated in the Moyni-

---

[135] Amiri Baraka (1934–2014, born Everett LeRoi Jones, also temporarily known as LeRoi Jones and Imamu Amear Baraka) was a major writer in the Black Arts Movement during the 1960s and 70s. He published *Home* as a collection of essays that tract his changing political worldviews. The *Autobiography of LeRoi Jones* was published in 1984 and republished in 1997 with additional passages cut from the earlier version by its editor.

han Report from 1965 as guilty of incapacitating Black men (Davis, "Reflections" 112–13; see, e.g., *Soledad Brother* 45, 48, 125). Yet, these sentiments are challenged in later correspondence with female members of the Soledad Defense Committee, such as Davis, "Joan," and Jackson's white lawyer Fay Stender, for all of whom George expresses deep affection and appreciation as women and activists.[136] It is probably especially those later letters, which, among other things, express affection to these women, as well as the explicitly autobiographical letter that introduces the collection, that caused white, formerly imprisoned French writer and activist Jean Genet to describe *Soledad Brother* as "a striking poem of love and combat" in his introduction to the collection. Thus, *Soledad Brother*, I argue, represents a complex combination of social analysis, life writing, and literary work in constant flux that situates especially young Black men as perpetual captives and fugitives in "neoslavery" (55, 236), while crowding out a more in-depth engagement with the genderedness of anti-blackness. Drawing, among others, on the tradition of slave narratives, Jackson further develops a notion of flight that involves the constant search for forms of refusal and self-defense both in the prison as plantation and in the larger American "neoslave" society while staying ultimately captive.

---

136 Davis took Jackson's gender convictions expressed in his early letters as an occasion to research the understudied role of Black women in slavery while she was in jail (James, "Introduction" 12). She published the results in her article "Reflections on the Black Woman's Role in the Community of Slaves" in 1971 and dedicated it to Jackson (Davis, "Reflections" 111). In a brief foreword to the article, Davis argues that had Jackson "*not been so precipitously and savagely extinguished, he would have surely accomplished a task he had already outlined some time ago: a systematic critique of his past misconceptions about black women and of their roots in the ideology of the established order*" (111). In fact, in the last letter of *Soledad Brother* written briefly after Jonathan's death, George asks Joan to "[g]o over all the letters I've sent you, any reference to Georgia being less than a perfect revolutionary's mama must be removed. Do it now! I want no possibility of anyone misunderstanding her as I did" (*Soledad Brother* 329). Because of his newly found respect for his mother's activism for her murdered son Jonathan and himself, he wishes to have any critique cut. This wish was apparently denied or at least not fully met, compromising Jackson's authority over the publication (cf. Perkins, ch. 3 and 6). Of course, Jackson's early gender convictions also reflect issues Black women faced in the Black Power movement. See Davis's and Shakur's autobiographies on their experiences with gender stereotypes in the movement (cf. also Perkins, ch. 5 and 6). For discussions of the roles of Black women in the movement from a historical perspective, see Gore, Theoharis, and Woodard. For a discussion of the impact of Davis's "Reflections" on Black feminism today, see Haley et al.

"I recall the very first kidnap"

Soledad Brother exhibits several formal and thematic characteristics that closely resemble the slave narrative tradition. Editorship, for instance, clearly influenced the compilation of the collection in important ways, creating a complex web of authorship, censorship, and mediation reminiscent of slave narratives. Editorship of *Soledad Brother* not only included the selection and ordering of letters and their copyediting but also the explicit request for an autobiographical letter to be written as an introduction to the collection (*Soledad Brother* 3). *Soledad Brother* and slave narratives also share a similar degree of self-censorship and an awareness of a potentially hostile readership that may use information against the author. Jackson frequently contemplates in his letters what he and his correspondents can write without censorship or interception of the letters through the prison authorities (see, e.g., 57, 74, 280, 305). His wish to keep information from potential readers other than the addressee clearly limits his writing about the activities of the Black Power movement, his mobilization inside the prison, and his and other people's involvement in criminalized activities. Consequently, the letters share more about the writer's political convictions with the addressee and – through the letter collection – also with an extended, presumably radical left readership than about the spectacle of uprisings, murders, and attempted kidnappings that might have sparked the attention of some of the collection's readers in the first place. Just as Douglass would not reveal details about his escape in his *Narrative* and Jacobs changed her own and other people's names in *Incidents*, Jackson's letters avoid certain topics and use indirect references, coded language, nicknames, or abbreviations to hide information from unintended readers (see, e.g., 196, 267, 316).

A further connection to slave narratives is the ambiguous representative-exceptional position that writers of slave narratives and writers of political prison life writing of the Black Power movement found themselves in. Even though Jackson explains "that he found difficulty in complying with the request to furnish a brief autobiography," Perkins observes that Jackson also "openly acknowledge[s] that he is both typical (in the sense that he shares the plight of other Black people under racist oppression) and atypical (in the sense that his resistance to oppression is both active and subversive)" (Perkins, ch. 1):

> I've been asked to explain myself, "briefly," before the world has done with me. It is difficult because I don't recognize uniqueness, not as it's applied to individualism, because it is too tightly tied into decadent capitalist culture. [...] But then how can I explain the runaway slave in terms that do not imply uniqueness? (*Soledad Brother* 4)

This recalls Douglass's and Jacobs's exceptional-representative status with respect to testimony of experiences of slavery and abolitionist activism of the nineteenth century. While the following close readings in this chapter will show that Shakur and Davis share with Jackson the explicit connection and integration of their individual stories with the experiences of the Black community at large, Jackson's self-representation oscillates not least due to the text type more strongly than Shakur's and Davis's. Jackson describes himself by turns as a lone, self-sufficient warrior-man who the prison system and anti-blackness cannot break ("I feel no pain of mind or body, and the harder it gets the better I like it" [*Soledad Brother* 38]), as a monster-like "subhuman" whose humanity was "destroyed" by the system of "abject slavery" in and outside of the prison (281), and as "born a slave in a captive society" (4).

The latter two, of course, clearly resonate with the slave narrative tradition. Especially the self-conceptualization of being "born a slave in a captive society," which appears on the first few pages of the collection, recalls the formulaic *ab ovo* beginnings of many slave narratives. The phrase introduces a fourteen-page-long "Autobiography" that forms an exemption with respect to the immediacy, the constantly changing narrating I, and the overall chronology of the letter collection at large. On request of the editor, the autobiographical letter presents a more or less coherent, closed coming-of-age story that shares a particularly large number of stylistic and thematic characteristics with both slave narratives and the Black Power autobiographies discussed below. Entitled with the date "June 10, 1970" and addressed to the editor "Greg" (3), it gives a concise overview of George's upbringing in urban, low-income neighborhoods and public housing projects in Chicago and Los Angeles as well as of his summers spent in the countryside of southern Illinois with family. The letter then narrates how George spent much time as a teenager on the streets of Chicago where racial profiling often limited his mobility until eventually criminalized activities led to his first imprisonment in a youth correctional facility at age 16 (10–15). The letter ends with the one year to life sentence two years later and returns to the time of narration when Jackson was being tried together "with two other brothers, John Clutchette and Fleeta Drumgo, for the alleged slaying of a prison guard" (16).

By claiming to have been "born a slave in a captive society" (4) at the beginning of the collection, Jackson immediately re-conceptualizes space as he extends the concept of captivity beyond physical imprisonment. He also introduces what Taija McDougall aptly calls the "non-passage of time" in *Soledad Brother* by broadening the concept of the enslaved beyond the historical end of slavery in the Americas (8). Jackson further connects his "slave" position to the womb of his mother, recalling "the fact that the condition of the child follows the condition of the mother and her own captive body in slavery, thereby making this womb a paradoxical space of confinement or preparation for a life of confinement as opposed to a place or space of certain nurture and protection" (G. Thomas, "'Neo-Slave Narratives'" 206).[137]

> It always starts with Mama, mine loved me. As testimony of her love, and her fear for the fate of the man-child all slave mothers hold, she attempted to press, hide, push, capture me in the womb. The conflicts and contradictions that will follow me to the tomb started right there in the womb. The feeling of being captured... this slave can never adjust to it, it's a thing that I just don't favor, then, now, never. (4)

According to McDougall, Jackson "works through capture as not only the state that he finds himself in, but also as that which forms Blackness as an ontological (non)position" (8). With a vocabulary that directly recalls Jackson's, Christina Sharpe argues that the birth canal – together with the Middle Passage and the coffle – "has functioned separately and collectively over time to dis/figure Black maternity, to turn the womb into a factory producing blackness as abjection much like the slave ship's hold and the prison" (*In the Wake* 74). The repetition of "womb" and "capture"/ "captured" in the passage and of "womb"/ "tomb" reinforce this notion of the inescapability of captivity on the stylistic level.

Also evocative of slave narratives' *ab ovo* beginnings and *partus sequitur ventrem*, George further argues that when he was born, he "felt free" (*Soledad Brother* 5), opening up at least two different levels of awareness: first, that of the new-born narrated I who perceived himself to be free, and second, that of an older narrating I who has gained a different consciousness and knows that he was born "captive," and that his teenage experiences with police surveillance, state violence, and physical imprisonment represent 'permutations' of his *a priori* captivity.[138] Writing to "Z." in March 1970, George – aged twenty-eight – notes, "[t]his is my tenth year of this [imprisonment], actually my twenty-eighth, but I was too numb to feel the first eighteen" (267). Thus, Jackson's conceptualization of captivity clearly exceeds an awareness of what Perkins describes as "the myriad ways in which he, like so many other Black men growing up in America," was "systematically prepared and mentally conditioned (by racism and the adverse material circumstances of his life) for eventual entrance into the criminal justice system" (Perkins, ch. 6). Instead, it embraces the broader definition of confine-

---

[137] As mentioned before in ch. 2.1., in the era of historical slavery in North America the legal doctrine *partus sequitur ventrem* held that the legal status of enslavement would be passed from enslaved woman to her children (Adrienne Davis 108).
[138] For "transatlantic slavery as the antecedent of contemporary surveillance technologies and practices" (Browne 546), see Simone Browne's work.

ment that Green proposes for literary analysis of African American literature ("Introduction" 2–4) and Dillon identifies in the essayistic work of Black Power activists Bukhari, Davis, and Shakur (see ch. 3 of Dillon's *Fugitive Life*). By drawing on the history of slavery and thereby 'queering' time and space in Dillon's sense, Jackson unmasks twentieth-century life for young Black men as still governed, or "possessed" by slavery and the Black border it established.

Later in the collection, a long letter to Stender from April 1970 again blends autobiographical references and political analysis with poetic language. Jackson describes slavery not merely as a legacy but as a visceral part of his memory that is in the first instance bodily:

> My recall is nearly perfect, time has faded nothing. I recall the very first kidnap. I've lived through the passage, died on the passage, laid in the unmarked, shallow graves of the millions who fertilized the Amerikan soil with their corpses; cotton and corn growing out of my chest, "unto the third and fourth generation," the tenth, the hundredth. My mind ranges back and forth through the uncounted generations, and I feel all that they ever felt, but double. (*Soledad Brother* 233–34)

With this vivid passage full of metaphors that reference the Middle Passage and plantation slavery in America through the crops the enslaved were forced to cultivate and also quotes from the Bible, Jackson not only shows the collective functioning of the Middle Passage and the birth canal that Sharpe posits. He also reveals "how the prison's connection to slavery reverses, compresses, and undoes the progress of time" (Dillon 91). George's "captive body is metaphorically infested" or "possessed" by slavery, a "possession" that has passed down, "modified and intensified" its "horrors" (91) from the Middle Passage, through plantation slavery to racialized mass incarceration.

From lessons in maroonage to flight as self-defense

Apart from enslavist notions of captivity that possess George and his generation of young African American men, *Soledad Brother* also uses the concept of flight to describe the ever-changing narrating I as a learned "escape artist" (*Soledad Brother* 4) and maroon revolutionary who strives for escape from the position of the prisoner condemned to death. According to his brief "Autobiography," his mother wanted to keep young George and his sister confined to their apartment in an attempt to protect them from life on the streets (5, 7–8). Yet, at an early age, George already finds ways to escape his parents' apartment to roam the streets of Chicago (5–8) where he then faces police surveillance and violence. Despite frequent police "pickups [...] for 'suspicion of' or because I was in the wrong part of town" (10), the narrating I describes narrated young George as 'uncapturable' because of his physical fitness and mental cleverness:

> There just wasn't any possibility of a policeman beating me in a footrace. A target that's really moving with evasive tactics is almost impossible to hit with a short-barreled revolver. Through a gangway with a gate that only a few can operate with speed [...] up a stairway through a door. Across roofs with seven- to ten-foot jumps in between (the pig is working mainly for money, bear in mind, I am running for my life). (10)

That George is literally running for his "life" already as a teenager quickly proves true. He is shot six times by a police officer, "standing with my hands in the air" (12), imprisoned in a youth facility (13), and eventually arrested on robbery charges (13). While the limitations to and deadly risks of his youthful fugitive mobility increase, he still escapes from jail by tying up "a brother," scheduled to be released on the next day, and leaving in his stead, successfully counting on the inability of the white officers to tell two Black male teenagers apart (15).

Greg Thomas also considers George's time in the countryside of Illinois as a formative training in flight ("'Neo-Slave Narratives'" 207). While George's mother perceived the trips as "removing [him] from harm's way" in the low-income neighborhoods of Chicago, George uses his stays to learn how to shoot guns, catch fish, and identify edible plants "that grow wild in most areas of the US" (*Soledad Brother* 8), unearthing "a new knowledge of survival in the 'wild,' and a newfound mobility" with which he "roams the country" (G. Thomas, "'Neo-Slave Narratives'" 207). This maroon knowledge recurs in a cluster of seven letters dated between 22 and 26 March 1970 and directed to Stender in the last third of the collection. Having argued for the need of an armed class revolution, he describes colorfully a fugitive love relationship and revolutionary comradeship he imagines to have with a woman:

> I want to hide, run, and look over my shoulder. The only woman that I could ever accept is one who would be willing to live out of a flight bag, sleep in a coal car, eat milk weed, bloodroot, wild greens, dandelions, a rabbit, a handful of rice. She would have to be willing to run and work all night and watch all day. She would bathe when we could, change clothes when we could. She would own nothing, not solely because she loved me, but because she loved the principle, the revolution, the people. (227)

As other letters show, Jackson develops this idealized notion of fugitivity with the increasing awareness that he might stay in prison all his life or die of a violent death rather than ever gain legal freedom (see, e.g., 250). Writing to Davis in May 1970, George explains somewhat condescendingly that even though he assumes "women [...] want at least a promise of brighter days, bright tomorrows," he has "no tomorrows at all" for consolation (298). While his youthful conception of flight emerges from a physical and geographical mobility that is increasingly hindered by state surveillance and state violence, Jackson's later conception of flight builds on the earlier youthful maroon notion but connects it to a fugitive love relationship from the perspective of physical imprisonment, isolation, and the feeling of being death-bound.[139]

The notion of death-boundedness contextualizes one of the most well-known quotes from *Soledad Brother*. The line "I may run, but all the time that I am, I'll be looking for a stick!" (328) is part of the penultimate letter of the collection, dated 28 July 1970 and addressed to Stender. As Michelle Koerner and Taija McDougall note in their respective articles on *Soledad Brother*, the white French poststructuralist philosopher Gilles Deleuze repeatedly misquoted the line in his work and ascribes it to Jackson without ever giving *Soledad Brother* as its source.[140] In the passage that follows the quote, Jackson questions agitatedly, as the many exclamation points suggest, any opposition between fight and flight and argues for the need of self-defense – possibly in opposition to non-violent protest as propagated by the Civil Rights advocate Martin Luther King Jr. who had been assassinated two years earlier:

> I may run, but all the time that I am, I'll be looking for a stick! A defensible position! It's never occurred to me to lie down and be kicked! It's silly! When I do that I'm depending on the kicker to grow tired. The better tactic is to twist his leg a little or pull it off if you can. An intellectual argument to an attacker against the logic of his violence – or one to myself concerning the wisdom of natural counterviolence – borders on, no, it overlaps the absurd!! (*Soledad Brother* 328)

In the rest of the letter, George addresses the question of self-defense with respect to cross-racial allyship between the non-Black free and Black fugitives

---

139 Abdul R. JanMohamed uses the term "death-bound" in *The Death-Bound-Subject: Richard Wright's Archaeology of Death* (2005) for his analysis of the role of violence in Wright's oeuvre.
140 For a detailed critique of the absence of a reference to *Soledad Brother* as the source of the quote in any of the French publications or English translations of Deleuze's work, also with Félix Guattari, see McDougall. It is telling that even Dillon, whose *Fugitive Life* explicitly addresses Black Power writing, fails to mention *Soledad Brother* as the origin of Deleuze und Guattari's "line of flight" (see ch. 4, esp. 124). In "Left Out: Notes on Absence, Nothingness and the Black Prisoner Theorist," McDougall convincingly argues that "Jackson spends the loose narrative of his collected letters ruminating on the meaning of an enforced nothingness," a "suspension between death and life, [...] that is illustrative of the prisoner generally, but of Jackson specifically as he theorizes the time between capture and escape," and that the missing citation of *Soledad Brother* in Deleuze's work and its reception "enforces his nothingness in the sense of the theoretical world" and the world of "institutional and academic practice" (8).

that recalls controversies between Black and white abolitionists of the nineteenth century:

> Don't mistake this as a message from George to Fay, it's a message from the hunted running blacks to those people of this society who profess to want to change the conditions that destroy life. These blacks are still in doubt as to whether those elements across the tracks want this change badly enough to accept the US being physically brought to its knees to attain it. (328)

De-individualizing his "defensible position" on the run to a level of abstraction that accounts for "the hunted running blacks" more generally, George argues for the need of violent self-defense and questions potential non-Black allies' commitment to support this endeavor.

Discussing the ways in which the "I may run" line and Jackson's writing on flight in *Soledad Brother* influenced the work of Deleuze, Koerner contends that the line invokes both Jackson's coming-of-age story of "a kid on the street running from a cop" and the history of "runaway slave[s]" (169). Koerner further argues that Jackson's writing can be understood as "thought on the run" (166) that "expresses the persistence of a capacity to reject – and reject absolutely – intolerable historical conditions of enslavement, imprisonment, and social death" and thereby "become[s] an affirmation of blackness as a condition of emancipated life" (157). In doing so, *Soledad Brother* "escapes what would otherwise be thought of as the historical conditions of its production," i.e., the text's – and as I assume by extension – Jackson's captivity (164), performing and theorizing fugitivity not only as a fugitive potential but as an active "counterattack" (161). According to Koerner, Jackson's "letters concretely assert writing as a freeing of life – of blackness – *from the terms of racist imprisonment*" (166).

Of course, what Koerner describes as "the connection between a process of rejection (escape) and the necessity to simultaneously construct a counterforce (the search for a weapon)" (169) recalls fugitivity as I have discussed it in relation to Tina Campt's work in chapter 1.4. Koerner's description of the "real force" of Jackson's writing as "a total refusal to adjust to existing conditions of capture, enslavement, and incarceration" (164) resonates with the "quotidian practice of refusal" which Campt defines "less by opposition or 'resistance,' and more by a refusal of the very premises that have reduced the lived experience of blackness to pathology and irreconcilability in the logic of white supremacy" (*Listening to Images* 32). Moreover, Campt actually cites Koerner's work on Jackson in the coda to *Listening to Images* in order to paraphrase her "notion of an anterior fugitivity" (113–14). What differentiates Campt's fugitive refusal from Koerner's is, however, Campt's reluctance to paraphrase flight and refusal as 'resistance' and to suggest – as Koerner does – that the writing actually escapes captivity.

Koerner rests her claim of the 'escapability' of captivity through fugitive writing, among other things, on the fact that Jackson professes to refuse to pass on captivity (157). After having introduced his aforementioned letter to Stender with the description of his body as possessed by the Middle Passage and plantation slavery (*Soledad Brother* 233), George, indeed, expresses the desire to rid himself of captivity through armed struggle. However, he also acknowledges – not least based on his experiences as a teenager on Chicago's streets and in prison – the chances of his brother and him dying in this attempt:

> I have a young couragous [sic] brother whom I love more than I love myself, but I have given him up to the revolution. I accept the possibility of his eventual death as I accept the possibility of my own. Some moment of weakness, a slip, a mistake, since we are the men who can make none, will bring the blow that kills. I accept this as a necessary part of our life. I don't want to raise any more black slaves. We have a determined enemy who will accept us only on a master-slave basis. When I revolt, slavery dies with me. I refuse to pass it down again. The terms of my existence are founded on that. (250)

I argue that Jackson links revolutionary activity so closely with death here that by inference he suggests that there cannot be an end to captivity apart from death unless revolution brings about the end of the classist anti-black system, what Wilderson perceives as the end of the world (see ch. 1.3.). Ultimately, Jackson understands his brother's armed court-room takeover as a performance of a freedom dream during which Jonathan, "black man-child with submachine gun in hand, [...] was free for a while" (329). Any potential 'escapability' of captivity lies in futurity, an unknown future after the end of the world. Until then Blackness as captivity "is that which cannot be escaped" (McDougall 8).

Koerner's discussion of Jackson as per Deleuze then lacks a more explicit reflection that while Jackson's fugitivity may indeed be understood not only as an enunciation of a fugitive potential but also as a literary performance of flight, it remains most literally a mental performance of flight in the unchanged confines of the prison that resulted in the author's death. Koerner's otherwise convincing and insightful article on the often unaccounted role of the Black radical tradition for late twentieth-century French philosophy[141] does not sufficiently acknowledge the physical limits to flight for Black people and George Jackson's *a priori* captivity which I have discussed above. Instead of understanding Jackson's writing as acting as "an autonomous force that *precedes* capture" (Koerner 168), it is

---

[141] For a similar investigation of the unaccounted role philosophies of the Black Panther Party played in white queer French theorist Michel Foucault's work, see Heiner. Heiner focusses particularly on relations of Jackson's and Davis's writing with Foucault's publications in the early to mid-1970s, not least his renowned *Discipline and Punish*.

the force of captivity and its permutations through 'Black time and space' in *Soledad Brother* and George Jackson's life that precedes his birth and already possesses him as a "prison-slave-in-waiting" (Wilderson, "Prison Slave" 18) in his mother's womb. As McDougall concludes by indirectly referencing Hartman, "Jackson cannot escape the weight of Blackness that presses in, breaks through the theorizing of Marxism, revolution, and carcerality. *Soledad Brother* labors to suggest escape/counterattack as the posture of resistance, of revolution, but 'scenes of subjection' break through the wall before he can" (8).

Jackson's fugitive refusal describes the impossibility and futurity of literal escape for African American men in late-twentieth-century America and the ongoing process of searching for a weapon in self-defense that may ultimately destroy the Black border as manifested in the walls of the prison to open up true lines of flight. As we saw in this close reading and will see in the following chapters as well, mental fugitivity, i.e. the search for forms of refusal within the bounds of literal captivity, cannot escape the different forms of captivity and (social) death that precede, exceed, and succeed escape. Yet, as Davis and Shakur focus their narratives more strongly on the community and the genderedness of anti-blackness, they also recast Jackson's male-connoted, individualized fugitivity as self-defense as Black feminist forms of familial reproduction and communal care.

Conjuring captive and fugitive communities in Angela Davis's autobiography

Angela Davis is probably the most well-known Black Power autobiographer discussed in this chapter. Today known not only as a social justice activist but also as an internationally renowned scholar, Davis was born in segregated Birmingham, Alabama, in 1944 where she grew up in a previously all-white neighborhood called 'Dynamite Hill' due to frequent Klan bombings of Black-owned homes. As a teenager, she became increasingly acquainted with communist thinking and the Civil Rights movement when she attended a private high school in Manhattan while living with a white Quaker family. After studying at Brandeis University and at universities in Germany and France, Davis pursued a Ph.D. with Herbert Marcuse at the University of California, San Diego, where she was banned from teaching because of her membership in the Communist Party USA (CPUSA). By the late 1960s, Davis had become a person of national interest who was kept under surveillance by the Federal Bureau of Investigation (FBI) because of her association with the Black Panther Party in Los Angeles and her involvement in the movement to free prisoners, such as the Soledad Brothers. In August 1970, Davis was charged with conspiracy, kidnapping, and murder, when Jonathan Jackson, George's younger brother, used a gun registered under her name in an armed take-over of a courtroom. Davis went underground and was added to the FBI's Ten Most Wanted list. She was caught two months later and spent more than a year in pretrial confinement in the states of New York and California until she was released on bail and finally acquitted of all charges.[142]

Angela Davis: An Autobiography was first published under the title *With My Mind on Freedom: An Autobiography* in 1974, about two years after Davis was released from jail. Davis published the text at a – for an autobiography – rather young age because during her imprisonment the campaign to 'Free Angela Davis and All Political Prisoners' had made her an internationally recognized political case and a stylized icon of the social justice movements at that time. Toni Morrison encouraged Davis to write her story and publish it with Random House (Davis in White). Morrison was working as the only Black trade editor at Random House in the early 1970s and "almost singlehandedly introduced black radical activists to mainstream American readers" by editing not only Davis's autobiography but also Jackson's *Blood in My Eye* (1971) and *To Die for the People: The Writings of Huey P. Newton* (1972) (James, "Toni Morrison"; cf. also G. Thomas, "'Neo-Slave Narratives'" 212).[143]

Davis followed Morrison's encouragement to use the autobiographical genre not tell a story of a "heroic individual" but of the movement (Davis in White) and wrote what she coined a "*political* autobiography" (*Angela Davis* xvi). Her auto-

---

142 For a brief summary of Davis's life until the publication of her autobiography, see James, *Imprisoned Intellectuals* ch. 3 and Perkins, ch. 1. For a critique of Davis's movements through her life as forms of flight, see James, "Black Revolutionary Icons" 137. James describes Davis's educational and career choices as flight from a confined Black community suffering from anti-black violence ('Dynamite Hill') to white privileged spaces (private high school, Brandeis University, and universities in Germany and France) before consciously entering African American activist circles at the University of California ("Black Revolutionary Icons" 137). James sees Davis's movement into Black radical spaces subverted again later in life by pursuing a liberal academic and activist career (137), pointing to the privilege of that choice in contrast to the ways in which others, such as Davis's codefendant Magee or Assata Shakur could or would not be rehabilitated in a similar way (137). Yet, this rehabilitation has only happened in liberal-leftist discourses, whereas conservative non-fiction publications continue to portray Davis as a fugitive criminal, as Gerund observes in her article "Angela Davis: the (Un)Making of a Political Fugitive in the Black Power Era."
143 Huey P. Newton (1942–1989) was co-founder of the Black Panther Party in 1966. For a brief biography, see James, *Imprisoned Intellectuals* ch. 4. At Random House, Morrison also edited the work of the African American writers Gayl Jones, June Jordan, and Toni Cade Bambara as well as of the African American boxer Muhammad Ali, among others (Davis and Griffin; Morrison in White).

biography thus shares characteristics with *If They Come in the Morning* (1971), a collection of essays and letters on and by political prisoners that Bettina Aptheker, Davis, and other members of the National United Committee to Free Angela Davis edited during her internment. In fact, Cherron Barnwell argues that the "style and purpose of *If* set the stage for" *Angela Davis* with the former functioning as "backdrop" and underlying "political treatise" and the latter representing the "dramatization thereof" (312, 310, 318).[144] Fitting a *political* autobiography, Chernekoff identifies the "battle rhetoric" of the era in *Angela Davis* that involves "name-calling and an openly and self-righteously biased perspective toward the enemy" as well as an emotionally charged language (49) that can also be discerned in the other two texts discussed in this chapter. Like them, *Angela Davis* is a literary contribution to the autobiographical tradition, socio-political history of the Black Power movement, and a theoretical treatise on Black Power, imprisonment, and liberation.

The autobiography details the conditions under which Davis fled, was caught, arrested, kept in jail, and tried. The text fuses these events with a coming-of-age story of her rising political awareness and increasing radicalization by interspersing the narrative with formative episodes of her childhood, education, and political activist career. The text is split into six parts, of which the first and the last two, tellingly entitled "Nests," "Walls," and "Bridges," narrate Davis's experiences of fugitivity, captivity, and the eventual release from jail, while part two to four ("Rocks," "Waters," and "Flames") narrate Davis's upbringing and coming of age chronologically in a flashback leading towards the narrated present. While the autobiography is, of course, written with much more retrospection than Jackson's letters, especially the events leading up to Davis's arrest and the jail and trial time are narrated with a strong immediacy which Chernekoff describes as "*in medias res*" writing (45) that issues from the short lapse of time between the experiences and their narration. This impression is reinforced by subheadings detailing the date of the events in varying frequency in part one, three, four, and five that may remind readers of Jackson's letter collection as well.

Due to the structure of the narrative, the plot is narrated not only in a non-linear fashion but, as Barnwell notes, also circular in that Davis's experiences as

---

144 *If They Come in the Morning* includes essays, letters, and interviews by Aptheker, Davis, Clutchette, Jackson, Magee, and Newton, among others. In representing what Barnwell describes as the "deindividuation" and "dehumanization" of prison life, *Angela Davis* not only enters into a dialogue with *If* but also anticipates, as Barnwell argues, Davis's "later political treatises on the racialization of crime and punishment" and her prison abolitionism (Barnwell 331; see, e.g., Davis, "From the Prison of Slavery," *Are Prisons Obsolete?*).

an African American fugitive from US law and as a captive in US jails frame her life story at its beginning and end (314). In this way, particularly the middle part offers insights into the lives of Black people in the segregated South and the North in the late 1940s, 50s, and early 60s that led to her political belief in an anti-racist, anti-sexist socialism and communism towards Black liberation as explicated at the beginning and ending of the narrative. Davis makes sense of "the criminalization of her political activities and ideological convictions" (Rodríguez, *Forced Passages* 124) by interpreting the increasing state repression of Black Power activists in the late 1960s and early 70s – together with her coming-of-age story – as collective experiences of a generation of Black political activists and the larger African American community. In her political autobiography, Davis insists that "her oppositional politics were more important than stories about her family, friends, and love life" and that "her thoughts and struggles are not unique but shared by many others" (Chernekoff 44).

Thus, Davis uses and reinterprets the autobiographical genre not only to document the history of the Black Power movement and of Black communities around that time (cf. James, "Black Revolutionary Icons" 143) but also to rally for support of the movement's efforts to liberate Black people. Drawing on the tradition of the slave narrative, Davis's autobiography interprets particularly her experiences of flight as indicative of the oppression of the African American community at large and as part of a larger history of enslavement, fugitivity, and captivity of African Americans since the Middle Passage. As we will see, the narrating I does so by imaginatively conjuring communities of formerly enslaved fugitives and by calling on as well as actively creating communities of (political) prisoners from within her jail cell. Ultimately, the circular focus on imprisonment emphasizes the subsistence of captivity beyond individual experiences as "an imprisoning cycle" of African American life that ultimately questions whether "Davis remains imprisoned" even after her release (Barnwell 314), held captive by the Black border of the anti-black world surrounding her.

Feeling the "very teeth of the dogs at their heels"

Angela Davis leaps into the first part of the narrative *in medias res* with Angela trying on a wig for disguise, as she prepares to leave a friend's apartment, where she has been hiding for a few days since she was charged with murder, kidnapping, and conspiracy. Her hands are "flounder[ing]" nervously "[l]ike broken wings" around her head, as she looks at her disguise in the mirror and hardly recognizes herself (4–5). Chernekoff finds "Davis's sense of being hunted and defenseless against forces far more powerful than she" reflected in this opening mirror scene (46). Barnwell interprets the fact that Angela does not instantly recognize herself in the mirror as showing how the "fugitive status was at least as threatening to her identity as it was to her life" (315–16). Gerund emphasizes how becoming a fugitive changes Angela not only in appearance but also mentally, the language with which Davis describes her experiences on the run highlighting her "vulnerability, fear, and loss of freedom" (288). Indeed, similes like the comparison of her hands with broken wings and the agents who "were swarming like angry wasps" around her friends (*Angela Davis* 6) in the first few pages of the narrative set a clear atmosphere of "fear" and "vulnerab[ility]" (5). While Angela acknowledges the need for "a quick, camouflaged move" (7) to hide from "agents," who she supposes to be either from the Los Angeles police or the FBI, she "hated" her "furtive, clandestine existence" (8). Yet, the section's title "Nests" can not only be associated with "'nets' used to capture someone or something" as well as "the (media) networks employed by her pursuers for her capture," but also with support networks for people on the run (Gerund 288), such as Angela's "true friend" Helen (*Angela Davis* 5) and by analogy the Underground Railroad of formerly enslaved fugitives in the nineteenth century.

When Angela leaves her friend's apartment to change her hideout because she is afraid to be found, she addresses fugitivity directly for the first time in the text, using the word 'fugitive' four times in only a few lines, and connects her "hunted state" to that of formerly enslaved fugitives as "perhaps a bit more complicated, but not all that different" (6). I cite this passage extensively before analyzing individual parts in the following:

> The route from Echo Park down to the Black neighborhood around West Adams was very familiar to me. I had driven it many times. But tonight the way seemed strange, full of the unknown perils of being a fugitive. And there was no getting around it – my life was now that of a fugitive, and fugitives are caressed every hour by paranoia. Every strange person I saw might be an agent in disguise, with bloodhounds waiting in the shrubbery for their masters' command. Living as a fugitive means resisting hysteria, distinguishing between the creations of a frightened imagination and the real signs that the enemy is near. I had to elude him, outsmart him. It would be difficult, but not impossible.
>
> Thousands of my ancestors had waited, as I had done, for nightfall to cover their steps, had leaned on one true friend to help them, had felt, as I did, the very teeth of the dogs at their heels.
>
> It was simple. I had to be worthy of them. (5–6)

In the last three lines of this quote, the narrator establishes a direct connection between herself as a fugitive and her fugitive enslaved "ancestors." Active agents in this paragraph are not only the narrated autobiographical I, the "thousands of" ancestors fleeing from slavery, the FBI (or police) "agent" as the narrator's "enemy" and "master" of the bloodhounds but also the implied 'master' (and probably also 'mistress') of fugitive enslaved people, and their dogs chasing the fugitives. A few lines before, the relation between the narrated present of fugitivity and the past of enslavement is already established, albeit less explicitly: The phrase "bloodhounds waiting in the shrubbery for their masters' command" (5) conveys a connection to "the very teeth of the dogs at their heels" (6) that fugitive enslaved people felt and the narrator feels as well. Even though the bloodhounds' master is actually the "agent in disguise" (5), an analogy to the legal owners of enslaved people and their dogs can obviously be drawn here.

Barnwell argues that in "analogizing her fugitive status to the ancestral history of fugitive slaves" Davis calls on "connotations of heroism" in "discourses that consider the social immorality of the historical circumstances that constructed the fugitive slave" (317). Following Barnwell, Gerund suggests in her analysis of the gendered and raced portrayal of Davis as a fugitive in her autobiography, contemporary media outlets, and recent non-scholarly books about fugitives, that Davis's identification with fugitive enslaved people seeking freedom from bondage serves "to counter the media image of the dangerous fugitive-criminal" (291).[145] I argue that Davis also creates a connection between her experience as a fugitive from US law and the experience of enslaved people fleeing from their enslavers to acknowledge the presence of enslavist forces in the lives of African Americans in the twentieth century more broadly, what Hartman describes as the "afterlife of slavery" (*Lose Your Mother* 234).

Indeed, Angela is very conscious of the 'Hartmanian' "fugitive's legacy" (234) handed down to her and other African Americans throughout the narrative. She, for example, describes Black jail officers as "prisoners themselves" who "[l]ike their predecessors, the Black overseers, [...] were guarding their sisters in exchange for a few bits of bread" (*Angela Davis* 43, cf. also Barnwell 323). Referencing James's redefinition of neo-slave narratives as abolitionist insurrectionary narratives that include Davis's autobiography, Neil Roberts argues that Davis in fact "adopt[ed] the experiential writing that fugitive and former slaves such as Douglass made prominent" (55). Gerund also observes several common characteristics between Davis's autobiography and the slave narrative tradition, such as "its political impetus and polemic style which utilizes self-fashioning as a stra-

---

[145] Barnwell and Gerund refer to the nineteenth-century abolitionist discourse and its conception of fugitive enslaved people as innocent, righteous agents seeking well-deserved freedom (see ch. 2.1.). Of course, the pro-slavery discourse of the time remains unacknowledged here as it painted a completely different picture in which fugitive enslaved people were criminalized and their (desire for) flight pathologized as a pseudo-illness (cf. Rushdy, *Neo-Slave Narratives* 257).

tegic rhetorical device, and its emphasis on the struggle for freedom and equality" (291). Gerund also lists strategic silences, the interpretation of individual experiences as representative for communal experiences, the isolation felt during flight, dependence on a network of supporters, as well as the need for disguise and clandestine movement at night as common characteristics on the level of content (291, 289).

Neil Roberts also points to an "eerie" historical parallel between the media reporting on the state's search for Davis during her time underground and the advertisements "that appeared after the passage of the 1850 Fugitive Slave Law offering rewards for chattel slaves in the South who took flight from a plantation in search of freedom in the North" (54). Yet, while formerly enslaved autobiographers, such as Jacobs, included their so-called fugitive slave advertisements mostly for authentication purposes in their narratives (see, e.g., *Incidents* 78–79), fugitive Black Power activists, such as Davis and Shakur, reference the media campaigns and the wanted posters in order to question the state's legitimization (Davis, *Angela Davis* 19; Shakur, *Assata* 3). "[B]y pitting official images" against "self-portraits" in their autobiographical writing, Chinosole argues, Jacobs, Davis, and Shakur produce an alternative lineage of "radical iconography" of Black fugitive women (*The African Diaspora* 97).

Apart from drawing parallels between the narrator's fugitive act and those of her enslaved ancestors as well as between the writing traditions that have shared those experiences with an interested public, in the above quotation we can also observe the complexity and ambiguity of fugitivity. It is described through the strong nouns "paranoia" and "unknown perils" that are qualified through a list of active verb combinations: "resisting hysteria," "distinguishing between the creations of a frightened imagination and the real signs that the enemy is near," and "to elude" and "outsmart" the "enemy" (*Angela Davis* 5). Moreover, a number of oppositions shape the paragraph. First, the route the narrator takes is usually "familiar" to her, as a fugitive it seems "strange" (5). Second, to elude the enemy as a fugitive is "difficult," while her attitude toward her identity as a fugitive seems "simple" to her – for she 'simply' "had to be worthy" of her ancestors (5) through mental ancestral self-positioning. Third, to elude the "enemy" you need "one true friend" (5). Last but not least, "paranoia" is described as "caressing" the fugitive (5), questioning an opposition between connotations of fear and caring. Familiar and strange, difficult and simple, enemy and friend, and a caressing paranoia are contrastive pairs that create and at the same time interrogate a range of meanings through which the ambiguity of the term fugitivity, so strongly dependent on context and perspective, is reflected in the text.

Clearly, flight for Angela is "portrayed as a trying experience that has little to do with freedom of mobility, and she defies tendencies to romanticize the fugitive's escape and elusive tactics" (Gerund 290). More importantly, recalling Dillon's conflation of prisoner and fugitive, being on the run is also confining for Angela as she has to hide in various apartments during her time underground, feeling "almost as much a prisoner as if I had been locked up in a jail" (*Angela Davis* 11, cf. Gerund 290). Linda's "literal seven-year confinement within a small attic space above her grandmother's living quarters, which she endures as a means of rejecting the confines of slavery, arguably emerges" in Davis's and Shakur's texts "revised as confinement underground" (D. Williams 33).

Ultimately however, despite reflecting on the confining, scary, and challenging aspects of flight with the strangely 'caressing paranoia' of being caught and feeling confined on the run looming large, fugitivity is also distinctly developed towards connotations of possibility and resistance against the agent/enemy/master in Davis's autobiography. Being fugitive raises Angela's awareness to her ancestral relation with fugitive enslaved people, a community she conjures in her imagination to refuse her isolation on the run. Angela's individual experience of fleeing, together with a cultural memory of African American fugitivity in slavery – most prominently shared in slave narratives – reaches a level of experience that is collective and imaginatively crosses time and space. As Dillon writes, slavery not only returns to the present as the prison and other forms of captivity, "for Shakur and Davis it also returns to drive resistance" (116).

Gerund critically notes that the analogies between (formerly) enslaved fugitives and fugitive Black Power activists and their texts, while discernable, also have their limits, not least with respect to the legal issue of "property and ownership" (292). While this may be true for Angela on the run, the difference, however, shrinks when Angela is caught and put in jail. After all, as Assata Shakur would later learn from a prison guard and share in her autobiography, the thirteenth amendment officially permits the enslavement of people in prisons (*Assata* 64, cf. Chinosole, *The African Diaspora* 116). The difference of property and ownership disappears even more so, when we consider not only legal definitions but the Black border and the structural positions of Black people theorized as outside of civil society by Wilderson. Davis ultimately draws on the slave narrative tradition to comment on the legacy of anti-black enslavement and its afterlife in the persecution and imprisonment of Black liberation activists and by extension the African American community in the twentieth century.

"The weight of imprisonment"

In contrast to its ambiguous conceptualization of fugitivity that ranges from resistance, physical escape, to perceived confinement, the legacy of slavery does not play such an overt role with respect to Angela's imprisonment in the autobiography. Instead, the text references contemporary strategies of refusal bar physical flight and theorizes indirectly any possibility of freedom outside of the prison walls as freedom dreams. When Angela is caught after two months underground, she is first incarcerated in the Women's House of Detention in New York City. Supporters, who formed the New York Committee to Free Angela Davis,[146] have arranged a rally in support of Angela's hunger strike against her initial solitary confinement so close to the jail that Angela can follow the demonstration and the speeches from the window in her cell. Reminded of past demonstrations in which she participated herself, Angela feels as if she were one of the demonstrators: "Looking down from my window, I became altogether engrossed in the speeches, sometimes *losing the sensation of captivity*, feeling myself down there on the street with them" (*Angela Davis* 46, emphasis mine). But when she hears her sister Fania make a speech, she is brought back to her personal reality in solitary confinement. I cite extensively again before discussing the passages in more detail in the following:

> Reflecting upon the impenetrability of this fortress, on all the things that kept me separated from my comrades barely a few hundred yards away, and reflecting on my solitary confinement – this prison within a prison that kept me separated from my sisters in captivity – I felt the weight of imprisonment perhaps more at that moment than at any time before.
>
> My frustration was immense. But before my thoughts led me further in the direction of self-pity, I brought them to a halt, reminding myself that this was precisely what solitary confinement was supposed to evoke. In such a state the keepers could control their victim. I would not let them conquer me. I transformed my frustration into raging energy for the fight.
>
> Against the background of the chants ringing up from the demonstration below, I took myself to task for having indulged in self-pity. What about George, John and Fleeta, and my codefendant, Ruchell Magee, who had endured far worse than I could ever expect to grap-

---

[146] As Childs points out, Herndon's and Davis's cases are connected in interesting ways through the New York Committee to Free Angela Davis and the activism of William and Louise Thompson Patterson. Not only did both Herndon and Davis succeed due to the "internationalization" of their political cases. Herndon's campaign was led by African American communist party leader William Patterson, "who headed the CPUSA's International Labor Defense," while about thirty years later the New York Committee to Free Angela Davis was headed by Louise Thompson Patterson, a social activist active in the Harlem Renaissance and Patterson's "comrade and wife" ("Angelo Herndon" 49).

ple with? What about Charles Jordan and his bout with that medieval strip-cell in Soledad Prison? What about those who had given their lives – Jonathan, McClain and Christmas? (46)

Particularly at the beginning of the paragraph, the text is fused with terms related to incarceration: reflecting on the "impenetrability of this fortress" in this "prison within a prison" lets Angela feel the "weight of imprisonment" in "solitary confinement." Again, key words are repeated: "solitary confinement" is mentioned twice, just as "self-pity" and "frustration" are, the two reactions the narrator suffers from while reflecting her situation as a captive. Remarkably, this reflection happens and the resulting "weight of imprisonment" is felt most intensely when the solitude of her solitary confinement is suddenly broken. The deprivation of most visual and aural stimuli is interrupted by expressions of solidarity from her "comrades" outside that are well visible and audible from Angela's cell.

But Angela eventually fights her self-pity by thinking about the many other captives in her jail and in different jails and prisons throughout the country who are incarcerated under worse conditions or have already died in acts of prison revolt. Angela 'roll calls' (cf. Perkins, ch. 1) the Soledad Brothers (Jackson, Clutchette, and Drumgo), her codefendant Magee, Charles Jordan,[147] as well as Jonathan Jackson, James McClain, and William A. Christmas who were killed during Jonathan's "Marin County Courthouse revolt" (*Angela Davis* 4). This time it is not the ancestral collective of fugitive enslaved people but a collectivity of living and dead 'political prisoners' or imprisoned prison activists that Angela reflects on and wants to live up to. The repeated terms "self-pity" and "frustration" are, thus, set against "raging energy" to be used "for the fight" and the refusal to "let them conquer me." As the first rather long sentence of the passage under examination makes clear, however, unlike the "comrades barely a few hundred yards away" on the street, the "sisters in captivity" in the neighboring cells are not visible and audible to Angela due to her isolation inside the jail, nor are the imprisoned and killed 'brothers' of course. Since physical and temporal proximity is out of reach, Angela mentally conjures up in more than one occasion a community of prisoners, both political and 'common,' performing a

---

147 Davis refers to Robert Charles Jordan, a prisoner of the Soledad Correctional Training Facility, who won a federal law suit that deemed his solitary confinement in a small 'strip cell' in the dark for twelve days 'cruel and unusual punishment' (see Stender 796; Davis, *Angela Davis* 28).

form of mental refusal similar to her efforts to be worthy of her enslaved ancestors in the first passage analyzed above.[148]

As Rodríguez points out, Davis "consistently struggled for physical proximity with fellow captive women" who she was often segregated from as the jail operators feared that she would radicalize them politically (*Forced Passages* 125). Yet, not only did authorities fear Angela's influence on other imprisoned women, they also feared her escape. Just as in the passage quoted above, the phrase "for the fight" could incidentally – with a slip of the tongue – shift from fight to *flight*, the prison authorities fear that Angela's fight *against* the prison system from within – supported by the campaign outside – might change into flight *from* the system in an instance. Consequently, Angela is isolated and kept under high security measures at all times and particularly so when she is transferred to a jail in California at the end of part one. The militarized measures during her transfer (70–73), like those on the way to a lawyers' conference, puzzle Angela who wonders: "Did they think I was going to run for it?" (290–91). Perkins argues that both "Shakur's and Davis's accounts reveal the extent to which Black women are victimized by the culture's myths of Black monstrosity," because "when shackled, each is treated as if she had some sort of superhuman powers that would enable her to violently overpower her captors and escape" (ch. 2). We can also see here again the unsoundness of the "'fight or flight' conundrum of Western European political thought" that Greg Thomas criticizes as "a false dichotomy that shields from sight" what he conceptualizes as "the Pan-African praxis of maroonage" ("'Neo-Slave Narratives'" 205). In other words, forms of refusal and flight from the position of social death are *neither* unmistakably fight *nor* flight but both, or none of the above. Consequentially, while the state fears the possibility of its prisoners' escape – a fear that Shakur would later prove justified –, the closest Davis's autobiography admits Angela to fleeing is in her dreams: "In my nighttime dream fantasies, I climbed through this skylight [of her cell] into freedom" (299–300). Again, her mental ability to refuse is crucial here, while the physical situation of incarceration remains.

The scene of solidary protest is doubled later on in the same part of the autobiography with an important shift in the set-up: another demonstration is taking place outside of the jail organized by the bail fund committee and the New

---

[148] On an earlier occasion during her second day in jail, Angela had already conjured up political prisoners in order to "fight the tendency to individualize my predicament" (28). She refers here to Jordan's case again and to "the brother who had painted a night sky on the ceiling of his cell, because it had been years since he had seen the moon and stars" (28). Davis is cross-referencing an anecdote here that she had read in Jackson's *Soledad Brother* in manuscript form (*Angela Davis* 27; cf. Jackson, *Soledad Brother* 313).

York Committee to Free Angela Davis before her transfer (cf. 64–66). As a reaction to the earlier rally and hunger strike (cf. 42–44), Angela is no longer kept in solitary confinement, but in a single cell on a corridor with the main jail population with whom Angela talks about imperialism and shares Jackson's *Soledad Brother* (*Angela Davis* 62). During the second protest then she can hear both the demonstrators outside and the imprisoned women in the neighboring cells of her corridor and on other floors. The reaction triggered by the demonstration is, thus, utterly different: individually felt frustration and self-pity in the first scene give way to collective feelings of empowerment in the second. Angela observes "powerful feelings of pride and confidence […] among the sisters in the jail" (65) that she shares with them. The chants from the demonstrators are met and joined by coordinated chants from the women in the cells, forming a dialogic exchange reminiscent of the African American call-and-response tradition, for instance, in "the work songs of the old chain gangs" (Barnwell 330). While during the first protest Angela has to face the jail authorities alone despite the support from outside, during the second protest expressions of solidarity and performances of resistance among the imprisoned and with the demonstrators are possible and very empowering for Angela and the other participating prisoners. In both cases, Angela consciously refuses to adopt the victim and criminal status assigned to her and the other prisoners by contextualizing her case in a larger framework and voicing the predicament of other political and 'common' prisoners who are literally isolated from each other through the walls of the jail and throughout other US jails and prisons (cf. 65).

According to Rodríguez, the second solidary protest "deconstructs the enforced invisibility of the prison while offering a subtle critique of the cult of personality that had already begun to manifest around her individual case" (*Forced Passages* 126). Moreover, including the names of prisoners throughout the country and of the female prisoners on Angela's cellblock in their chants, a wealth of captivity stories become embedded in Davis's narrative (Barnwell 312, 327), expanding not only the bounds of the autobiography but also epistemically suspending the walls of cells and prisons momentarily from within and without. As Rodríguez writes,

> Although neither the women inside nor the protesters outside could pretend to harbor a determining, definitive, or discrete political agency, they found in their collaboration – verbal, performative, and defiant in its simultaneous politicization of the Greenwich Avenue sidewalk and the looming jail – a form of resistance and radicalism that occupied a new political space while *reconstructing* it through acts of occupation and disruptive speech. (*Forced Passages* 126–27).

In the night after this second protest, Angela is clandestinely transferred to a jail in California, in other words the radical prison praxis Angela, her cell neighbors, and the protesters perform is immediately retaliated by the authorities (*Angela Davis* 67). The spatio-temporal enslavist-carceral continuum allows only momentary 'loopholes' of possibility in which Angela imagines and co-creates communities of enslaved and imprisoned people in and across time and space against the backdrop of what Davis, Jackson, and Shakur unmask in their writing as the prison as plantation and the so-called free world as a prison for Black and other minoritized communities.

Dreams of freedom and the "prison-slave-in-waiting"

Like *Soledad Brother*, *Angela Davis* clearly opts for what Alexandra Ganser, Katharina Gerund, and Heike Paul call a "discourse of liberation" (22) that depicts the fugitive and captive Angela as a "dissenting rebel" who constantly refuses to identify as "helpless victim," while the dominant public project onto her the image of the "dangerous outlaw" (22, cf. Gerund 279). Yet Angela does not accept the role of the "heroic *individual*" fleeing oppression and injustice (Ganser, Gerund, and Paul 22, emphasis mine). In fact, as Rodríguez points out, Davis argues in the introduction to her autobiography, "that no such thing as an 'individual' – in the sense of an autonomous, free-willed, fully self-conscious[,] and self-actualizing political subject – can exist under the American condition of normalized state violence" (*Forced Passages* 123). Instead, Davis calls on a collective of fugitives and captives as enslaved people and (political) prisoners, frustrating a main characteristic of traditional autobiographies that highlights the narrating subject (cf. Perkins, ch. 1).[149] As Gerund notes, in contrast to Shakur, Davis "rejects the position of an exile abroad" and instead locates "her place as a member of the imagined [African American] community" (296) against the state from within. Moreover, in her preface and epilogue Davis directly addresses her readers in order to "extend her political community [as reflected in the autobiography and] to include [the] readers" as well (Chernekoff 47–48). The narrative represents Angela as one Black fugitive and captive among many others and strongly works towards using the narrated personal experience to make a more general argument about a racist and capitalist system that holds Black people captive.

---

[149] By critiquing autobiography as a form from within that very form, Davis's and Shakur's autobiographies may be considered part of what Caren Kaplan has described as "out-law genres" (see also Smith and Watson 59).

Davis identifies the history of slavery as a core source of racism and classism in the twentieth-century United States. However, this history also figures as a source of power for the narrator by supplying her with a collective memory of how to refuse: Angela refuses the status of a victim, she refuses to understand her experiences as exceptional, and she ultimately refuses to accept the status quo. *Merriam-Webster* gives one of the less evident meanings of the term 'fugitive' that resonates with Campt's definition and that of the autobiography. The fugitive and captive Angela becomes "difficult to grasp or retain" for the authorities ("Fugitive"), not because they cannot literally keep her incarcerated – which they do pre-trial for more than one and a half years –, but because they cannot keep her from making use of every tool available to her to refuse "the very premises that have reduced the lived experience of blackness to pathology and irreconcilability in the logic of white supremacy" (Campt, *Listening to Images* 32).

Angela Davis demonstrates with the help of a personal example how a racist and classist system has systematically used anti-blackness against a large portion of the American population from the historical time of slavery until today. It criticizes racism and classism as being at the core of the US legal and political system. Solidarity among African Americans, but also with other ethnic minorities and white people, as well as resistance against the dominant system are understood as key strategies to improve the situation of Black and working class communities in the United States. Davis's political investment in writing her autobiography seems to have been to communicate this message further and to support the larger social justice movement on which she has had such an important impact.

Yet, Angela also acknowledges at moments that there is no safe, free place to inhabit before imprisonment or to reach after flight, only places of refuge. Just as the Northern States for fugitive enslaved people only provided constraint forms of legal freedom, Angela shows awareness of the *a priori* 'capturability' of Black people in the twentieth-century United States. When she describes her involvement in a campaign by relatives and friends of incarcerated African Americans, Davis uses the metaphor and physical reality of chains that connect both slavery and twentieth-century incarceration: "They didn't need to be educated or informed – they knew. The gray walls, the sound of chains had touched not only their lives, but *the lives of all Black people in the country*. Somewhere, at some time, they knew or knew of someone who wore those chains" (258, emphasis mine). From this point of view, Davis's autobiography not only anticipates racialized mass incarceration as it would evolve over the following decades, but also acknowledges what Afro-pessimist theory reveals as the structural positionality of Black unfreedom or 'social death.' Not unlike Jackson's "slave" position from his mother's womb to his death in prison, the chains of captivity in *Angela*

Davis prove flexible but unyielding, not a transitory experience on a clearly charted path to freedom but 'permutations' of captivity that include confinement underground and in prison as well as in society at large.

Thus, while the autobiography is undoubtedly not Afro-*pessimist* in its overall outlook, casting a look at the text that is informed by Afro-pessimism and Black feminist fugitive thought poses some challenging questions. Clearly, the autobiography shows resistance to be possible and to trigger change on the level of Angela's individual experience (and some of those around her). After all, she is released from jail on bail and finally acquitted of all charges. But does this resistance against the racist and classist society have any influence with respect to the structural positionality of Blackness in the United States as elaborated by Wilderson? Does her release from jail make Davis genuinely free or freer than Jackson who died in prison on a structural level? Wilderson would argue that Angela Davis – like all other people racialized as Black – continues to be a "prison-slave-in-waiting" ("Prison Slave" 18), who experiences freedom from state enforced incarceration, but remains structurally *a priori* captive or at least 'capturable' due to the prevailing history and presence of Black 'social death.'

Nonetheless, Davis's autobiography partakes in nothing short of the (further) development of what Hartman has called a "dream of freedom"; dreaming being the closest Black people can get to the concept of freedom when looking at it from the perspectives mapped out especially in Wilderson's work. From this point of view, Davis's autobiography that is also clearly indebted to Jackson's writing should be understood as an investment in a 'future reality of freedom' or, to use Campt's words, as an investment in futurity. Reading it from the perspective of Afro-pessimist and Black feminist fugitive thought, *Angela Davis* appears not only as a hopeful autobiographical narrative of captivity and fugitivity that grapples with the situation of Black people in the United States in the late 1960s and early 70s, but also and particularly as a personal expression of a collective and enduring dream of freedom elsewhere and 'else when.'

Analyzing passages of Davis's autobiography from such a perspective brings into focus the text's revolutionary rhetoric and the narrator's strategies to maintain a profound belief in liberatory change against all odds and the desolate backdrop of prevailing Black 'social death.' Clearly, the text stands as proof that social death and refusal, or the social life of the socially dead co-exist, actually relying on each other. As the analyses of Jackson's and Davis's autobiographical texts already show, there is an important differentiation to be made, however, between literal forms of captivity and the imagined, philosophical performances of mental flight that writing about literal forms of captivity from within the jail or prison seems to enable. This will become even clearer in the following analysis of fugitivity in Shakur's autobiography.

"You'll be in jail wherever you go": Confinement and flight in *Assata: An Autobiography*

Assata Shakur's autobiography shares a lot of themes and characteristics with Angela Davis's autobiography, while it has received less academic and public attention. Like Davis, in *Assata: An Autobiography* a formerly incarcerated Black Power activist shares experiences of imprisonment and flight in relation to the FBI's COINTELPRO in the United States in the late 1960s and early 70s and formulates Black resistance from the perspective of a political prisoner, Black revolutionary, and fugitive. Unlike Davis, however, after several trials of robbery, kidnapping, and murder Shakur[150] was sentenced to life in prison in 1977 for murdering a New Jersey state trooper even though forensic evidence showed that she was not physically able to shoot (cf. James, *Imprisoned Intellectuals* 114).[151] Shakur escaped prison in 1979 and received political asylum in Cuba, where she published her autobiography in 1987.[152] Besides narrating her experiences of surveillance, incarceration, and the trials, Shakur's autobiography

---

[150] Shakur's born name is Joanne Deborah Byron, her married name Chesimard. When she ponders the enslavist history of her husband's name during college, she rejects both and takes the name Assata Shakur (cf. *Assata* 185–86).

[151] Shakur was in the company of her comrades Zayd Malik Shakur and Sundiata Acoli on the day of the arrest. Zayd died during the shooting on the turnpike, while Sundiata escaped wounded. He was captured a few days later and also convicted of murder of the state trooper. He remains incarcerated.

Joy James contents that the charge of a Black person murdering a white police officer "functions today in a manner similar to the rape charge – the sexual assault of white females by black males – during the lynching era. This specific accusation, irrespective of evidence or facts, mobilises intense, punitive sentiment and racial rage that supports police, prosecutorial, and judicial misconduct in order to achieve swift and deadly retribution" ("Black Revolutionary Icons" 146). She notes, however, that Shakur and Mumia Abu-Jamal, who were both convicted of killing white policemen, received wide support from African American writers and scholars across the political spectrum, such as Henry Luis Gates Jr., Alice Walker, Manning Marable, and Cornel West, since they found that their "political trials where prosecutorial zeal and malfeasance [and] denied defendants due process" (146).

[152] Six years after the publication of *Assata*, Shakur's aunt Evelyn A. Williams published the memoir *Inadmissible Evidence: The Story of the African-American Trial Lawyer Who Defended the Black Liberation Army*. For an analysis of Williams writing about her work as Shakur's lawyer, see Chinosole, *The African Diaspora* ch. 7.

shares her understanding of the anti-black, capitalist structures of US society and illustrates the gendered role of Blackness and anti-blackness in the segregated US South and the urban US North of the 1950s, 60s, and 70s that clearly resonate with Davis's narration.

While *Assata* responds to the gendered anti-black spectacle of her capture, torture, incarceration, and her ultimate escape from prison, more mundane forms of anti-black violence and its refusal are reflected in the "retrospective oppositional narrative" (Chernekoff 50) in less obvious ways. Similar to Jackson's *Soledad Brother* and Green's broad definition of confinement, notions of imprisonment and flight include the ways in which Assata's life before the arrest and after her escape are subjected to other, more ambiguous forms of captivity that expose life for Black people around this time as an "incomplete project of freedom" (Hartman, "Venus in Two Acts" 4). Captivity and flight in *Assata* are tied to the history of enslavement and the literary tradition of slave narratives.[153] As Hames-García argues, Shakur's "autobiographical narrative [...] fuses autobiographical, literary, political, and philosophical elements so as to transform the genre of autobiography and to transcend its traditional limits" (95). Like Jackson, she refashions the autobiographical narrative into a "primer for revolutionary action" by drawing on the tradition of the slave narrative and Black political autobiography, such as Jacobs's, as a historical and literary archive (Hames-García 138, cf. also Perkins, ch. 2; Braxton). Similar to *Incidents*, family and motherhood play a central role in Shakur's conceptualization of freedom and its relation to revolution (cf. Hames-García 125–29, see also below). However, like Linda's attic space and the North, places of fugitivity and refuge in Shakur's "text of exile" (Paris 209) allow for freedom dreams but are at the same time still confined by the Black border and the gendered anti-black climate they emerge from.

In "The Vengeance of Vertigo," Wilderson analyzes the role of anti-black violence in the writing of Black Liberation Army members Bukhari and Kuwasi Balagoon.[154] While "Black paramilitary writings are to be commended for their pro-

---

[153] Shakur's prison and flight autobiography also intertextually speaks to fictional neo-slave narratives such as *Dessa Rose* by Sherley Anne Williams (1986) and Morrison's *Beloved* (1987). The novels were published around the same time as the autobiography and focus on fugitive enslaved women and mothers in North America during the historical era of slavery (G. Thomas, "'Neo-Slave Narratives'" 225). The latter will be discussed in chapter 2.3.

[154] Kuwasi Balagoon (1946–1986, birth name Donald Weems) was a bisexual member of the Black Panther Party and the Black Liberation Army. He escaped prison twice and was presumably involved in Shakur's liberation in 1979 (Umoja 214–15). He died in prison of an AIDS-related illness in 1986. Balagoon's early writing was published in the collective autobiography of the Panther 21 *Look for Me in the Whirlwind*. Later writings appeared in *A Soldier's Story: Revolutionary Writing by a New African Anarchist*. For a discussion of his life and writing, see Umoja.

clivity to subordinate the egoic individual to the collectivity of Black people on the move," Wilderson argues, "these rhetorical strategies are less attributable to conscious selection and combination decisions than they are to the quandary of a Black unconscious trapped by the disorientation of [...] a paradigm of violence which is too comprehensive for words" ("The Vengeance of Vertigo" par. 21). In other words, what may be perceived as a decentering of the Black imprisoned autobiographical self for the captive community that frustrates traditional autobiographical genre conventions can also be understood on a structural level as the lack of a Black 'Human' self that could own herself and her story instead of being captured by it. Wilderson further contends that the audiotaped message "To My People July 4th 1973," that Shakur had prepared in the first month of her incarceration and included in the autobiography (*Assata* 49–53),[155] unconsciously acknowledges that the Black Liberation Army and Black people in general are "void of relationality" ("Black Liberation Army" 175). Since for Wilderson Blackness exists outside of the time and space of civil society, Shakur's communiqué is unable to follow the narrative "arc of an emancipatory progression" that, as I summarized in chapter 1.5., would move from a prior state of equilibrium via momentary disequilibrium to equilibrium restored (178). Shakur is thus compelled to repeat the gratuitous violence that she and other Black people suffer from, while being unable to provide a "political strategy or therapeutic agency through which the violence which engulfs her flesh can be separated from the text's compulsion to repeat that violence" (180). Thus, like Jackson's understanding of his *a priori* captivity from birth, Shakur's writing exemplifies the "torturous clash between, on the one hand, an unconscious realization that structural violence has elaborated Blacks so as to make [their] existence void of analogy and, on the other hand, a plaintive yearning to be recognized and incorporated by analogy nonetheless" (178–79). As we will see below, gendered anti-black violence introduces *Assata*'s autobiographical I and permeates the autobiography, gesturing towards a non-human and *a priori* captive status that crosses boundaries from birth to Cuban exile, but cannot overcome the Black border.

---

155 The title of Shakur's speech clearly evokes Douglass's 1852 speech "What to the Slave Is the Fourth of July" (James, *Imprisoned Intellectuals* 17) in which Douglass "refuses even to entertain the question whether or not Blacks are fully human" (Chinosole, *The African Diaspora* 10), while criticizing the myth of the American Founding Fathers that excludes the (formerly) enslaved from the category of 'the Human' (Paul, *The Myths* 218–19).

Between social death and affirmations of life

Assata, originally published with the subtitle *The Autobiography of a Revolutionary* (Latner 464), is split into twenty-one chapters of varying length, sometimes accompanied by a poem. Chapters 1 to 12 alternate between two different plot lines: first, the plot of Assata's capture, hospitalization, incarceration, pregnancy, and trials, and second, the plot of her childhood, education, and political activism.[156] Similar to Davis's autobiography, *Assata* begins *in medias res* with Assata's arrest in 1973, steeping the beginning – as William Michael Paris aptly notes – both in "affirmation for life, birth, and return" through the poem "Affirmation" that precedes the first chapter (*Assata* 2), and in the "death and despair" of the violent arrest in the first chapter (210). The first six chapters with uneven numbers follow Assata from the arrest during which she was severely injured, through her time in hospital, and her experiences in jail, followed by the trials. In these jail and trial chapters, details about the daily life in prison, torture, and mistreatment by guards and doctors during her pregnancy as well as the course of events during the different trials dominate the narrative. Lennox S. Hinds, the African American editor of the autobiography and one of Shakur's attorneys, offers a "Trial Chronology" that lists the trials, their location, the charges, and the outcome as a peritext for reader orientation (*Assata* xiv). Hinds also provides a foreword that frames the autobiography as a truthful personal account (vi–xiii), functioning in surprisingly similar fashion to the prefaces of white abolitionist editors that introduced slave narratives for authenticating purposes (cf. Smith and Watson 101).

The first five chapters with even numbers share Assata's coming-of-age story. They trace her path chronologically and episodically in bildungsroman-fashion from her birth in 1947, her early childhood with her mother and sister in Jamaica, Queens, and with her grandparents in Wilmington, North Carolina, through her teenage days as a school dropout and runaway, to her time as a college student and emergent Black revolutionary in New York and California. After the violent *in medias res* beginning of the first chapter, the second chapter returns to Assata's birth by stating ironically that the "FBI cannot find any evidence that i was born" (18). As Paris notes, the state's inability to account for her life beyond her 'criminal record' shows how her status as "a full human subject" is called into question from the start (212). Stating boldly that "Anyway, I was born" (18), Sha-

---

[156] Joseph G. Ramsey describes the two plot lines that continue throughout the autobiography aptly as first "a narrative of Incarceration" and "a 'struggle for freedom'" and second "a narrative of Education" and "a 'struggle for consciousness'" (121).

kur "finds herself testifying in and against the court of the social imaginary in the United States" (Paris 211) that positions her as non-human, recalling the slave narrative writers' efforts to write themselves into being as 'Humans,' not least with their *ab ovo* beginnings. The narrative consequently struggles between affirming "life" first expressed in the aforementioned poem and the "trauma and violence" of social death that the first chapter introduces (Paris 210–11). As autodiegetic narrator and keen observer, Assata – again reminiscent of the slave narrative tradition – deals with these tensions by not merely focusing on herself but by including detailed descriptions of the urban environment she grows up in and of the people populating these places, paying close attention to race, class, and gender and how they intersect in people's lives. Similar to Angela, as a narrator Assata is highly selective, providing only limited information about her own life and focusing on the ways in which especially the challenges she encounters and the mistakes she makes can be interpreted as teachable lessons for understanding the main conflicts and antagonisms for Black people in the United States (cf. Hames-García 103).

The lucid, alternating narrative pattern interweaves the jail and trial plot with the coming-of-age plot and clearly connects Assata's experiences growing up with the events after her arrest, without ever allowing them to overlap. Paris contends that by keeping the two plots separately, Shakur "makes no attempt at unifying the two lives" and instead reinforces "the two tensions she [had] set out in the beginning of her work: affirmation/denial or life/death" (214). Chinosole compares the narrative movement back and forth between the two plots to a jazz composition in which the "chapters about her rebellion and its consequences are responses to those about her early life that explain how she arrived at such a radical political position," while Shakur's poems that close many chapters "accent the musical line as codas" (*The African Diaspora* 116). Connecting Chinosole's argument with the work of Moten and Campt discussed in chapter 1.4., Shakur's narrative can be seen as reflecting already on its macro-level the tension between the 'jazz' of Black social life and its structural subjection to gratuitous violence.

While chapters 13 to 18 abandon the strict alternating pattern, they are still fully dedicated to either of the two plot lines that come closer to their supposed climaxes, that is, on the one hand, the final Turnpike trial, the prison for life sentence, and the prison escape, and on the other, Assata joining the loosely organized underground resistance of the Black Liberation Army. Yet, both plot lines elide the potential climaxes, leaving the guilty verdict and Assata's time as a revolutionary underground in an ellipsis between chapter 18, which tells of the final preparations of the trial, and chapter 19, which describes Assata's life at New Jersey's Clinton Correctional Facility where she is moved to after the sentence.

These strategic gaps, of course, issue from the fact that Shakur – like Davis and Jackson as well as Douglass and Jacobs – was literally writing for her life and had to account for different audiences that were either "hostile" or "sympathetic" to her, her comrades, and their revolutionary project (Perkins, ch. 1 and 2).

As Chinosole argues in her study of eighteenth to twentieth-century US-based African American, Caribbean, and African autobiographical narratives, as a slave narrative writer Jacobs clearly "provides matrilineage" for Shakur. Yet, while "Jacobs depicts herself [as] a reformist, [...] Shakur proclaims herself as a revolutionary" so that Chinosole adds to Shakur's 'ancestry' Harriet Tubman as "Shakur's political foremother" (*The African Diaspora* 100).[157] For Chinosole, Shakur's text "is like a bold shout, answering Harriet Beecher Stowe's stereotypes of Black women in *Uncle Tom's Cabin* and advancing Jacobs'[s] portrayals of African American women" (115).

Hames-García also connects Shakur's writing and especially her concept of freedom as interdependence to *Incidents*, while describing Davis's autobiography that appeared about fifteen years earlier than *Assata* as a central intertext as well (98). Jacobs, Davis, and Shakur, Hames-García argues, develop a concept of freedom that not only relies on the physical freedom of the autobiographical subject but also on the struggle for future unconfined freedom of their families and communities that Hames-García describes with Ernst Bloch as "critical utopianism" (120–21). The autobiography reflects this stylistically not least through the consistent lowercasing of the autobiographical I that rejects a "perfected and masterful subject" (104) and redirects the attention away from the spectacle of her own captivity and flight towards the mundane plight of Black communities (cf. Perkins, ch. 1).[158] Similar to Davis's and Jacobs's texts, this attempt runs counter

---

[157] Harriet Tubman (1822–1913) was a formerly enslaved abolitionist involved in the Underground Railroad, who also fought and spied for the United States Army during the Civil War and later became an activist for women's suffrage (cf. Humez). Paul describes Tubman as a "Moses" figure, "because she repeatedly went back to the South after her own escape and led more than 70 slaves to escape" to the northern states and Canada West where she resided for a decade (*The Myths* 170; Henry, "Harriet Tubman").

[158] Chinosole remarks how the title and lower-case i invite readers to get to know the narrator/author intimately "on a first-name basis" (*The African Diaspora* 117). Of course, Shakur's orthographic practice of lower-casing the I is also reminiscent of "the grammar of the New Afrikan Independence movement" of the time, according to which the first person singular is consistently lower-cased and the first person plural capitalized to emphasize the community over the individual (Umoja 216).

Beside the lowercase "i," Paris and Perkins also address other orthographic peculiarities in Shakur's text that "translate[...] her resistance from the social sphere to the world of the text" (Perkins, ch. 4). Paris argues that Shakur "deploys resistance spelling such as 'kourt,' 'kourt-

to reader expectations and genre conventions of autobiographies, which Perkins defines broadly as "the narrative ordering of an individual's life that illuminates, in the process, his or her uniqueness" (ch. 1), epitomized in the American autobiography of Benjamin Franklin and its focus on a linear progress of the subject towards success and a complete self (Hames-García 102).[159] It also sets Davis's and Shakur's autobiographies apart from Jackson's *Soledad Brother* in which George expresses concern for the community, while conceptualizing flight as an individualized self-defense.

With Wilderson, the decentering of the self can also be understood as an unconscious reflection of the violent abjection of Blackness to the outside of civil society. Of course, if we cannot consider the fugitive life narrated as selfowned or 'Human,' Black Power autobiographies' unconscious acknowledgement of the non-existence of a Black autobiographical subject calls into question whether the writing under scrutiny here can or should be subsumed under the moniker of autobiography (from the Greek *autos* for self, *bios* for life, and *graphein* for 'to write'). After all, "[t]he emergence of autobiography as a literary genre and critical term [...] coincides with what has frequently been called the emergence of the modern subject around 1800" (Schwalm, par. 11), in other words with the emergence of the white modern subject and its "abjection" of Black people as chattel (cf. Broeck, *Gender*).[160] While feminist and postcolonial studies have requested the broadening of genre definitions and autobiographical canons to render autobiography studies more inclusive and diverse, acknowledging other non-heteronormative trajectories, traditions, and forms of narrating the self and lives (cf. Schwalm, par. 22), an Afro-pessimist analysis of Black auto-

---

room,' or 'amerika' in order to challenge the hegemonic imagination of these terms" and bring "to light the fundamental complicity between the language of justice in the United States and its racist history" (213). Perkins contents that the "consistent use of lowercase initials for proper names (e. g., amerika, new jersey, kourt, leavenworth) [...] and her deliberate alteration of conventional spellings (e. g., substituting the Germanic *k* for the letter c in the previous list)" allows Shakur to "better reflect[...] her own particular experience" (Perkins, ch. 4).

159 For a discussion of the ways in which slave narratives, such as Douglass's and Jacobs's, respond to "the Franklinian tradition of American revolutionary autobiography" (Levine 108), see Levine.

160 As Broeck argues, "the white modern subject (male or female) might be considered an abjector, that is, a motorizing force which needed Black thingification to 'know,' socially, culturally, politically, and epistemically, its subjectivity and its social being" (*Gender* 17).

biography would abandon the assumption of an autobiographical subject completely for the socially dead abject.[161]

From New York City runaway to revolutionary fugitive

After introducing the narrative with the violence of capture and of questioning the existence of her birth and her subjectivity in the eyes of the state in chapters 1 and 2, the coming-of-age plot quickly addresses the ways in which forms of captivity and flight further challenge Assata's conception of herself as a valuable human being. Before even becoming a teenager, arguments between her mother and her stepfather cause Assata to take long walks, cycling tours, and subway rides through "all kinds of neighborhoods – white, Black, Puerto Rican, Chinatown as well" (*Assata* 74). Assata describes the urge to escape the family's apartment as "some kind of disease" (74), indirectly referencing the ways in which enslavers and pseudo-scientists pathologized, disciplined, and regulated the movement of enslaved Black people in the nineteenth century as a pseudo-illness called 'drapetomania' (Rushdy, *Neo-Slave Narratives* 257). Not unlike Jackson, who "left home [as a teenager] a thousand times, never to return" (*Soledad Brother* 10), Assata too "was running away from home and [...] didn't even know it" (*Assata* 74). While staying vague about the conflicts with her family, in hindsight the narrating "i" understands the excursions not only as physical and "emotional distancing" from her family (Perkins, ch. 3) but as literal escapes from a childhood deeply troubled by anti-black racism and dominant white middle-class gender roles that affected her family and particularly her parents, who

---

[161] There has been no in-depth discussion of this question with respect to the tools and terms of autobiography studies to my knowledge so far. Yet, Taija McDougall gestures toward this point when she states in her discussion of *Soledad Brother* that Black life writing "both adheres to and obfuscates the autobiographical category" (McDougall 8). It may be worthwhile examining the degree to which notions of a 'hetero-(bio)-graphy' or 'autoethnography' (i.e. 'writing the othered self' or 'writing the cultural/ethnic/raced self') might better approximate the abjected autobiographical I in Black autobiography. For the conventional meaning of the term heterobiography that such an examination would necessarily depart from, see Schwalm, par. 7. On the term autoethnography, see Smith and Watson 157–60.

I thank Marius Henderson for sharing his thoughts with me on the question of the absence of an autobiographical subject in Black autobiography and the potential consequences for the use of genre terminology in a discussion of an earlier version of this chapter in the doctoral network "Perspectives in Cultural Analysis: Black Diaspora, Transnationalism, Decoloniality" at the University of Bremen, June 2019.

"were catching hell every day on their jobs, in society, and they took their frustrations out on each other" (*Assata* 73–74).

As soon as "the problems in [her] family intensified," running away from home becomes a more conscious choice for Assata (74–75). Aged 13, she runs away from home for an extended period of time, committing what might be perceived as "class suicide" (Chinosole, *The African Diaspora* 130) by 'choosing' to live temporarily among the most marginalized of the city, poor Black teenagers, gender nonconforming people, people with mental health issues and addictions as well as unhoused people and sex workers. She tries to make a living on her own, crashing on couches of acquaintances, sleeping on the subway, and in a cheap hotel in Manhattan's neighborhood Greenwich Village. She provides bare necessities first by learning shoplifting and fraud as survival strategies from street acquaintances, then by working odd jobs, all the while looking for a place of refuge in the anti-black urban environment: "At times, running away was fun and exciting. At other times it was miserable, cold, and lonely. [...] It was one hell of an education, and, when i think about it, i was one lucky chile. So many things could have happened to me, and almost did" (*Assata* 75).

Not unlike the training in maroonage/marronage that George receives while roaming the countryside of Illinois and the streets of Chicago, where he suffers from state surveillance and violence that specifically targets young Black men, the streets of New York City teach Assata her first lessons of flight, lessons that are more often than not structured by gendered anti-black violence. During her time on the streets, Assata learns quickly that the places of refuge available to her are not safe, particularly because men consider a Black female runaway teenager like her as always already sexually available to them (*Assata* 106). Assata is, among other things, sexually assaulted by a white cafeteria employer (102–103) and barely escapes gang rape. She is particularly shocked to realize that the group of African American boys around her own age who attempt to rape her, were "arguing and carryin on as if i wasn't even human, as if i was some kind of thing" (113–14). In fact, Assata escapes the rape not due to "any respect or consideration for her as a human being" but because of "the threat of property damage" she poses, when she threatens to destroy the apartment in self-defense (Perkins, ch. 3). As a Black teenage runaway, Assata experiences the invasion of gendered manifestations of anti-blackness on her own body framed – like Linda's before her – as 'non-human.' The narrating i further connects the anti-humanization of her being she experiences as a teenager by the white dominant society and in African American patriarchal communities to the history of slavery, tracing the issue "directly back to the plantation," where Black women were considered "less than a woman, [...] a cross between a whore and a workhorse" (*Assata* 116). By running away from a family and community that suffers from gendered anti-black racism which they also take "out on each other" (73–74), Assata re-counters those same challenges as immediate threats to her own body in the streets where she has to develop an escape strategy on the run, similar to Jackson's search for "a stick." In writing the autobiography, Shakur continues this search, drawing on the language and "grammar" of enslavement (Spillers, "Mama's Baby, Papa's Maybe") to process those experiences in hindsight and understand the captive structures that made those experiences possible.

Over the following coming-of-age chapters, the New York City teenage runaway grows into a Black Power activist and revolutionary on the run. Apart from the lessons learned on the street, different forms of education contribute to that development: a formal school education that Assata exposes as racist, not least with the help of the education she receives at the Manhattan Community College (*Assata* 32–34, 175–78); an informal cultural education her aunt gives her while they live together (135–36); as well as lessons in activism that she receives during and after graduation in activist communities in New York and California (181–91, 197–205). Assata shares the knowledge she gains on the history of the United States, slavery, and the role of racism generously with the readers throughout the text, explaining, for instance, the role of white male elitist economic interests in the American Revolutionary War and the Civil War (32–34).

After the completion of her activist education in California (ch. 13) and with the Black Panther Party in New York City (ch. 15), Assata eventually joins the loosely organized Black Liberation Army and by doing so becomes one of the "main targets" of the FBI (James, *Imprisoned Intellectuals* 114). As she explains, "[a]t the end of the sixties or the beginning of the seventies, people were going underground left and right" so that it almost seemed "as if the entire Black community was on the FBI's Most Wanted list" (*Assata* 234). Assata draws a close connection here between the Black community at large and its Black Power radicals that resurfaces when she narrates her first experience travelling as a disguised fugitive before dropping the plot strand and leaving her time underground completely out of the narrative. While Linda reveals her small and dark place of hiding that enabled physical escape, Angela's and Assata's 'garrets' underground are left mostly unknown.

When Assata vacates a friend's apartment in New York City to change her place of hiding, her subway ride is narrated in dramatic present tense. Assata scans the other passengers for potential threats to realize that her disguise as a so-called "maid" lets her blend in perfectly with the other Black women commuting early in the morning, most of whom Assata assumes to be working in the service sector. Almost all of them wear wigs to cover their natural hair just like Assata (and Angela on her escape before her), with "hair becom[ing] the [main] vehicle for masking" and trickster-ism[162] (Chinosole, *The African Diaspora* 119):

> Maybe we are all running and hiding. Maybe we are all running from something, all living a clandestine existence. Surely we are all being oppressed and persecuted. [...] A whole generation of Black women hiding under dead white people's hair. [...] I pray and struggle for the day when we can all come out from under these wigs. (239)

Drawing on the motif of masking that Sidonie Smith had identified in the Black autobiographical tradition since the slave narrative, Hames-García argues that Assata recognizes in her fellow commuters the risk of potentially "los[ing] herself in the mask" (107). Reminiscent of Barnwell's, Chernekoff's, and Gerund's interpretations of Davis's mirror and disguise scene as connecting fugitivity with vulnerability and a (fear of the) loss or change of her identity, Hames-García interprets Assata's disguise scene as pointing to the risk of "los[ing] herself to her disguise" (108). While Angela sees her reflection in the mirror, Assata sees herself mirrored in her co-commuters' faces, who – according to Hames-García – "have masked themselves for other reasons and have, perhaps, lost their sense of selfhood and succumbed to the psychological enslavement of sexist and racist oppression" (107–08). Whereas Assata's fugitive disguise "to preserve her physical freedom" may indeed gesture toward the risk of "losing the possibility for giving and receiving free recognition" (108), it actually suspends differences between her as a Black revolutionary fugitive from law enforcement and the daily practice of African American women dressing up for work in a white-dominated environment that forces them into underpaid service jobs and to strive for dominant beauty standards. The terms "running" and "hiding" are repeated twice and variations of "we are all" appear four times in this brief quotation on the "clandestine existence" of Black women in America, pointing to the mundane structural necessity to run and hide on an everyday basis and to the ways in which flight and camouflage continue to function as central survival practices of the majority of Black women in an anti-black world.

---

162 Apparently, a wig also played a central role in the events that resulted in George Jackson's death. According to James, Jackson hid a gun under an Afro wig, both of which had been smuggled into the prison and were used in the attempted prison takeover (James, *Imprisoned Intellectuals* 86).

From astro-travel to prison escape

In the jail and trial plot, Assata has to find ways to escape without disguise and while staying within the walls of her cell. Like *Angela Davis* and *Soledad Brother*, Shakur's autobiography documents how Black Power prisoners consciously exercise their mental abilities to refuse the pathologizing of their existence as prisoners by reading, writing, physical exercise, and most importantly active strategizing and participating in their court cases and political campaigns as forms of mental escape. While waiting for the Turnpike trial, Assata learns about another form of mental flight, i.e., of mentally dreaming yourself out of the prison. Eva, a Black woman considered "crazy" and a "hell raiser" (*Assata* 59), introduces Assata to "astro-travel":

> "[...]. You can go anywhere you want to," she told me. "You just have to project yourself."
> "Can you show me how to project myself the hell out of here?"
> "Oh, that's easy," she said, "I do that all the time. As a matter of fact, I'm not here now."
> "No," I said, "that's not good enough. I want to project my mind and body out of here."
> "You'll be in jail wherever you go," Eva said.
> "You have a point there," I told her, "but i'd rather be in a minimum security prison or on the streets than in the maximum security prison in here. The only difference between here and the streets is that one is maximum security and the other is minimum security. The police patrol our communities just like guards patrol here. I don't have the faintest idea how it feels to be free." (59–60)

Assata appreciates Eva's "astro-space projection" (59) as a mental form of escape and acknowledges that Eva often "made so much sense" it seemed that "it was really the world that was crazy" (60). Questioning notions of mental health and illness, Eva and Assata agree on the ways in which Black people are not only captives in prison but remain unfree outside of the prison walls as well, acknowledging what Jackson in *Soledad Brother* describes as an "inherited [...] neo-slave existence" (111) and what Hartman has theorized as "the precarious life of the ex-slave, a condition defined by the vulnerability to premature death and to gratuitous acts of violence" ("Venus in Two Acts" 4). Moreover, "Shakur argues that, because most people have never experienced true freedom, they cannot really know what it is from experience; yet they do have a sense of what it is not" (Hames-García 121). The passage points to the lack of difference between the streets and the prison and bends the spatial concept of the prison to describe other, more ambiguous forms of captivity of African American people. As Childs notes, in doing so, Shakur "joins many other Movement Era Black radical thinkers in expressing the degree to which official imprisonment is the zero degree of a generalized structure of social imprisonment for Black people in the United

States" ("Angelo Herndon" 52). Similar to Jacobs who questioned the freedom available to her in the North after her escape and Jackson's *a priori* captivity, Shakur emphasizes that in the United States in the 1970s Black people were not only captives in prison. She shows how she and a diverse cast of people she meets during her time as a teenage runaway remain unfree outside of the prison walls. Like Assata's, the lives of gender nonconforming sex worker "Miss Shirley" (*Assata* 108–09) and the girl "Lil Bit," who lives in roach-infested public housing (134), continue to be confined by poverty and gratuitous state and gendered anti-black violence.[163]

While appreciating Eva's critical analysis of society as captive, Assata considers Eva's mental fugitivity insufficient. She expresses dissatisfaction with an escape that allows her mind to travel while her body stays put, foreshadowing her prison escape two years later. Yet, how she escapes with the help of "comrades" remains unclear in the autobiography. Instead of providing information about the escape, the narrative voice goes to some length to insinuate and justify the escape through two prison visits of her mother, daughter, and grandmother. Again, like Jacobs, family and motherhood provide the reason for flight and potential self-confinement underground and on the run (cf. G. Thomas, "'Neo-Slave Narratives'" 222).

The first visit is from Assata's mother and Assata's four-year-old daughter, who was conceived and born during the trials and taken from Assata only a few days old.[164] The scene, narrated in dramatic present tense, with short sentences, and with a significant amount of direct dialogue, shows rather than tells and

---

[163] For a similar interpretation of Shakur's understanding of the captive status of Black women and women of color in the 'free' world as reflected in her essay "Women in Prison: How We Are" from 1978, see Dillon 113–17.

[164] During the Bronx bank robbery trial with codefendant Kamau Sadiki (birth name Fred Hilton), Assata and Kamau are repeatedly removed from the courtroom and sent to a separate room for their political interventions, which are considered disturbances (*Assata* 93–94). During the time they share the room privately, Assata becomes pregnant. After long legal fights about her medical care that risk her health and that of her baby, she gives birth to her daughter Kakuya Amala Olugbala Shakur in a hospital and only a few days later is returned to jail alone (143–45). As Greg Thomas points out, the "pregnancy on trial and in prison is a major scandal for neo-slavery," so that Shakur interprets bringing a child into this world as a revolutionary act that expresses her "possibilities of natural or supernatural reproduction" ("'Neo-Slave Narratives'" 222). Hames-García points to the ways in which revolution in *Assata* is conceptualized as motherhood (124–31) and the birth of her child therefore symbolizes not only the continued revolutionary struggle but also "the hopeful possibility that future generations of blacks will be free" (127, see also Rolle 168).

therefore makes more immediately comprehensible Assata's claim to a mother's need to be free for her innocent child.

> "You can get out of here, if you want to," she [Assata's daughter] screams. "You just don't want to."
> "No, i can't," i say, weakly. [...] I look hopelessly at my mother. [...] "Tell her to try to open the bars," she says in a whisper.
> "I can't open the door," i tell my daughter. "I can't get through the bars. You try and open the bars."
> My daughter goes over to the barred door that leads to the visiting room. She pulls and pushes. She yanks and she hits and she kicks the bars until she falls on the floor, a heap of exhaustion. I go over and pick her up. I hold and rock and kiss her. There is a look of resignation on her face that i can't stand. [...] When the guard says the visit is over, i cling to her for dear life. [...] She waves good-bye to me, her face clouded and worried, looking like a little adult. I go back to my cage and cry until i vomit. I decide that it is time to leave. (*Assata* 258)

Assata not only justifies her escape from prison with an unfair trial and verdict and the fact that she fears to be killed in prison (58–59), but also with the effect her own incarceration has on her daughter and her family who have to make ends meet in the unfree world outside of the prison without her. Like Jacobs's slave narrative, Shakur's autobiography indirectly reflects on the confining structures that systematically strain family relations inside and outside of the prison (see Knopf 221–24) while also offering potential loopholes for escape.

The indirect appeal to womanhood, motherhood, and the criminalization of whole Black communities as reasons to break the law and escape can also be found in Jacobs's narrative when Linda tries to convince her readers that she is a woman just like them who has a right to live legally free with her two children. Moreover, Hames-García identifies parallels between Linda's doubts about having her freedom bought "under an unjust and illegitimate slave law" and Assata's regrets at having legitimized the guilty verdict and the legal system that systematically persecuted Black Power activists through her active participation in her last trial (111). Ultimately, Linda's and Assata's appeals to freedom and free motherhood stand in stark contrast to contestations of their structural positions as (formerly) enslaved and captive fugitives outside of the realm of 'the Human,' onto whom legal rights could be bestowed. This unsolved tension of life in social death and fugitive forms of refusal permeates African American autobiographical genres throughout the ages.

During the second visit, Assata's grandmother tells Assata of a dream that predicts her escape:

> My grandmother came all the way from North Carolina. She came to tell me about her dream. My grandmother had been dreaming all of her life, and the dreams have come true. [...] "You're coming home soon," my grandmother told me, catching my eyes and staring down into them. "I don't know when it will be, but you're coming home. You're getting out of here." (260)

"[A] day or two before [... she] escape[s]" Assata speaks to her grandparents again over the phone, asking them about their "family's history, tracing the ties back to slavery" (261–62), thereby alluding to a connection between her imminent escape and the history of her enslaved ancestors. Fugitive enslaved people shared knowledge about escape routes and strategies for flight in coded form, in songs or stories for example (cf. H. B. Franklin, *Victim as Criminal* 90), and often kept some aspects if not the whole escape out of their public testimonies to allow others to use the same route or means of escape and not to endanger the lives of supporters (see 2.2.). Similarly, Assata appears to receive the news of her coming 'liberation'[165] through her grandmother's dream and, not unlike Davis, by conjuring up ancestral history, both of which remain the most explicit communication of the prison escape in the autobiography at large.

Still fugitive

The autobiography concludes with what Rodríguez describes as "an extended rumination on the ecstasy of her successful flight from prison (and the United States) and long-awaited reunion with her mother, daughter, and aunt" in Cuba in the "Postscript" (*Forced Passages* 64). In *deus ex machina* fashion, Assata's female family lineage are not only "major inspiration for her escape or liberation from prison" but also the reason for "her rather magical rematerialization as a self-identified 'Maroon woman' in revolutionary Cuba" (G. Thomas, "'Neo-Slave Narratives'" 222). The physical freedom Assata Shakur gains at the end of the autobiography, however, is considerably limited. As Hames-García contends, like Jacobs, "Assata remains only partly free at the end of her narrative both because others are still imprisoned and because the sexism, racism, and capitalist and imperialist exploitation of US society limits the freedom of people like Assata and all people" (112). After all, her comrade Zayd Malik Shakur dies

---

[165] As Wilderson contends in a question and answer session after a reading at San Francisco State College, Shakur did not escape but comrades liberated her. During the Q&A, he also explains that Shakur's autobiography, which he has been teaching frequently since 1988, is an important influence for his Afro-pessimist work ("Conversation").

during Assata's arrest, her codefendant Rema Olugbala is killed during a prison escape attempt and her original codefendant Sundiata Acoli receives a guilty verdict and remains in prison to this day (*Assata* 3–4, 163). Moreover, "[a]lthough not mentioned in Shakur's autobiography," several members of the Black Liberation Army were incarcerated for facilitating Shakur's liberation (Hames-García 118). What is more, while Shakur presumably still lives in Cuban exile today, due to the risk of kidnapping she is isolated from Cuban daily life to such an extent that Teishan A. Latner describes her life as "an exile within an exile" (471).[166] Thus, while the postscript of the autobiography paints a picture of a happy ending, particularly these extratextual circumstances fundamentally curtail any abstract philosophical argument about successful flight and unconfined freedom in Cuba (cf. P. Douglass 102; Hames-García 119; James, "Black Revolutionary Icons" 146).

After the publication of her autobiography and until the beginning of the twenty-first century, when she was still able to make public statements from Cuba, Shakur enforced the notion of a 'neo-slavery' present and consistently invoked what Rodríguez has termed "the fundamental endangerment of her status [as] an escaped and always fleeing black liberationist/prisoner/slave" (*Forced Passages* 64). In an open letter from 1998, Shakur describes herself as "a 20th century escaped slave" and pronounces Cuba not as a place of freedom but as "[o]ne of the Largest, Most Resistant and Most Courageous Palenques (Maroon Camps) that has ever existed on the Face of this Planet" ("An Open Letter"; see also Rodríguez, *Forced Passages* 64).[167] Shakur further describes efforts to have her extradited as a "re-incarnation of the Fugitive Slave Act" ("An Open Let-

---

[166] The risk of kidnapping and extradition increased with the repeated raise of Shakur's bounty, first to one million dollars in 2005 by the State of New Jersey and then to two million dollars in 2013, when the FBI added Shakur – at that point aged sixty-five – as the first woman and the second US American to the FBI most wanted terrorist list (Wilderson, "Black Liberation Army" 197; Latner 270–71).

[167] Of course, Shakur's celebration of Cuba as a safe haven for Black fugitives is an overstatement, not only with respect to the ways in which Shakur cannot live an unconfined life there, but also in terms of the position of Afro-Cubans in Cuban society. Even though Shakur was not the only African American radical who received political asylum in Cuba in the 1970s (e.g., Huey P. Newton also spent several years in Cuba to avoid prosecution of murder charges) and post-revolution Cuba has been consistently conceptualized as "multi-racial," Afro-Cubans have been addressing persisting anti-black racism on the island (Latner 456). The limits of Cuba as a safe haven for Black fugitives, of course, also recall the ways in which the US north and Canada provided only limited protection and refuge for fugitive enslaved people in the nineteenth century.

ter"; see also James, "Black Revolutionary Icons" 151).[168] In using the language of enslavement to describe her status in Cuban exile, Shakur emphasizes that – as Sidonie Smith has argued with respect to the travelling autobiographies of Langston Hughes and Claude McKay[169] – "[w]ith no geographical place of freedom, the exile may be forever plagued with a lack of at-home-ness" to such a degree that "the act of running becomes a new form of imprisonment" (75). In this sense, the escape dream of Assata's grandmother proved only partly true. While Assata Shakur escapes prison physically, she does not return "home" because there is no safe home to begin with. Instead, before and after her literal imprisonment she has to flee continuously and rely on makeshift places of refuge that assume imprisoning characteristics from New York streets to Cuban exile. Ultimately, Shakur confronts us with the paradox of literary manifestations of actual and imagined forms of flight and a reality of "creeping terror of nominal freedom for the black political exile" (Rodríguez, *Forced Passages* 64).

From Black Power to Black Lives Matter

As Rodríguez remarks, Shakur continues to function as "a venerated (if sometimes fetishized) signification of liberatory desire and possibility for many US radicals and revolutionaries" (*Forced Passages* 61). Indeed, narratives of fugitive Black Power revolutionaries, such as hers, continue to play a central role for scholars, artists, and activists attempting to understand the gendered manifestations of anti-blackness today, elaborate possible alternative conceptions of free-

---

[168] According to Greg Thomas, Shakur's definition of herself as a fugitive and rebellious "slave" is different from "Afro-pessimism's 'Slavery' or its 'Black/Slave' neatly bound up by the US state and a white ideology of Americanism" because Shakur "literally revolts against the idea" of enslaved people not resisting in her autobiography ("Afro-Blue Notes" 308). I consider Thomas's critique misguided here, however, because it conflates the Afro-pessimist differentiation between structure and experience that I discuss in chapter 1.3. Afro-pessimism does not deny a long history of enslaved people and their descendants finding forms to resist or refuse enslavement and anti-blackness (such as slave rebellions, marronage/maroonage, escape, or Cuban exile), it just locates these on the level of experience and performance that have failed to change the anti-black structure of US civil society.
[169] Claude McKay (1889–1948) and Langston Hughes (1902–1967) were both influential writers of the Harlem Renaissance. McKay was born in Jamaica and moved to the United States in 1912. He wrote several novels and two autobiographical books, *A Long Way from Home* (1937) and *My Green Hills of Jamaica*, which was published posthumously in 1979. African American Hughes also wrote, among others, two autobiographies *The Big Sea* (1940) and *I Wonder as I Wander* (1956), all of which deserve further investigation of their conceptualizations of captivity and fugitivity.

dom, and work towards future possibilities of Black liberation. With its focus on the emergence of the Black Lives Matter movement narrated through Patrisse Khan-Cullors's, her family's, and her communities' experiences since the 1980s until the election of Donald Trump, Khan-Cullors and asha bandele's *When They Call You a Terrorist: A Black Lives Matter Memoir* (2018) clearly stands in the tradition of Shakur's and Davis's Black Power autobiographies. Like Shakur, Khan-Cullors was branded a terrorist, a moniker which she uses in her political autobiography to describe the ways in which Black lives in the United States (and Canada) in the late twentieth and early twenty-first centuries are still de-humanized by gendered, homo- and transphobic, abelist anti-black violence, such as state-sanctioned surveillance, deprivation of health care and housing, criminalization, incarceration, and murder.

Davis's and Shakur's writing and activist work clearly form central intertexts for *When They Call You a Terrorist* which narrates Khan-Cullors's growth into political consciousness and the rise of the Black Lives Matter movement. Most noticeably, they are included as peritexts at the beginning (Davis provided a foreword to the memoir) and ending of the memoir, where Shakur and Davis are named in the acknowledgements as people who "another day" did the work that enabled Khan-Cullors and other BLM activists to do theirs (257). Moreover, an epigraph to the memoir taken from Shakur's "To My People" speech precedes Davis's foreword and is also included in chapter 14, when Patrisse travels together with organizers from all over the United States to Ferguson, Missouri, to protest the police killing of unarmed African American teenager Michael Brown in 2014:

> And before we take our leave, we offer one more piece, a chant I shout at the top of my lungs from another woman this nation meant to erase but who would not be erased, Assata Shakur. In the center of the room, I shout for the first time publicly these words as loudly as I can, with people echoing them after each line:
>
> IT IS OUR DUTY TO FIGHT FOR OUR FREEDOM!
> IT IS OUR DUTY TO WIN!
> WE MUST LOVE EACH OTHER AND SUPPORT EACH OTHER!
> WE HAVE NOTHING TO LOSE BUT OUR CHAINS! (222)

Citing Shakur's words from an audiotape that was smuggled from her jail cell in 1973 so that it reached 'her People' at this central point in the narrative locates Patrisse's coming-of-age story as an organizer and leader and the emergence of the Black Lives Matter movement in the Black feminist revolutionary legacy of the Black Power movement. As Angela Davis notes, in the Black Lives Matter movement Shakur's words "are often performed as a collective invocation [...] to accentuate the situatedness of the on-the-ground work of radical agents of change on a historical continuum that extends back to the 1960s" ("Assata's Message" 232).[170] In fact, Shakur situated her radial work in the legacy of revolutionary forbears as well when she drew with her rendition "on one of the concluding lines of *The Communist Manifesto* of 1848 by Karl Marx and Friedrich Engels" (Davis, "Assata's Message" 232), and re-appropriated the metaphor of the chains and their breaking to represent the legacy of African American captivity and fugitive forms of refusal that Davis also used in her autobiography to describe the captivity of African Americans inside and outside of prisons.

It may be less than surprising then that Jackson's writing, while oft-cited along with Davis's and Shakur's in the scholarly work of James and Rodríguez, for instance, is mostly absent in the writing of Black Lives Matter activists, such as Khan-Cullors, whose work is explicitly Black feminist and queer. Apart from the tendency of some of Jackson's early letters to reinforce negative stereotypes of Black women and his primary focus on the Black male revolutionary, his use of the Marxist slogan in *Soledad Brother* may provide reasons for this omission. Arguing in a letter from 1968 to his mother that prison guards "can't harm me, because the reality is that I have nothing to lose but my chains!" (184), Jackson propagates an explicitly self-centered, individualized search for forms of self-defense from the position of the male, *a priori* captive. His perspective, however, lacks Shakur's later call for communal care and love as a central part of revolutionary activity not only in the aforementioned speech but in her autobiography at large. In contrast, narratives of fugitive Black liberation activists and revolutionaries, such as Shakur and Jacobs, continue to play a central role for scholars and activists as they offer spaces of refuge in which to learn about century-old and ongoing gendered manifestations of anti-blackness as well as about potential loopholes in the anti-black world for freedom dreaming and to work towards future possibilities of liberation of all Black people. As narratives of "black revolutionary maternal" figures (James, "Afrarealism" 129), *Assata* and *Angela Davis* convey many covert and overt forms in which Black people, and especially Black women and queer people, continue to be held captive, most literally in the case of many Black Power activists who are growing old behind bars, such as Abu-Jamal and Acoli. They also draw attention to the ways in which they have found revolutionary ways to escape, care, and continue to run like Shakur.

---

[170] In chapter 1 of her study of the development of the Black Lives Matter movement, historian Barbara Ransby shows how organizing that emerged in the wake of the Black Power movement, such as the prison abolitionist group "Critical Resistance" founded by Angela Davis, Ruth Wilson Gilmore, and Rose Braz in 1998, represents important "antecedents" for Black Lives Matter organizing practices.

2.3 Fugitive Fictions of Slavery and Flight

> If the ghost of slavery still haunts our present, it is because we are still looking for an exit from the prison. Saidiya Hartman, *Lose Your Mother* 133.

> The undoing [of the plot of her undoing] begins with an escape to the woods, with perilous freedom, with *petit maroonage*, with wading in the water. Saidiya Hartman, "The Plot" 6.

In her article on "insurrectionist" Black Power autobiographies, with which I began the preceding chapter, Joy James extends her critique of slave narratives as ultimately reformist to late twentieth-century novels of slavery. These fictional neo-slave narratives, she argues, have "a similar national function" to slave narratives because in both "[r]edemption and safety continue to appear [...] as a variation of black success stories tied to integration, encoded in flight from enslavement [...] that bring closure to black rebellion" ("Black Revolutionary Icons" 136). While James finds the radical revolutionary stance she considers missing from slave narratives and neo-slave narratives in Black Power autobiographies, Greg Thomas proposes in the article "'Neo-slave Narratives' and Literacies of Maroonage" to juxtapose the analyses of African American historical novels of slavery with readings of the autobiographical writing of the Black Power movement. After all, both writing traditions show through different styles and forms as well as from various perspectives the ways in which African Americans may experience life in America in the twentieth century as a form of 'neo-slavery' (cf. ch. 1.5.). Similarly, in the study *Neo-Slave Narratives*, Ashraf H. A. Rushdy illustrates how the historiography of slavery since the 1950s, the Black Power movement, and its immediate backlash in the 1970s and 80s – not least in the form of the 'war on crime' and the rise of racialized mass incarceration (addressed in the previous chapter) – had a significant impact on the emergence of fictional neo-slave narratives.[171] As Rushdy points out, neo-slave narratives, such as Ishmael Reed's *Flight to Canada* and Sherley Anne Williams's *Dessa Rose*, emerged from increased attention to the history of US slavery from the perspectives of the enslaved and their descendants in academic and activist circles in the late 1960s and early 70s (*Neo-Slave Narratives*, ch. 2).[172] Focusing on late twentieth-century

---

[171] On the development of the historiography of slavery in the second half of the twentieth century, see also ch. 2.1. in this study. For a more elaborate discussion of critical prison studies' contribution to understanding the legacy of slavery in the rise of the prison industrial complex, see ch. 2.2.

[172] Besides *Flight to Canada* and *Dessa Rose*, Rushdy also analyzes Charles Johnson's novels *Oxherding Tale* (1982) and *Middle Passage* (1990) in his 1999 study. The first two chapters contextualize the emergence of neo-slave narratives within the political and historical discourses

neo-slave narratives by African American women in *Black Women Writers and the American Neo-Slave Narrative*, Elisabeth Beaulieu adds to Rushdy's contextualization the feminist movement, emphasizing its role in "prepar[ing] the way for the imaginative recreations later undertaken by black [female] writers" in neo-slave narratives (143), such as Octavia Butler's *Kindred*, Gayl Jones's *Corregidora*, Morrison's *Beloved*, and Williams's *Dessa Rose*.[173]

In "Failing to Make the Past Present," Stephen Best argues that the turn to slavery as a dominant topic in African American literature, with and after possibly the most well-known neo-slave narrative *Beloved*, brought about a "melancholic turn in recent African Americanist and African-diasporic cultural criticism" (456). According to Best, this turn is based on the premise of a "continuity between the slave past and our present" which "was forged at some historical nexus when slavery and race conjoined, and in the coupling of European colonial slavery and racial blackness a history both inevitable and determined proved the result" (454, see also 459–61).[174] In other words, Best criticizes the work of scholars, such as Hartman, who turn to the history of slavery, among others, as a primary lens to understanding African American cultural production, contemporary conceptions of Blackness, and the structures that continue to cause Black suffering today.[175] As I have already shown in chap-

---

of the 1960s, when Reed, Williams, and Johnson started to write, and the 1970s and 80s, when their neo-slave narratives were published. Rushdy also shows how the appropriation of the slave voice by white authors, such as in the case of the infamous novel *The Confessions of Nat Turner* by white US American writer William Styron (1967), caused African American writers to create their own versions of the history of slavery and counter racist stereotypes (89–90). Other famous examples of US American neo-slave narratives are, for example, Octavia Butler's *Kindred* (1979), Edward Jones's *The Known World* (2003), and Colson Whitehead's *Underground Railroad* (2016).
173 Octavia E. Butler (1947–2006) was an African American science fiction writer. Her speculative neo-slave narrative *Kindred* (1979) connects antebellum Maryland with 1970s-California through the African American protagonist Dana who is forced to travel between the two periods and places to save her white and Black ancestors and thereby her own twentieth-century existence. Gayl Jones's novel *Corregidora* (1975) also connects the twentieth-century with the historical time of slavery. Set in 1950s-Kentucky, it traces the story of the African American blues singer Ursa and her matrilineal ancestry back to enslaving Portuguese Brazil.
174 Best elaborates his critique of the melancholic turn in his monograph *None Like Us* (2018). In *What Was African American Literature?* (2011), Kenneth W. Warren outlines a similar critique in African American literature and its study in the post-Civil Rights era (*What Was*).
175 Best lists as examples of works that significantly contributed to the turn to slavery as a primary theme in African American literary discourse and as a dominant analytic lens for its study Spillers's essay "Mama's Baby, Papa's Maybe" (1987), Morrison's novel *Beloved* (1987), Paul Gilroy's *Black Atlantic* (1993), and Saidiya Hartman's *Lose Your Mother* (2007), among others. He also identifies a group of work "related" to the melancholic turn that "questions whether slavery

ter 2.2., what Best criticizes as an all-too ethically motivated "melancholic turn" to slavery, should not be understood primarily as a fictional creation of late twentieth-century African American literature and literary criticism. As (formerly) imprisoned and fugitive Black Power activists already argued in the 1970s, the "present day articulations" of the legacy of slavery, which Dylan Rodríguez summarizes as "apparatuses of policing, criminalization, widespread and state-sanctioned antiblack bodily violence, and ultimately massive imprisonment" ("Black Presidential Non-Slave" 18), reflect not only their individual experiences of imprisonment and flight as a form of neo-slavery but also those of the African American community at large.

James and Best point from different perspectives to two of the most contested issues in neo-slave narrative studies: on the one hand, the political and stylistic radicalism of neo-slave narratives' intertextual relationship to slave narratives and African American historiography and, on the other, the relationship the narratives supposedly create between the past and the post-slavery present, as well as the role this relationship may play in understanding Blackness and anti-blackness today. Best criticizes the interpretative trajectories that the majority of neo-slave narrative scholarship have followed, while James directs her critique towards neo-slave narratives' overt push to progressive reform instead of revolution. Yet, her critique applies even more fittingly to dominant lines of interpretation of neo-slave narratives rather than to the narratives themselves (cf. Broeck, "Trauma, Agency, Kitsch"). This chapter cannot and does not want to resolve these debates. Instead, it focusses on how concepts of captivity and flight – some of which emerged in the nineteenth-century slave narrative tradition and resurfaced in Black Power prison and flight autobiographies of the 1970s and 80s – contribute to the neo-slave narratives' complex fashioning of the relationship between the past and present and its implication for the concepts of Blackness and anti-blackness. Perpetual captivity and fugitivity, the question of refuge at risk of invasion, and the extending of the place and time of slavery which may lie in the past or "is always now" (*Beloved* 211), continue to appear in neo-slave narratives of the late twentieth and early twenty-first century. The narratives try to come to terms with the ethics and aesthetics of experiencing and memorializing the past in the post-Civil Rights era. As Best's and James's critiques already suggest and we will see in the following overview of neo-slave narrative scholarship, neo-slave narratives have often been read not

---

is over, whether it represents a historical object that has been overcome" without melancholy (*None Like Us* 140). Best names as examples Alexander; Childs, *Slaves of the State*; and McKittrick, *Demonic Grounds*, but fails to account for the role the writing of prisoners has played in this field. See also ch. 2.2. in this study.

only as trying to account for the history of slavery but also as bringing about the much-needed overcoming of cross-generational traumas of slavery that may involve healing and lead to better lives for Black Americans today (cf., e.g., Beaulieu, Keizer, Mitchell, Rody). While neo-slave narratives try to find a language and grammar to narrate slavery and often work towards narrative closure through different styles, I argue that upon closer inspection they also exhibit the limits of language and narrative to do so. The narratives clearly push for transformation, movement away from, overcoming of, and closure in and of the time and place of slavery, but on the level of their content and style stasis, confinement, and the perpetuity and circularity of confinement and flight undercut that push at critical instances.

In order to show the breadth of the neo-slave narrative tradition and its varying narrative strategies that covertly narrate perpetual captivity and fugitivity and their refusal, neo-slave narratives of different styles from the 1970s and 80s and from the early twenty-first century will be discussed in the following four subchapters. *Beloved*, the most influential US American neo-slave narrative and therefore almost unavoidable in a chapter on this writing tradition, approaches the tropes of captivity and fugitivity with a sincerity that hinges on pathos but deconstructs any assumptions of authenticity, easy memorialization, and healing – pushing postmodern and postcolonial narrative techniques to their limits. The seventh novel of the recently deceased African American Nobel laureate Toni Morrison (1931–2019) addresses the formerly enslaved and fugitive characters' struggle to remember and 'get over' the atrocities of slavery that were done to them and that the system made them do to each other. At the same time, the novel's complex style and structure calls into question whether, and if so, how the anti-black violence Sethe, Halle, their children, Sixo, and Paul D experience and perpetrate can be remembered, told, and overcome at all. *Beloved* tests the ability of narrative and the English language to tell ghost stories between the conventions of the historical novel, the gothic, and romance. It does so with clear intertextual relations to female-authored slave narratives, such as Jacobs's which was just gaining increasing attention as an authentic autobiographical narrative of a formerly enslaved woman and mother in the 1980s (see ch. 2.1.). *Beloved*'s female-centeredness and its addressing of forms of confinement and flight before, during, and after the Civil War can be also understood as a less obvious response to the Black Power and Black feminist autobiographical writing of the movement era, some of which – such as Angela Davis's autobiography – Morrison edited in the early 1970s.[176]

---

[176] Joy James and Greg Thomas point to this most literal connection between the emergence of

Ishmael Reed's *Flight to Canada* opts for an ironic and humorous approach to writing about slavery in the immediate aftermath of the Civil Rights and Black Power movement more than ten years before *Beloved* – an approach that was considered provocative and outright scandalous at the time. The novel represents a distinctly intertextual conversation with nineteenth-century North American literature about slavery, such as Josiah Henson's slave narrative (1849) and Harriet Beecher Stowe's infamous sentimental novel *Uncle Tom's Cabin* (1853), about the discursive appropriation of the voice of the enslaved and fugitive. In perfect postmodern fashion, the novel satirizes the life-and-death and captivity-and-freedom debates of fugitives, such as Henson, Frederick Douglass, and William Wells Brown, as the narrative collapses its nineteenth-century setting with anachronistic elements of popular culture and the liberation movements of the 1960s and 70s, in which Davis, Jackson, and Shakur were involved. In doing so, Reed's neo-slave narrative challenges the myth of Canada as a "land of promise far away beneath the light" (Henson, *Uncle Tom's Story* 78) as well as dualisms between resistance and assimilation or fight and flight. My analysis examines how the novel draws on the tradition of fugitive narration established in slave narratives, such as Henson's, and alienates familiar escape plots through anti-realistic satire in order to meditate on the history of slavery and flight, its narratibility, and the ways in which both shape post-Civil Rights North America.

While Reed's novel already broadens the dominant perspective on the history of slavery from the territory of the United States towards its northern border, Lawrence Hill adds a more specifically Black Canadian perspective which he expands to also include the northern Atlantic sphere between North America, Europe, and West Africa with his neo-slave narrative *Someone Knows My Name* published exactly twenty years after *Beloved* and more than thirty years after *Flight to Canada*. *Someone* represents an epic historical novel narrated by a strong, reliable Black female autodiegetic narrator who not only undergoes an unbelievable odyssey but also lives to tell about it and experiences a happy reunion with her daughter as legally free Black women in London. *Someone* adopts *Beloved*'s pathos and complex rounded characters but returns more directly to the realism and sentimentalism of nineteenth-century slavery literature that Reed had parodied so thoroughly, while including some of *Flight to Canada*'s self-referentiality. As my analysis will show, however, Hill's trust in a realist and sentimental narration of slavery and his concern with a truthful account ap-

---

fictional neo-slave narratives in the second half of the twentieth century and the writing of the Black Power movement (James, "Toni Morrison"; G. Thomas, "'Neo-Slave Narratives'" 212). As mentioned in ch. 2.2., Morrison published the writing by Angela Davis, George Jackson, and Huey P. Newton as editor with Random House while writing and publishing her first novels.

pears in stark contrast to Morrison's and Reed's earlier, more experimental neo-slave narratives whose form and stylistic choices had already thoroughly questioned the (realist) narratibility of the social and physical deaths and escapes of slavery. Yet, just as the protagonist Aminata needs to learn that she cannot return to the West African village from which she was captured, *Someone* shows cracks in the narrative construction of closure and a happy ending that *Beloved* had already burst open on the American literary scene decades earlier. Not only is return impossible, death, loss, and perpetual flight destabilize the overt plot drive for closure and coherence, pointing to the ways in which perpetual escape from enslavement, resettlement, homemaking, and the hope to return to the homeland are fugitive forms of refusal of those cast indefinitely as socially dead on the outside of the Black border. Ultimately, return to the slave narrative tradition and to the realism of the late nineteenth and early twentieth century proves limited, as the non-linear, circular, and perpetual forms of captive and fugitive narration continue to resurface.

My close reading of the "fugitive notes" of Teju Cole's 2011 novel *Open City* will close this chapter by broadening its outlook beyond historical novels of slavery. In contrast to the other three novels, *Open City* is not set in the historical time of slavery and its immediate aftermath but in the twenty-first century 'that slavery made' (Hartman, *Lose Your Mother* 133) and it is told from the perspective not of the (formerly) enslaved and their descendants but of a Nigerian German immigrant instead. In fact, the realist novel can be considered a neo-slave narrative only in its broadest sense (see definitions below) and has more frequently been considered a significant contribution to American literatures of migration, the recent literary renaissance of African literature, and the emergent transnational writing tradition of Afropolitanism. Yet, as my analysis of this novel shows, the novel includes in its episodic unreliable narration a neo-slave narrative fragment alongside other narrative fragments of the histories of violence that have affected Black people and people of color through the ages, such as migration, colonialism, genocide, imprisonment, and war. *Open City* addresses the afterlife of slavery specifically by literally and metaphorically burying slavery underneath the surface of New York City, the setting of the novel, together with other episodes of captivity and fugitivity in the novel's meandering. Moreover, fugitivity in this novel appears as a distinct narrative strategy of evasion, not least, of slavery and its afterlives. The protagonist represents himself as a cosmopolitan flâneur, who (re)visits sites of (forced) migration, enslavement, flight, and incarceration of Black people and people of color while leaving unaddressed the ways in which their experiences may reflect or affect his own. Cole's novel illustrates the inescapability of Black captivity and fugitivity in its composite literary traditions in the post-9/11 world.

Before I delve into close readings of the selected novels, let me give an overview of neo-slave narratives as an American writing tradition and its scholarly reception.[177] After attesting American fiction published in the last decade of the nineteenth century and the first half of the twentieth century a "displaced awareness of the legacy of slavery" (cf. also ch. 1.5.), Tim Armstrong first sees "a new era" inaugurated with Margaret Walker's *Jubilee* in 1966 (215). Many scholars of late twentieth- and early twenty-first-century narratives of slavery share Armstrong's observation. In the wake of the Black Power movement, novels like *Jubilee* and Ernest J. Gaines's *The Autobiography of Miss Jane Pittman* (1971) turned to experiences of slavery on "more self-conscious" and "historiographically complex and often quite specific terms" (Armstrong 215). In doing so, *Jubilee* and *The Autobiography of Miss Jane Pittman*, both based on oral history, can be considered important transitional texts that connect neo-slave narratives of the late twentieth century with earlier African American literary traditions, including, but not limited to slave narratives (see Dubey 332, 334; Rushdy, "Slavery and Historical Memory" 237–38, "Neo-Slave Narrative" 88, 92; Beaulieu xiv).[178]

---

[177] This chapter focusses on North American neo-slave narratives. Yet, neo-slave narratives have also been published and/or are set in the United Kingdom, the Anglo-, Franco-, or Hispanophone Caribbean, South Africa, and South America (cf. Anim-Addo and Lima, "Neo-Slave Narrative Genre"; Nehl).

[178] Walker (1915–1998), "one of the leading black woman writers of the mid-20th century" (Encyclopedia Britannica editors, "Margaret Walker"), based *Jubilee* on oral history of her great-grandmother and additional historical research. For an analysis of the novel, see Beaulieu 15–27. Gaines (1933–2019) wrote about the African American experience since the Civil War from the perspective of his home in rural Louisiana. He partially based *The Autobiography of Miss Jane Pittman* on oral history of family members (Encyclopedia Britannica editors, "Ernest J. Gaines").

Valerie Smith and Madhu Dubey also name Arna Bontemps's *Black Thunder* (1936) as an important antecedent (V. Smith, "Neo-Slave Narratives" 170–71; Dubey 332). This is particularly interesting as Bontemps's work is also a contemporary to the political autobiography of Angelo Herndon (cf. ch. 2.2.), suggesting a similar, but lesser known intertextual web of early twentieth-century captivity and fugitivity narratives on the African American left that I am observing at the end of the twentieth and beginning of twenty-first century in more detail in this study. Bontemps (1902–1973) was a poet and novelist of the Harlem Renaissance who later contributed to the scholarly re-discovery and re-evaluation of slave narratives in the late 1960s and early 70s, among other things, by editing the collection *Great Slave Narratives* (1969) and by publishing the biography *Free at Last: The Life of Frederick Douglass* (1971). For an analysis of *Black Thunder*, which represents a fictionalized account of a planned revolt of enslaved people led by the enslaved blacksmith Gabriel Prosser (1776–1800) in Richmond, Virginia in 1800, see Ryan 30–37.

With Walker's and Gaines's novels, slavery as a setting returns in fictionalized form on the American literary scene, increasingly "freed from [the slave narrative's] rigid nineteenth-century conventions and its obligation to flatter white audiences" (Beaulieu 143).[179] Beaulieu contends that late twentieth-century neo-slave narratives are "[i]maginative in ways its predecessors could not possibly be and yet factual in content and faithful to the spirit of the original slave narratives" (143). Like the Black Power autobiographies of Davis and Shakur, particularly female-authored neo-slave narratives function as "vehicle[s] directly responsible for revising how we perceive black women and black family relations and for exposing and repositioning the role that gender plays in narrativizing history" (4).[180] Late twentieth-century neo-slave narratives draw on recently re-issued slave narratives, oral history, the changing historiography, and earlier fictional narratives of slavery, and use these texts as templates and inspiration for their re-imaginations of the history of slavery and the emergence of (gendered) African American subjectivities.[181] In doing so, neo-slave narratives by Black women not only "revive[...] Harriet Jacobs'[s] early emphasis on the enslaved woman" and family (Beaulieu 2; cf. also Mitchell; Patton, *Women in Chains*). I would argue, they also reflect contemporary representations of political activists, such as Davis and Shakur, who emphasize the intersectionality of race and gender, to create strong female protagonists who are not only enslaved Black women and mothers, but also fugitives, survivors, and rebels.

Similar to their autobiographical forbears, neo-slave narratives deal with the experience of slavery in the seventeenth to nineteenth century primarily from the perspective of the (formerly) enslaved. Yet, "[h]aving attained a certain measure of [...] distance from the slave past sufficient to risk intimacy, along with in-

---

[179] For the complex structures of mediation and editorship that constrained slave narrative writers and narrators in sharing their views and experiences openly, see ch. 2.1.

[180] Beaulieu contends that female-authored neo-slave narratives of the late twentieth century countered the representation of Black women and men haunted by the racist and sexist 'matriarchy thesis' of the infamous Moynihan Report of 1965 (5) and its perpetuation in the form of stereotypical representation of the mass media (155). As mentioned in chapter 2.2., female Black Power activists were also targets of related stereotypes in the media and the movement, and used their autobiographies to counter them.

[181] For an analysis that accounts for autobiographical and fictional narratives of slavery from the mid-nineteenth to the late twentieth century, see Venetria K. Patton's *Women in Chains* (2000). Focusing on woman- and motherhood in slavery, Patton discusses Jacobs's slave narrative and Harriet E. Wilson's novel *Our Nig* (1859) in connection with the late nineteenth-century novels *Iola Leroy or Shadows Uplifted* (1892) by Frances E. W. Harper and *Contending Forces* (1900) by Pauline Hopkins as well as the late twentieth-century neo-slave narratives *Beloved*, *Dessa Rose*, and *Corregidora*.

creased access to publication and a growing mainstream audience" (Rody 22), novels, such as *Beloved*, "imagined [...] the interiority of enslaved and formerly enslaved people" after "[t]heir own [autobiographical] narratives had omitted their full selves for reasons of political exigency" (Brand, "Toni Morrison"). Moreover, while slave narratives propagated for the abolition of slavery and their content and style had to serve that purpose, neo-slave narratives tackle the politics, ethics, and aesthetics of experiencing and memorializing the past from today's perspective, reflecting, among other things, on Black identity formation and voice, the possibility of slave agency, resistance, reconciliation, and the healing of trans-generational trauma. The narratives grapple with the tropes of slave narratives and their writing tradition, calling attention to blank spaces in the historical records and memorialization of slavery, which some try to fill with fiction (Dubey 334). In this way, contemporary narratives of slavery in the twentieth and twenty-first century not only reimagine the past but indirectly comment on the past's relation to the post-Civil Rights present.

Neo-slave narratives narrate the inner lives of (formerly) enslaved people, many of which are women and children, in first or third-person voices, focalizing a large variety of characters set mostly in different moments and localities during the historical time of slavery (and its afterlives). In postmodern fashion, American neo-slave narratives offer innovative revisions of the American historical novel, the sentimental novel, romance, and the slave narrative as they oscillate between realist and "anti-realistic" representation (Goldberg 61), such as satire, the gothic, magical realism, and speculative fiction, producing "historiographic meta-fiction" (Mitchell 11) as "a distinctive brand of African American postmodernism" (Patton, "Black Subject Re-Forming" 880; cf. Spaulding). Narratives, such as *Beloved*, self-reflectively "eschew 'life-like' characters, the organization of incidents into a plot with a clearly defined beginning and ending, the chronological arrangement of events, and the use of reliable and omniscient narrators as being no longer sufficient to portray their perceptions of enslaved Black women and their worlds" and instead use "fragmentation, non-linearity, discontinuity, and cognitive disruptiveness" (Mitchell 10–11). Particularly, such stylistically innovative neo-slave narratives have received much scholarly attention because of how their innovative style contributes to the fashioning of a complicated connection between the postmodern present and the antebellum past.

The first full monographs dedicated to neo-slave narratives as a distinct African American writing tradition appeared in 1999 (Beaulieu; Rushdy, *Neo-Slave Narratives*). Many more literary studies works followed and diversified neo-slave narrative studies in the first decade of the twenty-first century. Studies focus, for example, on gender issues (Keizer; Mitchell; Patton, *Women in Chains*; Rody); the connection to earlier fictional and autobiographical African American writing traditions (Mitchell; Patton, *Women in Chains*; Ryan); the narratives' use of postmodern, non-realist, and speculative elements (Rushdy, *Remembering Generations*; Spaulding; Vint);[182] and on neo-slave narratives as part of broader writing traditions beyond African American literature, such as American narratives of slavery (Ryan) and narratives of the Black diaspora, especially in the United States and the Anglophone Caribbean (Keizer, Rody). Several introductory essays about the emerging field ensued (Dubey; Rushdy, "Slavery and Historical Memory," "Neo-Slave Narrative"; V. Smith, "Neo-slave Narratives"). Scholarship on neo-slave narratives at the end of the second decade of the twenty-first century broadened the scope further by focusing on hitherto underappreciated genres, such as short stories, poetry, and drama, but also audio-visual and performing arts (Anim-Addo and Lima, "Neo-Slave Narrative Genre"; Tawil, *Slavery in American Literature*). Neo-slave narrative scholarship has also increasingly paid attention to texts that reveal "a hinterland of new writing in several languages" other than English (Anim-Addo and Lima, "The Power" 11; see Anim-Addo and Lima, "Neo-Slave Narrative Genre"). Last but not least, Best's critique of the turn to "the slave past" in African American literature, and his call to examine the "primacy" of this turn in "black critical thought" ("On Failing" 453) has launched a welcome self-reflective discussion of the relationship between the past and present that neo-slave scholarship from the last twenty years proposes and its underlying assumptions, desires, and fears (Best, *None Like Us*; Colbert, Patterson, and Levy-Hussen; Crawford, "Turn to Melancholy"; Levy-Hussen, *How to Read*; Levy-Hussen, "Trauma").

The term 'neo-slave narratives' was first used as a critical side note by Ishmael Reed in an interview in 1984 (Reed and Martin) and then by the scholar Bernard W. Bell to describe a "modern narrative of escape from bondage to freedom" in his study *The Afro-American Novel and its Tradition* in 1987 (in Beaulieu 25). Since its earliest use, the meaning, scope, and appropriateness of the term have been debated. Rushdy outlined one of the most cited narrower definitions of neo-slave narratives as late twentieth-century first-person narratives of slavery in his first study *Neo-Slave Narratives* from 1999. In the same year, Beaulieu discussed the portrayal of gender roles, gender relations, and especially enslaved motherhood in female-authored neo-slave narratives of the 1970s to 90s and emphasized their emancipatory power for Black female identities and their entanglements with the history of slavery "with a spirit of celebration" (xiv). She there-

---

[182] For an overview of speculative fiction and the neo-slave narrative tradition, see Tucker. For more in-depth analyses of neo-slave narratives that use non-realist and speculative elements, such as *Beloved* and *Kindred*, see, for instance, Vint or Spaulding.

fore suggests describing late-twentieth-century neo-slave narratives written by African American women as "freedom narratives," as they "celebrate [...] freedom from commodification, and freedom from the invisibility that has historically enshrouded the enslaved mother" (14). Similarly, in focusing on the Black female self and her liberation, Angelyn Mitchell calls late twentieth-century African American women's narratives of slavery in her study *Freedom to Remember* (2002) "liberatory narratives" (3). According to Mitchell, liberatory narratives surpass slave narratives, or "emancipatory narratives" as she calls them (xi), because the former "engages the historical period of chattel slavery in order to provide new models of liberation by problematizing the concept of freedom" (3). A liberatory narrative "reveals the unspeakable – indeed, the unacknowledged – residuals of slavery" (xii) that illuminate "the enduring effects of our racist and sexist American history in today's society" (3). In order to include Black diasporic narratives from the Anglophone Caribbean less closely connected to the slave narrative tradition, Arlene Keizer prefers the term "contemporary narratives of slavery" in her monograph *Black Subjects: Identity Formation in the Contemporary Narrative of Slavery* (2004). Tim Ryan adjusts Keizer's terminology with respect to the US American twentieth-century context of his study *Calls and Responses: The American Novel of Slavery Since 'Gone with the Wind'* (2008) that also provides a tabular overview of the "general historical development of the cultural conversation about slavery" in historiography, fiction, drama, and film of the twentieth century (215, see 217–26). As Ryan's and Keizer's studies suggest, the term 'narrative of slavery' may provide the broadest conceptualization of a cultural tradition that includes prose and film, especially but not exclusively in the modern and postmodern era (cf. Patton, *Women in Chains*).[183]

As the writing tradition keeps growing, a flourishing debate also continues to identify potential subcategories and new developments. In his more recent publications, Rushdy differentiates between three main forms: "the historical novel, the pseudo-autobiographical slave narrative, and the novel of remem-

---

[183] On slavery as a thematic focus in American literature, music, and film, see Ezra F. Tawil's *Cambridge Companion to Slavery in American Literature* from 2016. For a discussion of textual and non-textual forms, such as visual arts, as part of the neo-slave narrative tradition across the Black diaspora, see the *Callaloo* special issue on the neo-slave narrative genre edited by Joan Anim-Addo and Maria Helena Lima in 2017. It includes analyses of poetry (Campa), sculptures (Iasiello), visual arts (McCoy) as well as fiction and drama from the UK (Iromuanya, Pérez-Fernandéz), Brazil (Maddox), and the francophone Caribbean (Puertas). Childs's analysis of Angelo Herndon's autobiography as a neo-slave narrative and antecedent of the political autobiographies of the Civil Rights and Black Power movements discussed in ch. 2.2. is also part of this issue ("Angelo Herndon").

bered generations" ("Neo-Slave Narrative" 90), the latter of which he describes as "palimpsest narratives" in his second book-length study of the tradition *Remembering Generations* from 2000 (8–9).[184] Timothy A. Spaulding uses the term "postmodern slave narrative" specifically to describe neo-slave narratives that adopt non-realist, gothic, or speculative elements, such as *Beloved* and *Flight to Canada*. Giving an outline of neo-slave narratives as a writing tradition from the perspective of the second decade of the new century, Markus Nehl offers the concept of "generations" to differentiate between earlier contributions of the mid- to late twentieth century (including *Beloved*) and the more recent publications of the first decade of the new century (23–30). According to Nehl, neo-slave narratives of the "second generation," such as *Someone*, tend to mix genres, including fiction and non-fiction, and are more transnationally oriented (32, 30–32), while speaking more directly to Hartman's notion of "the afterlife of slavery" in the United States (and beyond) (12). Instead of assuming a linear progression of neo-slave narrative styles from one generation to the next, Margo Natalie Crawford defines "post-neo-slave narratives" as narratives that address the "architecture of the unknown" of slavery both in the twentieth and twenty-first century ("Post-Neo-Slave Narrative" 72). Rejecting revisionist approaches that draw on the slave narrative tradition and try to fill historical gaps with fiction, post-neo-slave narratives, Crawford argues, dwell on "the unknown" and stylistically improvise literary forms that require readers to learn a "counter-literacy" (71) in the twentieth and twenty-first century.[185]

---

184 Rushdy dedicated his first study *Neo-Slave Narratives* exclusively to "pseudo-autobiographical slave narratives," which he describes as "fictional autobiographies" and "contemporary novels that assume the form, adopt the conventions, and take on the first-person voice of the ante-bellum slave narrative" (22, 3). *Remembering Generations* focusses on "palimpsest narratives" which Rushdy defines as "novel[s] representing late-twentieth-century African American subjects who confront familial secrets attesting to the ongoing effects of slavery" in the form of "an antebellum ancestor" (8, 5). Both definitions are narrower than the ones Rushdy proposes in his overviews of neo-slave narratives for *The Cambridge Companion to Slavery in American Literature* ("Slavery and Historical Memory") and *The Cambridge Companion to the African American Novel* ("Neo-Slave Narrative").
185 Crawford describes Morrison's novel *A Mercy* (2008), Edward Jones's *The Known World* (2003), Monifa Love's *Freedom in the Dismal* (1998), and Amiri Baraka's play *The Slave* (1964) as post-neo slave narratives. Nehl includes *A Mercy* as well as Hill's *Someone*, Hartman's *Lose Your Mother*, Yvette Christiansë's *Unconfessed* (2006), and Marlon James's *The Book of Night Women* (2009) as part of a second generation of neo-slave narratives that carry out a transnational dialogue. He observes differences in these second-generation approaches to the "aesthetic and ethical challenge of how to re-imagine slavery" and the difficult task of narrating the experiences of enslaved women after the watershed of *Beloved*. Nehl thus identifies a divide in his corpus of second-generation neo-slave narratives between the male and female writers. Morri-

As the following close readings will show, both *Flight to Canada* and *Beloved* oscillate between Rushdy's definitions of the historical novel and the 'pseudo-autobiographical slave narrative,' while *Someone* can be considered a 'pseudo-autobiographical slave narrative' that takes on the proportions of a historical novel. Cole's *Open City* surpasses any of the definitions mentioned, while including elements of many of them in a truly palimpsest fashion that, however, also goes beyond Rushdy's definition of the 'palimpsest narrative.' Yet, while insightful with respect to the development and diversification of neo-slave narrative studies, the terminological genre debates do not concern me here as much as the conceptualization of fugitivity within and across the texts broadly joined by the term neo-slave narrative that I use as an umbrella term for North American late twentieth- and early twenty-first-century novels of slavery. The term neo-slave narrative explicitly marks the clear connection to the slave narrative tradition but also allows me to draw connections to the political 'neo-abolitionist' autobiographies discussed in chapter 2.2. While the Black Power movement plays an important role in Rushdy's understanding of the neo-slave narratives' emergence, he does not consider the close literary relationship between neo-slave narratives and Black Power autobiographies. I agree with Thomas that Black Power prison narratives that lie beyond Rushdy's definitions deal with captivity and flight in the afterlives of slavery in a way that is connected to the fictional processing of the history of slavery. To avoid terminological confusion and the blurring of stylistic and genre distinctions, however, I refrain from broadening the term's definition to Thomas's or James's suggested proportions. Moreover, while accepting Keizer's and Mitchell's critiques of the term neo-slave narrative, I refrain from using their suggested alternatives because they not only exclusively refer to narratives by women but also overemphasize the narratives' strive for freedom and liberation. Assuming an instructive effect of neo-slave narratives on readers, Mitchell, for instance, argues that Black women's "liberatory narratives" are "liberatory not only in content and form, but in their projected and ideal reception as well" (21). My critique of this notion goes beyond Ryan's argument who cautions against all too celebratory readings of neo-slave narratives, because they risk reducing the narratives to "didactic protest fiction in which literary complexity is subsumed to a simplistic ideological agenda"

---

son, Hartman, and Christiansë, Nehl contends, "warn against an easy appropriation of black history and draw attention to the impossibility of working through the past in order to heal the wounds of slavery" (21) by adopting narrative strategies such as non-linearity, multi-perspectivity, and narrative fragmentation and by resisting "the temptation to fill in the gaps and silences of the archive" (36). James and Hill, according to Nehl, fail to account for this. I discuss Nehl's study in more detail in my review of his monograph (von Gleich, "Markus Nehl").

and thereby obscure the innovative stylistic contributions those texts also make (12). More importantly, this study insists, the narratives cannot escape the antiblack captivity from the historical era of slavery through its afterlives that makes the unconfined narration of Black life, liberation, and freedom impossible.[186]

"Do you think it'll be any different in Canada?": Fugitive narration from Josiah Henson's slave narrative to Ishmael Reed's *Flight to Canada*

In *Flight to Canada* (1976), Reed stages a pronounced intertextual conversation with nineteenth-century narratives of slavery, such as *The Life of Josiah Henson, Formerly a Slave, Now an Inhabitant of Canada, as Narrated by Himself* (1849) and the novel *Uncle Tom's Cabin* (1853), about the discursive appropriation of the voices of the enslaved and fugitive, and their twentieth-century descendants in North America. Using prominent postmodern aesthetics, the novel takes on the political debates among abolitionists, politicians, the (formerly) enslaved, and plantation owners in North America before, during, and after the US American Civil War. It collapses these nineteenth-century North American discourses with anachronistic elements of popular culture – especially technology related to movement and media communication, such as telephones, television, yachts, and cars, but also late twentieth-century colloquialisms and cultural references (Juan-Navarro 92) – and with discourses of the political movements of the 1960s for civil, Indigenous, and women's rights as well as Black Power. *Flight to Canada* thus approaches American literature about slavery with biting irony and humour, which Black feminist scholars, such as Sondra O'Neale and Michelle Wallace considered provocative if not outright problematic in the immediate aftermath of the movement era, not least with respect to Reed's archetypical or stereotypical characters, particularly women of color (cf. Mielke 4, 7). Ultimately, Reed's "nineteenth-century narrative in theme and plot" that is "overla[id ...] with a twentieth-century cornucopia of in-your-face ridicule, rebuke and sarcasm" (Beck 133) challenges dualistic notions of resistance or assimilation, fight or flight, and the myth of Canada as a safe haven for African American fugitive men.

---

[186] I do not adopt the broader term of 'narratives of slavery' because of my focus on African American and Black diasporic narratives in North America whose origin lies, at least in part, in its nineteenth-century slave narrative tradition, while I propose to understand neo-slave narratives as part of the larger Black diasporic writing tradition of fugitive narration that also includes a novel such as *Open City*.

Flight to Canada consists of twenty-nine short chapters, separated in three parts, and two poems by fugitive writer and protagonist Raven Quickskill. The first poem "Flight to Canada" opens the novel. A first-person narration in italics follows and deliberates what Janet Kemper Beck calls the "most well-known theft of a slave's story" (135), i.e., Stowe's appropriation of Henson's life story for her novel.[187] Afterwards, the narrative switches to a figural narrative perspective that starts with Raven sitting at the desk of his former legal owner Arthur Swille III about to write a slave narrative about his formerly enslaved friend "Uncle Robin" (and indirectly himself). In fact, the whole novel appears to be the very communal meta-slave narrative Raven writes (Juan-Navarro 93). The macrostructure of *Flight to Canada* is, thus, an interesting twist on the structures of slave narratives and Black Power prison autobiographies that also seems to serve as a foil for later neo-slave narratives, such as Lawrence Hill's *Someone Knows My Name*, as we will see in another section of this chapter. Reed discards both slave narratives' *ab ovo* and Black Power autobiographies' *in medias res* beginnings and instead begins *in ultimas res* and introduces a "proleptic *mise en abyme*" (Juan-Navarro 93) to reflect on the mediation processes of slave narratives and other African American writing traditions.

The main part of the narrative then tells the story that led up to Raven writing the slave narrative chronologically by switching between two parallel plot lines. The first plot line concerns the events taking place at Swille's plantation in Virginia and focusses mostly on the overdrawn characters Robin, a seemingly docile servant along the lines of Stowe's Uncle Tom, the enslaved woman ruling the plantation household "Mammy" Barracuda, Swille, and his wife. The second plot line narrates Raven, Stray Leechfield, and 40s' "search for freedom" (Velikova 37) after their escapes to the northern US states and Canada. As Roumania Velikova notes, in the novel the plantation in Virginia is a parody of Edgar Allan Poe's "southern gothic [...] subverted by trickster slaves drawn in the tradition of Charles Chesnutt and William Wells Brown" (37), while the experiences "in the North derive from the narratives of such runaway slaves-turned-abolitionists as Douglass and Brown" (38) as well as Henson. I will pay attention mostly to

---

[187] Henson's slave narrative gained popularity in the nineteenth century not least due to prevailing assumptions that Stowe had used the narrative as inspiration for the main character "Uncle Tom" of her commercially successful novel *Uncle Tom's Cabin* (Sinanan 75). As Sinanan explains, Stowe's *Key to Uncle Tom's Cabin* (1853) suggests a connection between Henson and Stowe's protagonist that was further reinforced "when Stowe herself provided the preface to Henson's second autobiography, *Truth Stranger than Fiction* a few years later. Although in the preface to this text, Stowe made no claims that Henson was a model for Uncle Tom, the title alone invited the reader to read the narrative as a real basis for the fictional character's life" (75).

this second plot line in my analysis of the novel's conceptualization of flight and how the two plot lines are interwoven in the novel's ending.

The "cartoonesque" characters of *Flight to Canada* (Levecq 284) are characterized through self-characterization and much commentary by other characters as well as their actions, telling names, and individualized speech patterns. Robin, Stray, and Barracuda are excellent examples where name, vernacular, and code switching as well as description of their outer appearance clearly contribute to their characterizations which reference, play with, and potentially break racist stereotypes of the "Uncle Tom," the "minstrel-porn-star-pimp," and "the mammy" (Carpio 572–80, cf. Ryan 126–27).[188] The narrative style of the main part is mostly dramatic and scenic, imitating many conventions of the historical novel (Rushdy, *Neo-Slave Narratives* 124) as well as nineteenth-century narratives and drama, for example, in detailed realist representations of spatial settings in present tense and the use of dialogue. The end of the novel provides closure in two ways. The narrative comes not only full circle to where it began in terms of time and place. It also provides a form of poetic justice when the 'main villain' Swille dies after the ghost of his beloved sister – or more likely the enslaved Pompey impersonating her (Rushdy, *Neo-Slave Narrative* 107) – pushes him into the fire, and the formerly fugitive and enslaved find a place of refuge at the plantation which Robin inherits after changing Swille's will.[189] Yet, as Hortense Spillers points out, "to rehearse the plot of *Flight to Canada* is ultimately dissatisfying [...] because the novel does not actually 'concern' the plot very much at all" ("Changing the Letter" 195). Instead, the text's focus constitutes its "clever juxtaposition of prior texts, plots, landscapes, icons, historical and other invented identities, poems, songs, puns, and sight gags, in a dizzying shift of time and textual perspective" that reminds Spillers of "an evening of American television" (195).

---

188 The names of Robin, Raven, and his lover Princess Tralarala Quaw Quaw, of course, also contribute to their characterizations as they refer to types of birds or in Quaw Quaw's case appear like "a transcription of a bird call" and thereby reference the flight and communication of birds as central elements to their characters (Mielke 12, see also Crisu 111).

189 Rushdy convincingly argues that Pompey committed the murder disguised as the ghost of Swille's sister and discerns in this scene riddle-like elements of an "Edgar Allan Poe mystery and a plain old-fashioned whodunit" plot which Reed transforms into a "Neo-HooDoo whodunit" narrative (*Neo-Slave Narratives* 106–07). Rushdy also points to the ways in which "[b]y inscribing into *Flight to Canada* the violent resistance of Pompey Reed reinscribes into Henson's own narrative that dialectic of material and spiritual resistance Stowe had removed. Even the original Uncle Tom[, i.e., Henson] was willing to consider violence [against his legal owner], Reed reminds us" (108).

Flight to Canada and Reed's idiosyncratic novelistic writing style have received considerable attention in African American literary studies over the decades (cf. Dick). The majority of research was conducted in the 1980s and 90s and primarily focusses on his first five novels published between 1967 and 1976, with special attention given to Reed's most well-known and praised novel *Mumbo Jumbo* (1972). As Spillers's description of *Flight to Canada* suggests, Reed's dense intertextual aesthetics and polemical representation of African Americans, white Americans, Indigenous people, and their political struggles challenges readers and critics alike (cf. Mvuyekure 203–04). The challenges Reed's writing pose may also explain why many critics who have worked on Reed have used "neo-hoodooism" as a main line of interpretation that Reed himself has proposed (see, e. g., Jordan; Lindroth; Lock; Mikics; Mvuyekure; Rushdy, *Neo-Slave Narratives* 125–28). In his novels, poems, and interviews of the 1970s to 90s, Reed developed 'neo-hoodooism' as a syncretic writing style which he traces back to Haitian voodoo and Black North American hoodoo traditions, while combining them with various other cultural traditions and postmodern aesthetics (cf. Carpio, Crisu).

A central part of 'neo-hoodooism' as per Reed is, for instance, the reoccurring trickster figure which he connects both to Black American and Indigenous cultural traditions in works such as *Flight to Canada*, in which characters like Robin, "are viewed as poseurs, or mask-wearers, who are not really as they seem" (Harris 469, see also Lindroth). Just as with Reed's larger oeuvre, many analyses of *Flight to Canada* have thus considered its use of voodoo aesthetics and themes, and the ways in which they are combined with postmodernism and the Africa American literary tradition (Carpio, Juan-Navarro, Lock). In introductions to the neo-slave narrative tradition and literature about slavery more generally, *Flight to Canada*, like Charles Johnson's *Oxherding Tale* (1982), is often referenced as an example for a humorous and satiric neo-slave narrative (Dubey 338–39; Rushdy, *Neo-Slave Narratives* ch. 4; V. Smith, "Neo-slave Narratives" 172; Rody 23–24; Ryan 125–29). Scholars have also examined the novel's various intertextual relations, for example, to Brown's novel *Clotel or, The President's Daughter: A Narrative of Slave Life in the United States* (1853, see Crisu); Stowe's *Uncle Tom's Cabin* (Crisu; Rushdy, *Neo-Slave Narratives* ch. 4; Spillers "Changing the Letter"); Douglass's, Henson's, and Brown's slave narratives (Harris; Moraru; Rushdy, *Neo-Slave Narratives*; Velikova); and to the representation of the American South by Edgar Allan Poe and other white writers (Moraru). Other themes in *Flight to Canada* that scholars scrutinized are, among others, the representation of Native American women as using, but also going beyond the 'Pocahontas' script (Mielke); the role of the nation, (cultural) nationalism, and postnationalism (Jordan, Levecq); the rewriting of the local history of Buffalo

(Velikova); and the relation of what in the novel is described as "material" versus "cultural slavery" (Reed, *Flight to Canada* 67; see Walsh).[190]

Flight to Canada draws on a tradition of fugitive narration that slave narratives established and the autobiographical writing of Black Power activists further developed at Reed's time of writing. I argue that particularly Henson's own rewriting of his slave narrative in his later autobiographies and Reed's further rewriting of those narratives in his novel suggest a perpetuity of flight inside and outside of territories of legal slavery and during and after the historical time period – a perpetuity that I already examined in Douglass's and Jacobs's slave narratives in chapter 2.1. Thus, drawing on this discussion of slave narratives, in the following, I first briefly focus on Henson's conceptualization of flight to Canada as a freedom dream caught up in the perpetuity of flight. I show how Henson's slave narrative and his later autobiographies – just like Douglass's and Jacobs's narratives – subtly question the possibility of unconfined freedom in nineteenth-century North America. Second, I focus on Reed's satirical and parodist rewriting of Henson's concepts in *Flight to Canada* in more detail. Reed's novel disrupts the realism and linearity of historiography (cf. Dubey 338) and sentimental nineteenth-century novels of slavery to connect the mid-nineteenth century with the post-Civil Rights era, both of which appear politically and culturally "not that different" for Black Americans and other racialized communities, such as Indigenous people and white Jewish immigrants (Velikova 39). More importantly, by using the slave narrative tradition and Henson's writing in particular, the novel argues that Black American writing at the bicentenary of US American Independence still suffers from similar confines and forms of appropriation, mediation, and critical scrutiny of its validity, authenticity, and artfulness exactly one hundred years after the publication of Henson's third autobiographical text (Beck 133).

Flight and the freedom dream of Canada in Henson's slave narrative

As Heike Paul notes, "Josiah Henson has probably been as foundational a figure for the Canadian context as Frederick Douglass has been for the US-American"

---

190 In a dispute with Raven about their experiences of oppression, racism, genocide, and enslavement, the side character Mel Leer claims that white Jews like himself suffer from "cultural slavery" (67). Richard Walsh understands Mel's differentiation between "material" and "cultural slavery" (67) as a comment on "the monoculture's institutionalized subordination of all other cultures: the institutional structure of slavery [that] remains in sublimated form, as the machinery of a state of oppression" (63) in places without 'material' or legal slavery.

context ("Remembering the Fugitive" 265–66) which makes looking at his writing in preparation for an analysis of Reed's, and also later Hill's neo-slave narratives more than worthwhile. Henson's writing broadens this study's initially exclusive perspective on fugitivity from the United States (cf. ch. 2.1. and 2.2.) to a more hemispheric North American approach and accounts for the fact that fugitives have and continue to cross national borders. Henson (1785–1883) has become part of a "Canadian foundational myth" about the Underground Railroad that contrasts "Canada as a safe haven" from the United States with its history of slavery and the slave trade (265, 264). Paralleled with "the biblical figure" of "Moses leading his people to Canaan/Canada" (261), Henson is considered representative for tens of thousands of fugitives who in the nineteenth century, especially after the passing of the Fugitive Slave Act, fled from the legal reach of slavery in the United States and "settled throughout British North America in small towns and larger city centres" that would later make up Canada West (261; cf. Henry, "Fugitive Slave Act," "Underground Railroad").

While considered representative of nineteenth-century African American fugitives from US America to Canada and widely read in the late nineteenth century, introductions into the field of African American slave narrative studies often mention Henson's writing only marginally (see, e.g., Ernest, *The Oxford Handbook*; Fisch, *The Cambridge Companion*). Moreover, collections and analyses of Black Canadian literature that have appeared in increasing numbers since the early 2000s frequently mention Henson's writing as a central contribution but fail to provide in-depth analyses (see Siemerling, *Black Atlantic Reconsidered*, esp. 90–93; Clarke, *Odysseys Home*; Clarke, *Directions Home*). Reasons for this oversight may be found in the fact that Henson's narratives were actually not written by himself but by amanuenses,[191] and that they tend to be located at the margins rather than the center of various national (US and Canadian), regional, and cultural (African American, Black Canadian, Black North American, and Black diasporic) writing traditions.[192]

---

[191] Apparently, Henson first told his story to the abolitionist Samuel A. Eliot who wrote Henson's 1849 slave narrative "as narrated by himself" (Siemerling, *Black Atlantic Reconsidered* 90), while he is said to have written his later autobiographies himself despite the fact that "today it is assumed that Henson was never more than semi-literate" (Paul, "Remembering the Fugitive" 266).

[192] Exceptions to this oversight are studies that approach Henson's work primarily from a historical perspective and/or retell his life story to an interested public (see, e.g., Brock, Murray, Troiano, Winks). For a brief summary of Henson's life and writing, see, e.g., Simson. On the challenges to locate slave narratives in national, regional, and cultural writing traditions, see R. R. Thomas.

2.3 Fugitive Fictions of Slavery and Flight — 173

Like Douglass's, Henson's slave narrative starts in the typical formulaic *ab ovo* fashion and with a brief recapitulation of his family relations and childhood in Maryland affected by violence. It retells how he grows into working as an overseer on plantations in Maryland and Kentucky as a teenager and later becomes a preacher. In contrast to Frederick, Josiah starts to desire escape only as a grown man when he fears to be sold away from his wife and children. In typical slave narrative fashion, contributing to and aligning with the aforementioned myth, Henson describes Canada in religious terms as the Promised Land when contemplating his escape:

> Canada was often spoken of as the only sure refuge to pursuit, and that blessed land was now the desire of my longing heart. ... I knew the North Star – blessed be god for setting it in the heavens! Like the Star of Bethlehem, it announced where my salvation lay. Could I follow it through forest, and stream, and field, it would guide my feet in the way of hope. I thought of it as my God-given guide to the land of promise far away beneath the light. I knew that it had let thousands of my poor, hunted brethren to freedom and blessedness. (*Uncle Tom's Story* 78–79)[193]

After deliberating the dangers of flight, such as death due to deprivation, during recapture, or of punishment, and the risks of staying enslaved (79), Josiah persuades his wife to escape with him and they flee from slavery in Kentucky, mostly on foot, in 1830 to the north, with Josiah carrying two of their three children on his back. The physical "burden" on his "raw" shoulders and the "fearful dread of detection ever pursued" Josiah and his family on their trek as they fear that "the dogs and slave-hunters" would follow them (83). Luckily, Josiah's family remains undetected; they survive the trek, and arrive in Canada by boat. Henson describes the scene of arrival vividly, comparing the happy arrival and newly gained legal freedom comically with a form of madness:

> It was the 28th of October, 1830, in the morning, when my feet first touched the Canada shore. I threw myself on the ground, rolled in the sand, seized handfuls of it and kissed them, and danced around, till, in the eyes of several who were present, I passed for a madman. "He's some crazy fellow," said a Colonel Warren, who happened to be there. "Oh no, master! Don't you know? I'm free!" He burst into a shout of laughter. "Well, I never knew freedom make a man roll in the sand in such a fashion." Still I could not control myself. I hugged and kissed my wife and children, and, until the first exuberant burst of feeling was over, went on as before. (96)

---

[193] All citations of Henson's narrative are taken from the 1876 autobiography.

After his arrival, the Henson family "settled on the first spot in Canada" (103) where Josiah co-founded the Dawn community (Dresden, Ontario) at the end of the narrative.

Similar to Douglass, Henson published several autobiographical texts apart from his slave narrative: *Truth Stranger Than Fiction: Father Henson's Story of His Own Life* (1858) and *Uncle Tom's Story of His Life: An Autobiography of the Rev. Josiah Henson* (1876), among other versions edited by John Lobb. The later autobiographies reproduce the slave narrative in revised form, while adding chapters on Henson's life as a preacher, community leader, Underground Railroad conductor, entrepreneur, and abolitionist orator after his escape.[194] They, thus, continue to narrate Henson's life after "the 'happy ending' of the fugitive's arrival in freedom" (Paul, "Remembering the Fugitive" 262). Whereas slave narratives "typically focus on the atrocities of slavery and the perils of escape" and "Canada tends to surface primarily through anticipatory references to freedom rather than as a concrete and fully developed place" (67), *Uncle Tom's Story of His Life*, which I focus on here, dedicates the majority of its thirty-one chapters to Josiah's life in Canada after his successful flight. It thereby not only reproduces the myth of Canada as a safe haven but also subtly provides reasons to question it.

While the later narratives certainly show much appreciation of the newfound legal freedom and independence Josiah slowly acquires in terms of income, housing, and literacy, they also address significant challenges for fugitives in Canada. Henson explicitly argues that fugitives need to "be able to defend their natural and inalienable rights after they became freeman and citizens of Canada" (172) because, "[t]hough Canada was the land of freedom to the fugitive slaves, [...] they met with so much prejudice at first, on account of their colour, that it was with difficulty they could procure the common comforts of life" (174). Henson seems to point to a variety of anti-black constraints to legal freedom in Canada that complicate life for formerly enslaved Black people, such as the difficulty to settle and build a sustainable Black community that could financially provide for itself in a dominantly white society, geographical mobility or "diasporic restlessness" (Siemerling, *Black Atlantic Reconsidered* 92), and the fugi-

---

194 Henson's first narrative has a different chapter structure than the later narratives, names of legal owners and overseers are not spelled out due to the immediacy of the events described, and it ends after what is chapter 18 in the 1876 autobiography briefly after the successful flight. The 1854 edition adds chapters 19 to 22 that narrate Henson settling in Canada as a legally free man and closes with a chapter on the difficult situation of fugitive enslaved people in Canada. The 1876 edition adds nine more chapters.

tives' inability to express themselves freely. In Henson's narratives, these constraints are shown to be inextricably linked.

Many of the chapters added to the original slave narrative are concerned with Henson's efforts to raise money from white abolitionists in the US North and the UK to buy land for the Dawn settlement and build a school and two mills. Thus, the later narratives try to establish the autobiographical subject and by extension his community as respectable, trustworthy Christians, self-made and independent people, and free Black citizens of Canada, for instance, with citations of page-long recommendation letters as proof (see, e. g., ch. 30). Yet, conflicting interests with white community leaders and debt frequently interfere. In other words, anti-blackness in the law, in business transactions, and in people's minds work against the community's efforts.[195] Thus, Henson's orating tours and the publications of his narrative in various versions can be considered part of his larger effort to survive and articulate survival as a form of life, arrival, and settlement.

Reinforcing the rumored connection between Henson and Stowe's fictional character through the inclusion of a preface by Stowe in the second autobiography and the merging of Uncle Tom with Henson in the titles of the third autobiography, of course, made the later autobiographies not only comprehensible and acceptable to a white liberal readership but also more profitable. John Ernest, however, argues that in this way, Henson's "life had become so identified with that of Uncle Tom that any hope of understanding the actual man was lost in the fame of the fictional character" ("Introduction" 9), which may have posed another reason why Henson's writing has not received as much scholarly attention. Yet, Ernest's argument seems to take for granted that a true story of Henson's life was at some point identifiable. He underestimates the general role fictionalization plays in any autobiographical writing process and the particularly complex mediation processes of slave narratives between writer or narrator, editor, amanuensis, publisher, and intended readership that affected Jacobs's and Douglass's much more appreciated writing as well (see ch. 2.1.). In any case, Henson strategically uses "his prominence and popularity both nationally and internationally" within the bounds of his limited possibilities as a fugitive Black man in an anti-black environment "to encourage and promote black settlement in

---

195 For the details of Henson's entrepreneurial ventures and the complex reasons of their failure, see Winks.

Canada and thus assists in the creation of his own myth" (Paul, "Remembering the Fugitive" 260).[196]

Yet, despite Henson's dedication to the Dawn settlement, the narratives also show how settlement in Canada did not mean definite homemaking. For many fugitives it not only involved struggle for survival but also return to rescue family, friends, and other enslaved people through the Underground Railroad, frequent travel across the Canadian-US border and to other parts of the world, as well as resettlement in the US northern states and elsewhere.[197] As Paul concludes, for many free and fugitive Black people Canada represented "a safe place for pondering all available future possibilities and for engaging in political activism, but it was a place of transit rather than a place of belonging" ("Remembering the Fugitive" 272). Josiah returns to the American South repeatedly and is proud to document 118 people in whose escape he assisted (*Uncle Tom's Story* 120). He also frequently travels to the northern United States and to the UK to preach, raise money to support his settlement in Canada, lecture, and promote his publications. While Henson clearly chose Canada as his main place of residence, the narrative shows that Josiah nonetheless remains restless until he becomes too old to travel and desires to finish his third speaker and book tour through Great Britain and return to Canada for good.

Surely, the Canadian settlement provides refuge from the risks of recapture under the Fugitive Slave Act. Yet, as I argue in chapter 2.1., the slave narrative writers' and narrators' geographical and literary restlessness – i.e., their ongoing physical movements and their rewritings of their lives – can also be understood as subtle signs of their acknowledgement that slavery and escape from enslavement do not represent singular events that can be easily overcome. The Black border between social death and unconfined freedom continues to be insurmountable even after crossing national borders and entering territories outside of the legal reach of slavery. Thus, the point *after* the event of enslavism still has to occur at the moment of writing and rewriting for Henson, Douglass, and Jacobs. After having successfully escaped the geographies of slavery in the American South, the extended reach of enslavism makes the US American

---

[196] As Paul notes, the blurring of the fictional character of Uncle Tom with Henson continues today. At Henson's place of settlement in Ontario, his story is epitomized in the so-called 'Uncle Tom's Cabin Historic Site' where paradoxically Henson's story is remembered for his symbolic role in the history of Black presences in Upper Canada and Black Canadian history in general through a cabin named after Stowe's fictional character ("Remembering the Fugitive" 262–63).
[197] For further discussion of the diverse corpus of autobiographical writing by fugitive Black Americans in the 1850s that addresses their multidirectional movement across the Canadian-US border, see Sawallisch.

North, Canada, and the larger Atlantic world merely a place of refuge at constant risk of invasion by enslavist forces. Flight for Henson does not end in the American North nor Canada but takes on a perpetuity. In other words, rewriting his autobiographical narratives of flight and arrival challenges the assumed possibility of arrival, settlement, and homemaking for fugitives in nineteenth-century North America. As we will see in the following, it is this notion of the perpetuity of slavery and flight and its complex forms of mediation in Henson's texts that Reed's own rewriting of the slave narrative tradition picks up.

Fugitive tracks, traces, and a loophole of retreat in *Flight to Canada*

Similar to Henson's narratives, the flights of Reed's fugitive characters do not end in the northern US states because of the danger of recapture and re-enslavement. Particularly Raven and 40s understand that Swille, the superrich plantation owner with a liking for sadomasochistic whippings and a lavish life style, is obsessed with recapturing them even after their legal emancipation (Reed, *Flight to Canada* 177). The third fugitive from Swille's plantation Stray Leechfield, however, believes he can buy himself free after his escape (73) and therefore underestimates the risk of recapture. He ends up being the only fugitive who is forcefully returned to the plantation by Swille's "tracers" (177). As Robin tells him upon his repatriation (luckily after Swille's death and Robin's takeover of the plantation): "He didn't want money. He wanted the slave in you" (177). The different strategies Raven, 40s, Stray, and Robin adopt to escape slavery and recapture as well as the forms of refusal available to them in and outside of the territories of legal slavery form a central thread that runs through the novel's rewriting of the slave narrative tradition.

This thread first starts to unfold when Raven's poem "Flight to Canada" comes into Swille's hands, provides information about Raven's whereabouts in the fictional town of "Emancipation City," and causes Swille to send his "tracers" after the fugitives (52). As critics have noted, Raven's poem functions both as a trace that endangers Raven's successful escape and as a catalyzer that moves his flight onward (Lock 73; see also, e.g., Rushdy, *Neo-Slave Narratives* 122–24). But even before he is chased by the tracers because of his poem, Raven does not have a permanent residence in Emancipation City and instead moves from place to place as a house sitter for "sympathizers to the cause" (Reed, *Flight to Canada* 61; see Levecq 288). Whereas "[h]is housesitting functions for various abolitionists [already] signal his aimless wandering through the country, even through its so-called progressive spaces, without any sense of home or belonging" (Levecq 291), the arrival of the 'tracers' accelerates Raven's flight further. In other words, when "Raven Quickskill finds himself pursued by the written and print texts of his own refusal" of slavery (McCoy 196), the novel reveals the ways in which Raven's initial escape from the plantation has not ended because it failed to bring Raven unconfined freedom or a complete detachment of enslavism and its world-structuring forces. In this way, *Flight to Canada* clearly pays "tribute to those antebellum slave narratives," such as Douglass's, Jacobs's, and Henson's, "that had earlier recorded this same dilemma" of a text that shall bring freedom but endangers any potential freedom at the same time (Rushdy, *Neo-Slave Narratives* 122–23).

Importantly, while paying heed to slave narratives through the poem as an escape trace and accelerator, the novel starts after the initial escapes of the fugitives that are usually at the center of slave narratives and only relays some of them mediated in dialogue after the fact.[198] Moreover, the tracers' attempted recapture of Raven and his onwards escape are significantly defamiliarized in Reed's novel. When two tracers dressed in "blazers" and carrying "a briefcase" arrive at the house Raven was "'watching' for a few months" with "orders to repossess" him, Raven invites them into the house and engages in formal conversation, before escaping through a window (Reed, *Flight to Canada* 61). Reacquiring human property appears here as a formal business transaction that involves paperwork rather than the physical force, manhunts, and bloodhounds known from slave narratives and which Josiah fears.[199] What is more, after Raven's rather easy and banal escape through a window that may recall George Jackson's escapes from home as a boy, Raven's status as a fugitive is described with dramatic, cliché images of flight from crime fiction:

> Quickskill walked the streets. He kept seeing license plates with VIRGINIA on them; they seemed to be following him. He put his collar up around his neck. He put his hands in his pocket. He kept walking against the shop windows, sliding around the corners. He was a fugitive. He was what you'd call a spare fugitive instead of a busy fugitive: he didn't have the hundreds of wigs, the make-up, the quick changes busy fugitives had to go through; he was a fugitive, but there was no way he could disguise himself. (76)

---

[198] While the novel does not explain how 40s escaped from the plantation, Stray describes and criticizes Raven's escape strategy as sly and servile: "You were the first one to hat, but you did it in a sneaky way. What kind of way was that, we thought. Just like a house slave. Tipping away. Following a white man and his wife onto the boat with a trunk on your shoulder, and when the guard ax you where you going, you say you with them" (Reed, *Flight to Canada* 74). For a discussion of Stray's initial escape, see below.

[199] As Ryan notes, the majority of physical violence in the novel is directed towards white characters, and white women in particular (127).

Christine Levecq argues that when the tracers are on Raven's trail, they "make him move furtively from house to hotel to street, [and] his newfound mobility becomes the symptom of an errant diaspora rather than a sign of free movement" (290–91). Importantly, while using familiar language from crime and detective novels, Reed recalls the relevance of disguise for fugitives from slavery as addressed in slave narratives, but emphasizes its futility because formerly enslaved fugitives and their descendants are always already marked by the racialization of their bodies as "Black" which equaled – and according to Wilderson – still equals "Slave" (*Red, White, and Black* 7).

The quoted passage not only references slave narratives but also recalls Black Power activists' autobiographical writing about experiences of FBI surveillance, disguise, and the feeling of being hunted. However, while Angela Davis's and Assata Shakur's autobiographies readily adopt the motif of formerly enslaved fugitives fleeing from legal owners and their bloodhounds to describe these experiences, Reed refuses to reproduce this familiar, violent image in his narrative. As Corina A. Crisu argues, Raven's escape in *Flight to Canada* is really "a succession of escapes" built after Douglass's "repeated border crossings and sense of endangered liberty" as documented in his slave narrative and his two following autobiographies (114). As I showed in the first part of this subchapter this holds true for Henson's narratives too as they – like Raven's poem – provide traces of perpetual fugitivity that Reed follows, while making central changes to their familiar plotlines. Published a few years after Davis's text, Reed's novel appears to use disguise and adaptation (of stereotypes, intertexts, tropes, and discourses from the nineteenth and twentieth century) as a main approach to telling a story of escape from slavery. While he uses the heavily mediated literary form of the slave narrative from Henson, Douglass, and Brown, and the sentimental novel from Stowe, Reed also tries on other literary forms and intertexts, such as crime fiction elements found in Black Power autobiographies' scenes of flight in order to see which ones allow him to tell a story of Black North American flight from a post-Civil Rights perspective.

After the imminent danger of recapture becomes apparent with the arrival of the tracers, Raven wants to warn his fellow fugitives Stray and 40s. As readers learn from various conversations among them and particularly through figural commentary *in praesentia* and *absentia*, Raven, 40s, and Stray had fled individually from their legal owner's plantation in Virginia to Emancipation City before the beginning of the narrative and have taken on very different escape and survival strategies since then. In her examination of "Vodoo, Fetishism, and Stereotype" in *Flight to Canada*, Glenda R. Carpio describes Stray as an "ambivalent figure of the minstrel performer" (566), "whose name insinuates his deviant (astray) and parasitic (Stray) characteristics," and who "does not believe he can redefine the power relations that render him property, insisting instead that the only possibility left for him is to manipulate the rules of the market" (577). As Swille recalls, Stray's escape was "very dramatic" (Reed, *Flight to Canada* 36): Stray "stabbed the man" who uncovered his illegal poultry business and then "fled on a white horse, his cape furling in the wind" (36). Afterwards – as 40s tells Raven in Stray's absence – Stray had financed his onward-escape by having his business partner, an "Indentured Servant friend, the [white Jewish] Immigrant, Mel Leer" (67), pretend to be his owner, sell, and then "kidnap" him back (80). Now in Emancipation City, Mel and Stray sell pornographic pictures of Stray having sex with white women and offer to rent Stray out as "Your Slave for One Day" (71, 80) because, as Stray explains to Raven, "[i]f anybody is going to buy and sell me, it's going to be me" (73). Levecq considers Stray's strategic use of pornography and prostitution as an integrationist survival strategy (288, 291). Working consciously within the capitalist and enslavist system of the US American North, Stray's freedom is clearly circumscribed by the ongoing commercialization of his body as property. Yet, he also points to the ways in which alternative strategies, such as Raven's work as an abolitionist writer and lecturer (like Douglass and Brown), are also complicit (Reed, *Flight to Canada* 72–73). After all, Raven "also performs as a racialized being to market a consciously crafted image of himself" (Carpio 579–80), despite the fact that he refuses to buy himself from Swille out of principle (Reed, *Flight to Canada* 73–74).

While there is hardly any information about 40s' initial escape provided, his hiding and defense strategies in Emancipation City are the most clearly articulated of the three. 40s lives on a houseboat "down near the river" (76) armed to the teeth and prepared to hide in the mountains "for twenty years" to escape recapture if necessary (78). Importantly, his self-proclaimed "home" (76) is not only a place of refuge but also a means of transportation and thus a potential escape vehicle. 40s argues that Raven and Stray are fools for trying to buy, sell, and lecture themselves out of slavery into freedom because it clearly makes them traceable to "Swille […] [and] his dogs, both the four-legged and the two-legged ones" (78). According to Crisu, 40s "criticizes both Leechfield's pragmatism and Raven's idealism," and instead isolates himself, striving for invisibility (in contrast to Stray's sexualized hypervisibility) in order to survive and stay out of the enslaving South (113). In Crisu's words, "[l]ike Ralph Ellison's invisible man, 40s has his own hole, where he prepares for camouflaged fight" if necessary (113). His approach to escape is thus a militant and individualistic strategy of hiding, survival, and self-defense.

Taken as a tripartite, all of the fugitives' strategies of hidden militancy and eremitism (40s), hypervisible sexualization and commercialization (Stray), and Raven covering the middle ground between the two with political activism and art (Crisu 113) illustrate the contradictions and conflicts involved not only in the antebellum nineteenth century but in the twentieth-century movement era as well. Together with Robin's attempt to find refuge and refuse the system from within slavery, the novel ironizes all strategies and questions any assumed clear-cut differences between them across the centuries. In fact, Robin, who does not flee the plantation and instead disguises himself as a servile servant, compares himself to the rebel Nat Turner at the very end of the novel, pointing out that while Turner attempted resurrection and was now dead, he was alive and managed to inherit a large plantation (Reed, *Flight to Canada* 178). *Flight to Canada* thus not only shows that the (formerly) enslaved had overt and subtle forms of refusal at their perusal that break the mold of assimilation, resistance, and flight. It also rejects judgment of the Black Power movement as to which strategy deserves more respect (cf. Rushdy, *Neo-Slave Narratives* 102–04).

Apart from the different forms of disguise, flight, and refusal as adopted by 40s, Stray, Robin, and Raven, the eponymous flight to Canada represents another important concept that Reed rewrites from nineteenth-century literature of slavery. Similar to Henson's slave narrative, for most of the novel Canada represents an ambiguous dream of freedom rather than a distinct geographical place that shall free Raven from the restless fugitive existence in the northern US states. While Raven only briefly stays north of the Niagara River, Canada looms large as an imagined place of refuge for Raven, just as it had for his historical forbear Henson. Yet, Reed pushes the ambiguity of Canada as a projection surface to a new extreme:

> While others had their tarot cards, their Ouija boards, their I-Ching, their cowrie shells, he had his 'writings.' They were his bows and arrows. He was so much against slavery that he had begun to include prose and poetry in the same book, so that there would be no arbitrary boundaries between them. He preferred Canada to slavery, whether Canada was exile, death, art, liberation, or a woman. Each man his own Canada. There was much avian imagery in the poetry of slaves. Poetry about dreams and flight. They wanted to cross that Black Rock Ferry to freedom even though they had different notions as to what freedom was. (88)

Rachel Adams argues that in "Quickskill's formulation, Canada is slavery's idealized antithesis" (66) and the novel examines how "slave narratives helped to crystallize the mythic equation of the north with freedom" (66). Given that as Glen Anthony Harris points out, "Quickskill seems always just short of Canada; his goal is forever just beyond reach. The harder he struggles, the more his object retreats before him" (470), I suggest that Raven's literary and physical restlessness in the US North rather than slavery actually represents the idealized antithesis of Canada.

After risking and accelerating Raven's flight, the poem finally brings Raven to Canada when he uses the honorarium he receives for its publication to finance his journey with a "yacht" across the Niagara River (Reed, *Flight to Canada* 87, 145). Upon landing on the Canadian shore, Raven is described as "too tired and depressed to greet this prospect with joyful exclamation of former slaves who reached this moment of Jubilation" (154). Again, the novel refuses to reproduce a familiar motif of nineteenth-century literature of slavery and instead uses techniques of alienation when the novel cites a Henson-like jubilant arrival directly afterwards set off in italics and quotation marks (155). The direct quote is taken from Eber Pettit's *Sketches in the History of the Underground Railroad: Comprising Many Thrilling Incidents of the Escape of Fugitives from Slavery, and the Perils of Those who Aided Them* from 1879 (66–67), in which the white abolitionist and Underground Railroad conductor reports on the arrival of fugitives in Canada. Tim Ryan argues that "the moving testaments of escaped slaves contain a powerful emotional core that is necessarily absent from the determinedly anti-realistic *Flight to Canada*" (128). Ryan contends that while offering stylistic innovation Reed fails to address "the often-horrific realities of slavery" (129) in the juxtaposition of Raven's uneventful arrival and the quoted jubilant scene. While Ryan's critique of Reed's comic approach to narrating escape from slavery may be understandable particularly in comparison to other neo-slave narratives he examines, such as Butler's *Kindred* and Morrison's *Beloved*, Ryan fails to consider the narrative perspective of the quoted excerpt in more detail.

> "While they were on my vessel I felt little interest in them, and had no idea that the love of liberty as a part of man's nature was in the least possible degree felt or understood by them. Before entering Buffalo harbor, I ran in near the Canada shore, manned the boat and landed them on the beach ... They said, 'Is this Canada?' I said, 'Yes, there are no slaves in this country'; then I witnessed a scene I shall never forget. They seemed to be transformed; a new light shone in their eyes, their tongues were loosed, they laughed and cried, prayed and sang praises, fell upon the ground and kissed it, hugged and kissed each other, crying 'Bress de Lord! Oh! I'se free before I die!'" (Reed, *Flight to Canada* 155)

In the excerpt, an uninvolved, presumably white observer first admits his initial disinterest in the fugitives, who he had deemed incapable of the 'Human' "*love of liberty*" and then provides an eyewitness report of the fugitives' euphoric arrival on the Canadian shore. With the juxtaposition of Raven's somber arrival narrated from a figural narrative perspective with an eyewitness report that necessarily lacks insight into and empathy for the fugitives' feelings and perceptions, *Flight to Canada* again points readers to the mediation and appropriation processes of narratives of slavery in the nineteenth century, disallowing readers to sentimentally feel for the fugitives. In doing so, I argue the novel questions the possibility of authenticity, narrative agency, and of telling a coherent, authoritative narrative of flight and arrival from the 1850s to the 1970s. As we will see, in the following close readings of Morrison's and Hill's texts, later neo-slave narratives raise these questions as well.

Fitting to this focus on mediation and appropriation of voices and perspectives, in contrast to Josiah, Raven does not in fact experience life in Canada after his arrival. Instead, he relies on the judgment of his friend, the free Black "Carpenter" who had moved to Canada a little while ago. The two men accidentally meet on the Canadian side of the Niagara Falls at a hotel. Raven had just arrived reunited with his lover, the Native American dancer "Princess Tralarara Quaw Quaw," and Carpenter is on his way back to the United States. Carpenter, who says he was beaten up by "mobocrats" (Reed, *Flight to Canada* 159), describes Canada as an extension of the capitalist and racist US America (Crisu 115). According to him, the place appears "like any American strip near any American airport" where "[v]igilantes" worse than the "Klan" harass fugitive slaves, "and the slaves have to send their children to schools where their presence is subject to catcalls and harassment" (Reed, *Flight to Canada* 160). Thus, Raven learns from Carpenter that Canada is not the paradise for Black people and people of color of which he had dreamed. Instead, Canada falls short of the idealized freedom dream that fails, however, less because of Raven's experiences in Canada and more so because Raven immediately retreats from the counter-image of Canada as an extension of the United States Carpenter presents to him. Again, the novel addresses the problem of taking mediated information for face value here and indirectly encourages its readers to acknowledge and try to see through the layers of mediation. After hearing Carpenter's report of racism and violence in Canada, Raven thinks he has lost something he was assured of all his life, i.e., "whatever went wrong I would always have Canada to go to" (161). The presumed endpoint of perpetual fugitivity proves to be no endpoint at all but a mere discursive reproduction of the US American north. When Carpenter also relays the news of Swille's death (161), Raven consequently leaves Canada for Emancipation City without seeing anything from Canada but the hotel with a view of the Niagara Falls. After corresponding with Robin, Raven makes his return journey to the Virginian plantation, where Robin in the meantime has slowly poisoned Swille (174), changed his will (171), and then delayed help when Swille is pushed into the fire (136–37), in order to inherit the estate and create a boarding house for Black artists and workers after Swille's death (11–14).

Upon the "dematerializing of Canada" as an endpoint of flight (Levecq 293), many critics follow Robin's suggestion that when even Canada proves not to be the guarantor for freedom "*Canada, like freedom is a state of mind*" (178). In this reading, the novel suggests disconnecting freedom from a specific locality and instead propagates freedom as an inner pose, which Raven finds in his writing (Reed, *Flight to Canada* 88, Juan-Navarro 93). For as Santiago Juan-Navarro contends in his analysis of "Self-Reflexivity and Historical Revisionism in Ishmael Reed's Neo-Hoodoo Aesthetics" in the novel,

> The spatial and temporal coordinates soon acquire an overtly metafictional quality. Quickskill's flight to Canada allegorizes the search for an aesthetic utopia represented by writing in a state of liberty, an ideal for which Reed has fought in all his works. The novel ends with its protagonist coming to an understanding of his own condition: Canada is only a state of mind, a desirable ideal the black writer has to strive for, no matter where he may be physically. (93)

Yet, the passage set off in italics in which Robin describes Canada as mental freedom at the novel's ending is more ambiguous than that, because Robin introduces the thought by wondering whether Raven found "*what he was looking for in Canada? Probably all that freedom gets to you. Too much freedom makes you lazy. Nothing to fight. Well, I guess Canada, like freedom, is a state of mind*" (178). Before he compares himself favorably with Turner, he continues,

> *I couldn't do for no Canada. Not me. I'm too old. I done had my Canadas. I'm like the fellow who, when they asked why he sent for a helicopter to get him out of prison, answered, "I was too old to go over the wall." That's how I feel. Too old to go over the wall. Somebody had to stay. Might as well have been me and* [his wife] *Judy.* (178)

Thus, while the novel surely allows for the equation of freedom with an inner pose, it also immediately calls that notion into question. After all, Raven does not enjoy the 'freedom of the mind' anywhere else but the plantation. Moreover, return for Raven is a reaction to the ever retreating of an alternative elsewhere that he can project his hopes towards as well as a retreat on his part from the potential proliferation of the geographies of slavery from the American South and the northern US states to Canada. Thus, his and Quaw Quaw's "chronologically and regionally backward movement[s]" (Mielke 16) upon the disillusionment of Canada also represents a retreat from alternative models of freedom not bound to a place and beyond the US nation state (Levecq 293, Mielke 16). As Christian Moraru contends,

> Obviously, Reed distrusts the utopia of a cultural or political space outside structures. On the contrary, subjects as self-emancipating (re)writers can reposition themselves only within textuality, within the realm of structure and discursive practices. One should not toil to flee structures – one should struggle to restructure, rewrite them. As *Flight to Canada* suggests, evasion – in all senses – is a poor option as well as a dangerous illusion. (111)

Using his ability to read and write as well as to camouflage his wits and true intentions, Robin, who had stayed enslaved and superficially seems closest to the infamous character of Uncle Tom in Stowe's novel, rids the enslaved community on the plantation off their legal owner. Inheriting the property, he then creates what Jacobs calls a "loophole of retreat," i.e., a place of refuge, an enclosed form of freedom in the center of the enslavist system. While Jacobs's loophole was a tiny attic space on her legal owner's plantation, where she hid from his sexualized violence for seven years before her escape to the US northern states – endangering her mental and physical health –, however, Robin's loophole takes the comfortable form of a large plantation estate "six times as big as Monaco" (Reed, *Flight to Canada* 107).[200] Paradoxically, at the end of Reed's neo-slave narrative the fugitives seem to find the safe haven so dearly sought after not in the US northern states nor Canada but in the enslavist plantation geography of the American South.

Rushdy interprets the murder of the plantation owner Swille, who is stylized in the novel as representative of the enslavist, colonialist, imperialist, and capitalist system, as the fall or at least destabilization of that system (*Neo-Slave Narratives* 112–13). I, however, would question the insurrectionist power of the plantation in the hands of the formerly enslaved. In the end, the narrative does not explicate how the boarding house will continue to exist within the larger anti-black society of the American South that Robin describes as "the devil's country home" and "the land of the hunted and haunted" even after taking over the plantation (173). Moreover, when Raven visits 40s at his houseboat and contends that his weapons are excessive because "[w]e are not in Virginia anymore", 40s replies: "Shit. Virginia everywhere. Virginia outside. You might be Virginia'" (Reed, *Flight to Canada* 76). With this statement, 40s actually anticipates Raven's disillusionment with Canada as a safe haven and points to the ways in which slavery and anti-blackness can reach into all kinds of geographical and interpersonal spaces assumed safe. Raven, too, ponders the possibility of never-ending

---

[200] This is not the first indirect reference to Jacobs's slave narrative. As Crisu points out, Raven's poem "Flight to Canada," like Linda's faked letters to her legal owner from her garret, shall trick Swille into believing that he has already reached Canada and is therefore out of his legal reach (112). Moreover, while *Flight to Canada* does not address the sexualized violence against enslaved women that Jacobs's slave narrative hints at and Black female writers of neo-slave narratives address in more detail, the American South of Reed's novel is nonetheless clearly portrayed as dominated by "sexual deviance" (Levecq 290), such as sadomasochism and incestuous relationships primarily among the white population. In this way, the novel shows "the connections between chattel slavery's reification of people and bodies and the fetishistic notions of race and sexuality it has produced" (Carpio 563).

enslavement before his flight to Canada, also in relation to the control enslaved people may force onto one another as they too still exist within the larger anti-black, enslavist system:

> Slaves judged other slaves like the auctioneer and his clients judged them. Was there no end to slavery? Was a slave condemned to serve another Master as soon as he got rid of one? Were overseers to be replaced by new overseers? Was this some game, some fickle punishment for sins committed in former lives? Slavery on top of slavery? Would he ever be free to do what he pleased as long as he didn't interfere with another man's rights? Slaves held each other in bondage; a hostile stare from one slave criticizing the behavior of another slave could be just as painful as a spike collar – a gesture as fettering as a cage. (144)

Raven and 40s' insights into the reach of anti-blackness and enslavism across geographies and into the minds of the (formerly) enslaved may provide hints that Robin's boarding house may prove vulnerable as well. Thus, I agree with Spillers who describes Reed's work as "sufficiently interstitial that 'flight to Canada,' indeed 'flight,' remains entirely possible" ("Changing the Letter" 183). Yet, in contrast to Beck's claim that Raven's return to the plantation has "brought the slave and his narrative, literally and figuratively, home" (139), I argue that Reed's novel indirectly exposes that – while there is flight – there is no arrival in freedom or return home for formerly enslaved and fugitive men in the transatlantic world of slavery, neither in the fugitive dream of Canada nor in any loopholes of retreat available. O'Neale comes to a similar conclusion in her 1978 review of the novel when she writes that Reed "manipulates time and history to provide a cyclical pattern of that tired truth: the black man's condition has never changed. Whether in 1855 or 1975, the black man still longs for a non-existent flight to Canada in quest of a freedom that is unobtainable at home. Catch 22" (175).

In sum, by spanning a rich intertextual net that blends nineteenth-century discourses of slavery and the abolitionist movement with the discourses of late-twentieth-century popular culture and the Black Power movement, Reed produces a hybrid narrative that reveals continuities of anti-blackness, captivity, fugitivity, and unfreedom for Black people from the historical era of slavery until the postmodern, post-Civil Rights time and across and beyond geographies of legal slavery in North America. Reed also showcases diverse strategies of enslaved people to refuse both the legal captivity of the antebellum South and other forms of confinement and constraint.

Reed's novel joins the ranks with an increasing number of neo-slave narratives that question the myth of Canada as a safe haven for African American fugitives. However, *Flight to Canada* does not so much fill blatant gaps in the United States and Canada's memorialization of slavery and Black North American history, as later neo-slave narratives, such as Hill's *Someone Knows My Name* from 2007 would do, as we will see below. Instead, *Flight to Canada* primarily exposes the myth from distinctly Black male perspectives south of the Canadian-US border and uses innovative stylistic revisions that later neo-slave narratives, such as *Beloved* and *Someone Knows My Name*, would fall back on in order to account for Black female fugitivity and diasporic refusal in the eighteenth and nineteenth century and comment on their late twentieth-century reincarnations.

"Rememories" of captivity and flight in Toni Morrison's *Beloved*

In commemoration of her death in 2019, Dionne Brand described Toni Morrison as "the greatest writer in English of the 20th century and the 21st" ("Toni Morrison"). Indeed, Morrison was the first African American woman to win the Nobel Prize in literature in 1993 and scholars consider the publication of *Beloved* as a watershed moment not only in the neo-slave narrative tradition, the American literary scene, and in African American studies, but also in the public discourse on the history of slavery in North America (Best, "On Failing"; Broeck, "Enslavement as Regime" par. 2; Misrahi-Barak; Nehl). As Angela Davis contends, because of Morrison's *Beloved* "[w]e can never think about slavery in the same way" (qtd. in Greenfield-Sanders). The prize-winning novel narrates the story of enslaved Sethe who escapes slavery on a plantation named "Sweet Home" in Kentucky, pregnant and on foot, across the Ohio River to the 'Free' North. When faced with recapture shortly after arriving in Cincinnati, Ohio, Sethe kills her first baby daughter, who remains unnamed in the novel, in an attempt to save her children from imminent re-enslavement. The deceased daughter haunts Sethe's family home on Bluestone Road as a baby ghost and then reappears as a young woman named Beloved after Paul D, another fugitive and formerly enslaved man from Sweet Home, arrives on Sethe's porch.[201]

For Carmen Gillespie, the main themes of the novel are encoded in the title's particles: the question of *Being* (i.e., being human, free, and dead as another "state of being"); the theme of *love* (i.e., mother love and the love story between

---

201 Dubey points to a variety of "interpretive possibilities entertained in the novel" with respect to the character of Beloved from "ghost or a hallucination" to "symbolic projection" or "a flesh and blood reincarnation of Sethe's dead daughter," the last of which, of course, has been established as a dominant reading that requires readers to "suspend rational notions of mundane reality" (342).

Paul D and Sethe);[202] and the role of the past as reflected in the title ending on the letter *d* (29–33; cf. also V. Smith, "Neo-slave Narratives" 179). Inspired by a newspaper clipping about the historical figure Margaret Garner, which Morrison encountered during her work for *The Black Book* at Random House,[203] Morrison put the underappreciated perspective of enslaved and fugitive women as mothers at the center of her 1987 novel. Clearly, many of *Beloved*'s themes also resonate with Jacobs's writing and Black Power autobiographies, such as Shakur's. Around the time *Beloved* was published, Jacobs's slave narrative was receiving increasing attention in the scholarly community as an authentic slave narrative (see ch. 2.1.), while other novels such as *Dessa Rose* (1986) addressed similar issues in the post-Civil Rights and post-Black Power era. After Jacobs had admitted that her self-confinement in the attic space left physical and psychological effects that will stay with her for the rest of her life, Morrison provides fictional insight into the kinds of marks enslavement and escape from enslavement left on the bodies and the souls of the formerly enslaved (cf. Heinert 81–82).

Beloved is split into three "books" of different sizes, which are made up of numerous untitled chapters and smaller narrative units the majority of which are anachronistically told, and sometimes retold from a figural narrative perspective that focalizes the protagonists Sethe and Paul D as well as several other characters. At instances, free indirect discourse also switches to interior monologue and stream of consciousness. The novel uses "fragmented narrative and flashbacks to tell multiple stories simultaneously" (Vint 243), "depriving [...] [readers] of the usual coordinates in time and space" (Best, "On Failing" 469). Thus, the major narrative planes only become gradually graspable over the course of the narrative with substantial effort from the readers, which – as is the case with *Flight to Canada* – does not lend itself "easily to summary" (V. Smith, "Neo-slave Narratives" 175). The planes are on the one hand the narrated present of post-Civil War Cincinnati in 1873, and on the other Sethe's and other characters' mostly unmarked "rememories" of what happened almost twenty years earlier in the 1850s on the Sweet Home plantation, during flight, and in

---

[202] The novel also hints at other potential love stories, such as the consensual relationship in which Sethe was conceived or the relationship "Baby Suggs holy" had with her husband. Yet, Greg Thomas argues that these forms of "erotic maroonage" may be "overshadowed by Morrison's accentuated, controversial, and indeed problematic imagery of black men 'fucking cows' in the absence of women – a reductive cattle equation that could undermine the antichattel politics of maroonage in this particular narrative of slavery" ("'Neo-Slave Narratives'" 225).

[203] *The Black Book* (1974) is a collection of diverse archival material that Middleton A. Harris et al. compiled and Morrison edited as an examination of African American life and history through the ages (Gillespie 237).

the first twenty-eight days on Bluestone Road. Dennis Childs points out how the "rememories," a term Sethe uses as both subject and verb, represent "the haunted experience of Sethe, Paul D, and other characters, suggesting the undead nature of chattel slavery within the context of de jure freedom" ("Angelo Herndon" 52). Thus, *Beloved*'s 'rememories' actually function similar to Douglass's and Henson's reiterations of arrival, settlement, and survival and also resonate with Black Power autobiographies' insistence on anti-blackness and confinement as a form of 'neo-slavery' after emancipation (cf. Childs, "'You Ain't Seen Nothin'" 275).[204] As Valerie Smith puts it in an introductory essay on neo-slave narratives,

> even though set after the Civil War in the North *Beloved* is nevertheless a novel about slavery. The characters have been so profoundly affected by the experience of slavery that time cannot separate them from its horrors or undo its effects. Indeed, by setting the novel during Reconstruction, Morrison invokes the inescapability of slavery, for the very name assigned to the period calls to mind the havoc and destruction wrought during both the antebellum era and the Civil War years. ("Neo-slave Narratives" 175)

Scholars have read and taught *Beloved* as a stylistically brilliant and thematically challenging American masterpiece under all kinds of auspices in African American studies, American studies, and world literature since the early 1990s. By the early twenty-first century, countless monographs, edited volumes, and companions on Morrison's oeuvre (see, e. g., Gillespie; Kubitschek; Middleton; Montgomery; Roynon; Tally, *Toni Morrison*) and the novel *Beloved* (see, e. g., Andrews and McKay; Bloom; Tally, *Beloved*) have appeared.[205] Examples of topics addressed in the area of neo-slave narrative studies include the trope of "the returning daughter" (Rody); the novel's complex combination of the slave narrative tradition, the historical novel, and the gothic (Heinert) as well as its drive towards "liberation" (Mitchell); and the role of "musical improvisation" in the novel (Keizer). As Harriet Mullen shows in her comparative analysis of "Resistant Orality in *Uncle Tom's Cabin, Our Nig, Incidents in the Life of a Slave Girl,* and *Beloved*," "Morrison's literate ghost story has as much, or more, in common with the Gothic as the sentimental tradition" (273). According to Ryan, *Beloved* is also indebted to *Kindred* in much of its style and themes and in contrast to *Flight to Canada*, which Ryan reads as an exclusively discursive approach to the history of slavery, addresses slavery "both as discursive field *and* horrifying reality" (145, emphasis

---

[204] For a more detailed summary of the content and its main themes, see, e.g., Gillespie 19–28 and Roynon 45–55.
[205] For an overview of research on Morrison's oeuvre in the 1970s, 80s, and 90s, see Roynon 114–25.

mine). Sherryl Vint describes *Beloved* together with *Kindred* and *Flight to Canada* as "fantastic neo-slave narratives" that "critique the limitations of realist forms and 'objective' history to convey African-American history" and "revise and resist the tropes of nineteenth-century slave narratives, particularly their increased emphasis, influenced by Civil Rights and feminist struggle, on embodiment and embodied experience for understanding slavery" (241–42, 255). In a pointed rereading of the critical reception of *Beloved* at the turn of the century, Sabine Broeck convincingly argues that much scholarship has adopted a "kitsch" reading of the novel as being about the successful overcoming of the traumas of chattel slavery, healing, and redemption despite the narrative's much more contradictory plot and narrative form ("Trauma, Agency, Kitsch"). This is true for the majority of earlier works listed above while more recent studies, such as Tessa Roynon's 2013 *Cambridge Introduction to Toni Morrison*, agree that "Morrison [indeed] conveys Sethe's erroneous belief that Beloved's reappearance atones for past suffering, while the author implicitly questions whether any kind of atonement or recompense is possible at all" (52).

Despite the abundance of *Beloved* research and the fact that forms of flight and border crossings are plenty in the novel (G. Thomas, "'Neo-Slave Narratives'" 214), studies that focus on the notions of flight, borders, confinement, and refuge in *Beloved* are still rare. Notable exceptions that both refuse dominant redemptive readings and address fugitivity and confinement are, for instance, Dennis Childs's and Greg Thomas's articles. The latter, which I already discussed in chapters 2.1. and 2.2., proposes to read neo-slave narratives, such as *Beloved*, alongside Black Power autobiographies for their common conceptualization of forms of "maroonage." In the former, "'You Ain't Seen Nothin' Yet': *Beloved*, the American Chain Gang, and the Middle Passage Remix" from 2009, Childs discusses Paul D's "rememories" of captivity in the chain gang as "forward-haunting" (274). He considers the inclusion of the chain gang, "normally associated with the late nineteenth to mid-twentieth century into the context of the late antebellum period" as a "strategic anachronism" (285) that addresses the proliferation of African American confinement from the Middle Passage through slavery into the post-emancipation era and also indirectly foreshadows the twentieth-century rise of the prison industrial complex.

In the following necessarily selective close reading, I focus on some of the novel's narrative units that narrate flight and the parallel proliferation of confinement in the liminal period between the pre- and post-Civil War periods in US America in the novel. As examinations of Sixo's failed escape and Sethe's escape, or rather Denver's birth story will show, the novel associates flight with a fleeting possibility of reproduction as futurity which, however, is immediately questioned through their degree of mediation that calls into question the reliable narratibility of escape as such. Moreover, Sixo's death, Sethe's risk of recapture, and the murder of her daughter further question the existence of places of refuge for the formerly enslaved and fugitive. In fact, Sethe's form of self- and 'child-defense' together with Beloved's return as a young woman from what appears to be an eternal Middle Passage cast into question whether physical death or the passing from one state of being into another can be a form of sanctuary for the formerly enslaved and fugitive. *Beloved* represents home and death in the immediate aftermath of slavery as a proliferation of forms of confinement – from Paul D's repeated escapes and captures, through death as the ongoing Middle Passage, to the house on Bluestone House as a claustrophobic place of confinement.

Stealing away

While Sethe is the most well-known fugitive in the novel, whose escape story I will examine in more detail in the following, many more fugitives populate the novel's world who have received less scholarly attention, such as Paul D and Sixo. Greg Thomas aptly describes Sethe's family and the other enslaved people on the Sweet Home plantation as "a community of runners" ("'Neo-Slave Narratives'" 214), i.e., a collective of potential maroons, "runaways likely to run off on any given day, at any given minute" (223). Yet, as Sethe admits in a first-person narration that addresses a "you" who is presumably Beloved, Sethe and her husband Halle did not have running on their "minds" for most of their time on the plantation because they thought they would be able to buy themselves free (Morrison 191, 197). But this changes when their legal owner Mr. Garner dies and is replaced by Mrs. Garner's relative "schoolteacher" who forbids Halle to work for money on Sundays, introduces physical punishment, and also starts to study the enslaved people on the plantation to satisfy his "ethnographic interest" (Mitchell 92, cf. Morrison 191, 193, 219–20). As Sethe explains, Sixo introduces them to the idea of flight after Mrs. Garner sells the first enslaved man, Paul F, from their midst (197) and thereby transforms the community of enslaved into Thomas's "community of runners" ("'Neo-Slave Narratives'" 214). As a result, Sixo "started watching the sky," "not the high part, the low part where it touched the trees," and frequently "crept at night" (197). When he eventually "learned about the train," Sethe "could tell his mind was gone from Sweet Home" already (197). Morrison describes Sixo's search for the North Star and the Underground Railroad through indirection, only gesturing towards the familiar motifs from slave narratives, such as Henson's. Moreover, she makes Sixo key to the escape plan that the enslaved people on Sweet Home work out (222), while refusing to provide insight into his own perspective. In fact, the perspectives of those who fail to escape Sweet Home and go mad or die, like Halle and Sixo, are not accessible to the readers throughout the novel. Instead, it is only Sethe, Paul D, and Halle's mother "Baby Suggs," the Sweet Home survivors, who may try to, or cannot help but 'rememory.'

The fact that readers access information about Sixo only through Sethe's and Paul D's recollections contributes to Sixo's mysteriousness. As readers learn, Sixo speaks an unidentified language besides English, his complexion is "indigo" (21, 222), and he has a special connection to the natural world which Mitchell describes as a "cultural and spiritual rootedness" all off which leads her to suggest that Sixo was African, a Middle Passage survivor (103). Sixo is also already a short-term run-away before he finds out about the North Star and the Underground Railroad. Using his ability to "melt[...] into the woods," Sixo leaves the bounds of the plantation regularly to walk thirty-four miles to meet his fourteen-year old girlfriend "Patsy the thirty-mile woman" and return before his absence is noticed (24–25, cf. G. Thomas, "'Neo-Slave Narratives'" 224). In other words, he is frequently 'stealing himself away' for a night, "literally and figuratively carried away by [his] desire" during his trips whose historical precedents functioned, according to Hartman, as a temporary "way of redressing the natal alienation or enforced 'kinlessness' of the enslaved" (*Scenes of Subjection* 66, 67). During these "short lived flights from captivity" (65–66), Sixo moves along the Black border and presses against it with everyday practices of refusal which allows social life of the socially dead. However, instead of shedding the status of social death by crossing over, these practices are "undertaken with the acknowledgment that conditions will most likely remain the same" or worsen (51) due to "the sheer incommensurability of the force that [slavery] deploys in response to the small challenges waged against it" (55). And worsen they will in the novel when the escape plan fails, as we will see in the following.

As readers find out when Sethe's partial memory of the escape plan is later complemented with Paul D's perspective (Morrison 219–29), after months of planning, hardly anything goes according to plan when they receive the signal to meet up and take a wagon to the North together. Schoolteacher catches Sixo and Paul D at the meeting point, whereas Patsy escapes alone. Sethe manages to send her three children Howard, Buglar, and the unnamed nine-month-old girl in the wagon ahead to her legally free mother-in-law Baby Suggs in Cincinnati, Ohio. When she cannot find Halle, who has gone presumably mad after witnessing schoolteacher's nephews abusing and whipping Sethe, Sethe runs off alone to follow her children (228). Caught and tied to a tree, Sixo laughs and sings, so that schoolteacher deems him lost as a usable property and therefore burns and kills him while Paul D is watching (226).[206] Although Sixo's escape fails, he is confident that his girlfriend escaped pregnant. The two enslaved pregnant women Sethe and Patsy thus steal away not just themselves but also the freedom dreams embodied in their children and unborn babies (225, 227, 229). Yet, as the novel also shows, carrying the Black female futurity of the promise of successful flight and newborn life that Sixo dreams up when he laughs, sings, and shouts out "Seven-O" (226) also means carrying the potentiality of reproducing death and enslavement because of the law of *partus sequitur ventrem* and the Fugitive Slave Act. While the novel never provides insight into what happens to Patsy and her unborn baby Seven-O, in the narrative fragments that share Sethe's escape and arrival the limits of escape and arrival become clear. After all, as Paul D overhears schoolteacher saying, the legal owner is set on securing a profitable enslavist future by recapturing "the breeding one, her three pickaninnies and whatever the foal might be" (227).

Sethe's escape is narrated in three main narrative fragments shared first in two separate parts from the perspective of Denver, the daughter born on the run (29–35, 76–85), and then from Sethe's perspective (90–93). When we recollect those fragments, the following summary of the escape can be constructed. Sethe makes her way alone on foot towards the north, six-months pregnant, barefooted, and with raw whipping wounds on her back. Close to the Ohio River, which separates the enslaving American South from the so-called free North, physical exhaustion lets Sethe lie down in the woods and consider waiting for death. In this moment, the white teenager Amy Denver appears, looking for edibles (32). As the talkative Amy explains to Sethe, Amy is the daughter of an immigrant mother who worked as an indentured servant for "Mr. Buddy" to pay off her passage (33). After her mother's death, Mr. Buddy forces Amy to work for him under dire circumstances occasionally with the use of physical punishment (79). Now she is on the run to Boston to start a new life and acquire the best "velvet" (32). During their brief chance encounter, Amy helps Sethe find shelter, massages her swollen, numb feet (34–35), treats the wounds on her back, helps her walk to the river, and assists Sethe in delivering her second daughter Denver (named after Amy) prematurely in a broken boat on the riverbank (83–85). After Amy leaves Sethe out of fear of being identified with a fugitive Black girl, Sethe finds the Underground Railroad conductor Stamp Paid who rows

---

206 As a rebellious Middle Passage survivor, Sixo seems also connected to Sethe's mother. As a "terribly elliptical passage in *Beloved*" suggests (G. Thomas, "'Neo-Slave Narratives'" 225), "Ma'am" may have also survived the Middle Passage and was presumably involved in rebellious activity for which she was brutalized and hanged with several others (Morrison 61–62).

her to the other side of the river, where Ella picks her up and takes her to Baby Suggs's house (90–93). Sethe's endurance, her unlikely encounter with Amy, and the help of the Underground Railroad conductors and members of the Black community in Cincinnati Stamp Paid and Ella ensure Sethe's survival and successful escape.[207]

Hames-García argues that Morrison – like Shakur and Jacobs – develops a concept of freedom out of the particular perspective of the formerly enslaved or captive mother that is "by definition one's responsibility to the freedom of others" (130). Sethe, Assata, and Linda indeed connect their escapes from imprisonment and enslavement directly to their desires to be with and care for their children. Yet, as many critics have noted, while Assata and Linda also express a desire for freedom for themselves, Sethe escapes almost exclusively because she wants to save her children from being sold away and to be free to love and care for them (Rody 216, Vint 247, Mitchell 93). After sending her children to the North ahead, her desire to provide for her children epitomized first in the need to bring her toddler nursing milk (Morrison 17) and then in her fear of dying while the unborn baby "lived on – an hour? a day? a day and a night? – in her lifeless body" (31) drives her along the ordeal of the lonely, risky trek. Sethe later emphasizes in a conversation with Paul D that escaping and rescuing her children from enslavement was the first and "only thing I did on my own" (162).[208] She admits she had help, but it was she who told herself "Go on, and Now" when it was necessary (162). Rody describes Sethe's escape not only as "a significant feminization of the archetypal slave escape narrative" (29) but as a "maternalized version" because it is intermittently interrupted by Denver's birth and her reasons for escape are "entirely maternal" (216).

---

[207] Scholars have criticized the inclusion of a white savior/helper/midwife in this scene. Chinosole, for example, criticizes that both "*Beloved* and *Dessa Rose* pay homage to the idolized privileged European American women inversely" through the inclusion of "a kindly poor or abandoned [white woman] who plays a decisive role in the African's quest for freedom" (*The African Diaspora* 104). Rody too points out that the escape plot would have worked without Amy, a fugitive white girl who exhibits several similarities to Sethe (both are "poor and alone, running North, away from a violent father-owner figure, recalling [...] [their] lost mother[s]"). The inclusion of Amy as facilitator of both Denver's birth and Sethe's escape, Rody argues, contributes to the "myth of American female interracial bonding" (217, 30).

[208] According to Rody, Paul D "brings to the text a voice of tribal griot-cum-historical eyewitness," whose presence evokes Sethe's attempts at "female autobiography" and also "offers hope of futurity at the telling's end" (Rody 27, 30). I propose a different reading of the novel's futurity in its narrative fragments of escape in this chapter, while I agree that Paul D functions as a trigger for some of the 'rememories' that the novel tries to tell.

On the level of narrative transmission, I argue, Sethe's escape story is not only maternalized but also significantly shaped by the perspectives of the daughters. The majority of the escape story is highly mediated through Denver's memory of her mother's story about "the magic of her birth" (29). Signs of this mediation appear only irregularly in the first fragment and completely disappear in the second so that readers may think they experience the escape and encounter with Amy directly from Sethe's perspective. Yet, the degree of mediation becomes clear when the second fragment is introduced by Beloved demanding from Denver to tell her "how Sethe made you in the boat" and Denver admits that Sethe "never told me all of it," but wants "to construct out of the strings she had heard all her life a net to hold Beloved" (77). Denver reiterates the first part of the story from the earlier memory fragment (29–35) as a brief report, but "[n]ow, watching Beloved's face alert and hungry, how she took in every word, asking questions about the color of things and their sizes, her downright craving to know" (77), embellishes the story for Beloved's entertainment. Among the embellishments are familiar images of escape in slave narratives ("Behind her dogs, perhaps; guns probably" [78]) which, as we saw in the preceding close readings, are also cited in Reed's novel and have been used to great effect in Black Power autobiographies, such as Davis's. Moreover, Denver starts to conceive of the story not as a "monologue," but "a duet" in which she and Beloved "did their best they could to create what really happened, how it really was" (78) when Sethe escaped slavery pregnant with Denver, risking death to bring nursing milk to her nine-month-old daughter who was waiting for her in Ohio. In effect, Beloved and Denver's 'duet' calls attention not only to the mediatedness of the escape/birth story but also to the ways in which the novel at large interweaves characters' fragmented 'rememories' in an effort to create a coherent narrative of slavery, flight, and their aftermaths. Yet, despite this effort, the novel does not provide, maybe cannot even tell, "what really happened, how it really was" (78). Instead, there are only highly mediated, biased narrative fragments of a "miracle birth" with the help of Amy who Ella describes as a ghost-like midwife.[209] Magical realist or *deus ex machina* elements, such as the miraculous appearance and help from Amy, are juxtaposed with all-too real and at the same time very lyrical

---

[209] Upon receiving Sethe on the northern riverside and hearing of a white girl who assisted in Denver's birth, Ella first makes sure to leave the scene as quickly as possible out of fear that Amy might have told them off (92). Years later Ella suggests in a conversation with Stamp Paid that it was more likely that Amy was a ghost rather than a friendly, helpful white girl: "A *white*woman? Well, I know what kind of white that was" (187). The chances of a ghost in white appearing before Sethe during her escape, Ella seems to argue, are far higher than a white girl helping a Black fugitive.

descriptions of Sethe's physical suffering on the run, the wounds on her back and her feet (see, e.g. 79–80). These elements form narrative fragments that function as placeholders, in a way, for an unlikely escape story that refuses narration and proves not to be a successful escape over twenty-eight days after Sethe's arrival on Bluestone Road. "Storytelling becomes the text's self-conscious task" and its "distinctive tone arises from the very difficulty of telling," Rody contends, because "[a]ny recuperations arise against a background of storylessness" (25, 27).

The "box," the house, and the hold

Before I discuss the events after Sethe's arrival, I want to look at Paul D's 'rememories' after he was captured while trying to escape from Sweet Home. As four separate narrative fragments from Paul D's perspective show, after the failed escape attempt Paul D becomes a recurrent prisoner and fugitive. Captured and tortured for their communal escape attempt from the plantation by having to carry an iron bit in his mouth (69, 71), Paul D is sold to Brandywine who leads him and ten other enslaved people in a coffle from Kentucky to Virginia. Because he attempts to kill Brandywine (106), Paul D is sent to a chain gang in Alfred, Georgia, where he is exposed to rape by the white guards in the morning, hard physical labor during the day, and imprisonment in a small "box" in the ground at night (106–09). While working and singing work songs, the prisoners "killed the flirt whom folks called life for leading them on. Making them think the next sunrise would be worth it; that another stroke of time would do it at last" (109). In the rhythm of their work songs, which is reflected stylistically in the repeated succession of short phrases separated by semicolons, they beat down "the women they knew; the children they had been; the animals they had tamed themselves or seen others tame" (108). In other words, they crushed the potential futurity that they, like Sixo, dreamed of, their hopes for a better tomorrow. As Childs observes, Paul D's experience of captivity in "the box" is explicitly connected to the Middle Passage, suggesting a "Middle Passage carceral model" that proliferates across the fragmented border between pre- and post-emancipation and the enslaving South and the 'free' North in the novel (274). Both Paul D's "trauma of living burial within the prison camp 'box'" and an unnamed man's trauma "in the underwater tomb of the slave ship hold" referenced in Beloved's monologue are met with the same "somatic response" of physical trembling ("'You Ain't Seen Nothin'" 281). Childs further argues that Paul D's repeated experience of captivity across time and space "suggests how 'rememory' goes much deeper than the realm of survivor psychology – that chattel slavery haunted its way into the experience of Black freedom at the level of penal practice and other forms of racial capitalist violence from 1865 to the prison industrial complex" ("Angelo Herndon" 52).

The "chain leader" of the chain gang, who Paul D names "Hi Man," gives the calls "Hiiii! at dawn" and "Hoooo!" at night to start and stop their work (108). When Paul D notices that Hi Man does not receive orders from the white guards to give these signals, Paul D imagines that Hi Man assumes this "responsibility [...] because he alone knew what was enough, what was too much, when things were over, when the time had come" (108). Drawing on Hortense Spillers's notion of the "intramural," Christina Sharpe describes the daily interventions of Hi Man as an "intramural" form of care (Terrefe and Sharpe, par. 86). This care is not "the 'care' of the state, which is care as prison cell, as grave, as mental institution, etc.," nor does it "change the circumstances of their being imprisoned" (par. 86). Instead, it stops "that particular violence of the white men demanding fellatio" (par. 86) and "redirect[s] the violence of rape to the violence of other captive labor" ("'And to Survive'" 173). This care therefore "sounds and holds the note that keeps the men with whom he [Hi Man] is chained from the brink" (Sharpe, *In the Wake* 133). Yet, in doing so, it also ensures the functioning of the chain gang for the time being as it necessarily operates within the bounds of the 'state-care' of captivity. As Paul D notes, the prisoners who "had been there enough years to have maimed, mutilated, maybe even buried" hope for life and a tomorrow "kept watch over the others," because "if one pitched and ran – all, all forty-six, would be yanked by the chain that bound them and no telling who or how many would survive. A man could risk his own life, but not his brothers" (109). Hi Man's daily "shout" of "mercy" (Morrison 108) can be understood along similar lines as Sixo's short-term flights, i.e., as an expression of social life within the bounds of social death.

Also, like Sixo's short-term flights, the prisoners' "ordinary note[s] of care" (Sharpe, *In the Wake* 132) together with the happenstance of extreme weather create the circumstances that facilitate an escape attempt. The forty-six chain gang members' silent care for each other establishes the ability to communicate without words through their chains that binds them together and that forces them to move as one (Morrison 110–11) and the necessary "trust" to escape from the hold-like "cages-cum-graves" (Sharpe, "'And to Survive'" 173). During a flooding caused by heavy rain the captives emerge as a group from their boxes through the mud that almost drowns them "[l]ike the unshriven dead, zombies on the loose, holding the chains in their hand" (Morrison 109). As Childs points out,

> Morrison's description of the newly freed men as 'zombies' registers both the paradoxical capacities of attempted agency and collective action on the part of those facing the most abject forms of dominance and how – within the context of continued collective subjection – the 'success' of such acts of rebellion cannot most often be described as a clean break from unfreedom. ("'You Ain't Seen Nothin'" 290)

The captives do not gain unconfined freedom through their care and escape but transform themselves first into a make-shift community of captives and then into a make-shift collective of socially dead fugitives who chance upon another community of outcasts after their escape. A camp of Cherokee, who are sick from an infectious disease, provide food and news as another form of care and literally free the men "with Buffalo hair" from their shackles (Morrison 111–12). Thus, for a brief moment Morrison juxtaposes the abjection of the formerly enslaved with the violence of settler colonialism against Native Americans "who chose a fugitive life rather than Oklahoma" (111). Yet, she does so without making any claims to relationality beyond a bare form of what might be called 'intermural' care that takes place in acknowledgement of and despite the Black border that separates the two communities structurally.

Afraid of "tracking dogs" and unsure where to go, Paul D is reluctant to leave the Cherokee but eventually asks for the way to the "Free North. Magical North. Welcoming benevolent North" (112), an ironic, hyperbolic description of the trope of the 'free North' reminiscent of Reed's novel that seems to indirectly gesture to the fact that it provides no unconfined form of freedom – neither for African American captives on the run nor Native American fugitives, displaced, dispersed, and sickened from the white settlers. Acting on the advice of the Cherokee, Paul D does not follow the North Star – as might be expected – but tree blossoms (112). As signs of spring moving northward into colder climates, the intermural care of the Cherokee, if you will, lead Paul D to the northern states. Just like Sixo and writers/narrators of slave narratives, Paul D's escape requires and (re)creates a close relationship to the natural environment and some of its inhabitants. Yet, Morrison refuses to fall back on the familiar trope of the North Star, while also rewriting encounters between formerly enslaved fugitives and Native Americans. Whereas slave narratives, such as Henson's, frequently include scenes of chance encounters with Native Americans as helpers who support the fugitives' escape from the narrators/writers perspective (*Uncle Tom's Story* 89–90),[210] Morrison offers an elliptical glimpse (of three paragraphs' length)

---

[210] Another slave narrative that includes Native Americans is, for example, *Twenty-Two Years a Slave and Forty Years a Freeman* (1857) by Austin Steward. For a discussion of the narrative, see

into a "[d]ecimated but stubborn" Indigenous community and their particular "calamit[ies]" by focalizing on an unspecified "they" (Morrison 111).

After about six months, the trees stop to bloom and Paul D, "[c]rawling out of the woods, cross-eyed with hunger and loneliness," asks for help from a "weaver lady" in Wilmington, Delaware who takes him in and with whom he stays for eighteen months until he starts to follow tree blossoms again (131, 113). From then on until his arrival on Sethe's porch, phases of fugitivity and captivity in the northern and the southern states alternate for Paul D (267–70), again blurring the lines between enslaving and non-enslaving territory, between the time before, during, and after the Civil War, and between captivity and flight. Paul D summarizes his futile escapes from captivity that last for eighteen years with only intermitted times of refuge before he arrives in Sethe's yard:

> In five tries he had not had one permanent success. Every one of his escapes (from Sweet Home, from Brandywine, from Alfred, Georgia, from Wilmington, from Northpoint) had been frustrated. Alone, undisguised, with visible skin, memorable hair and no whiteman to protect him, he never stayed uncaught. [...] And in all those escapes he could not help being astonished by the beauty of this land that was not his. (Morrison 268)

While reiterating the importance of the natural environment and the fact that the land was stolen from the Indigenous population, Morrison also points here to both the hypervisibility of Blackness and the perpetuity of flight, which recurs in many of the texts under scrutiny. Referencing George Jackson's writing, Childs suggests that Jackson's coining of the term "(neo)slavery" adequately reflects this proliferation of captivity ("'You Ain't Seen Nothin'" 290) even during flight which Morrison aptly describes as phases not of freedom but of being intermittently "uncaught" (268).

In contrast to Paul D's oscillation between captivity and fugitive 'uncaughtness,' Sethe settles in the house on Bluestone Road after her escape across the Ohio River. As she explains to Paul D upon his arrival, "[n]o more running – from nothing. I will never run from another thing on this earth. I took one journey and I paid for the ticket, but let me tell you something, Paul D Garner: it cost too much!" (15). In Sethe's "rememories" of the "twenty-eight days" after her arrival on Bluestone Road, the house is described as a "cheerful, buzzing" haven ruled by Baby Suggs, an "unchurched preacher" and spiritual leader of the Black community (86–87, 89). Baby Suggs preaches a "healing philosophy of Black flesh" which Greg Thomas describes as "erotic maroonage as well as spiritual maroon-

---

Sawallisch, ch. 5 and especially her discussion of its representation of Indigenous characters (179–82).

age" ("'Neo-Slave Narratives'" 227). By drawing on Sharpe's notion of care "in the wake," Calvin Warren describes Baby Suggs's spiritual work for and with the community in "the clearing" – as represented in the 1998 movie adaptation of the novel – as exhibiting "a deep understanding of black care" ("Black Care 46).

> Laughing, crying, and dancing provide form for an indecipherable violation – one that language cannot adequately address [...]. The captives, then, must rely on the 'non-sense' sign (laughing, crying, and dancing), as institutional care would describe it, to give form to an affective dimension [of physical and metaphysical injury]. ("Black Care" 46)

While resonating with Thomas's interpretation of Baby Suggs's role, Warren "hesitate[s] to use the word 'healing,' since this is often the 'sign' of metaphysical overcoming and domination, and anti-black violence continues without end and can never be overcome" (45–46). After all, the 'haven' is quickly exposed as vulnerable to invasion by enslavist and anti-black forces, exposed as merely a place of momentary refuge in a makeshift community of 'uncaught' fugitives, "[w]hen the four horsemen [...] – schoolteacher, one nephew, one slave catcher and a sheriff" arrive to recapture Sethe and her children (Morrison 148). Baby Suggs anticipates the arrival of schoolteacher as a foreboding that smells "Dark and coming" (138). When Sethe recognizes schoolteacher, she rushes to the cold house with her children and attempts to kill them, planning to commit suicide afterwards. As she "rememories" and presumably does not speak out loud during another conversation with Paul D, in order to be with her children in death safe from re-enslavement, she "collected every bit of life she had made, all the parts of her that were precious and fine and beautiful and carried, pushed, dragged them through the veil, out, away, over there where no one could hurt them" (163).[211] Sethe succeeds in killing her nine-month old daughter, before Baby Suggs and Stamp Paid intervene and save the other three children. When schoolteacher arrives on the scene, he judges Sethe – like Sixo before her – as no longer usable property and leaves her to the sheriff who sends Sethe with Denver to jail (149).[212]

---

[211] In this chapter, Sethe evades telling Paul D of the infanticide by circling around the topic for a long time (Morrison 159–65). This narrative strategy of evasion will be discussed in more detail in the close reading of Teju Cole's *Open City* below.

[212] Like Paul D and Sethe, the side characters Stamp Paid and Ella also illustrate how captivity and flight have affected the whole community, especially with respect to sexualized violence. As Stamp Paid tells Paul D, he fled from his legal owner and deserted his wife Vashti after being forced to hand her "over [...] to his master's son" (184–85). Ella too had escaped from a house "where she was shared by father and son" (256), held captive for their sexual pleasures

After the invasion, the house becomes a place of confinement, stasis, haunting, and death for its inhabitants. As Mitchell notes, the description of the house without a backdoor and no visitors, enclosed by a fence, and the upstairs windows facing the sky appears somewhat "[t]omb-like" (91). It exhibits a "claustrophobic containment from which all the characters suffer" that is reminiscent of Linda's garret and further laden with a "stagnant and oppressive nature of the past, unexamined and uninterrogated" (91). The house haunted by the ghost of the dead baby, "the gothic of the system of slavery" if you will (Patton, "Black Subject Re-Forming" 882), represents the present characterized by stasis and confinement as well as haunted by the past (Levy-Hussen, "Trauma" 198). In this way, even though Sethe and Baby Suggs have a 'home of their own,' which Linda so desires at the end of her narrative, this home is haunted and remains confining for its inhabitants. In fact, Morrison shows that Robin's description of the American South as "the land of the hunted and haunted" in Reed's novel (173) also applies to the postbellum American north – or as Baby Suggs notes, there is "[n]ot a house in the country ain't packed to its rafters with some dead Negro's grief" (5).

In the aftermath of "the Misery" (Morrison 177), Baby Suggs abandons her work as preacher, retreats from the community, confines herself to bed, and starts "ponder[ing] color" (201) until she dies. Baby Suggs is frustrated with the anti-black world in which "there was no bad luck but white-people" (104) because – unlike Hi Man whose "ordinary note of care" (Sharpe, *In the Wake* 132) halts or at least redirects the anti-black violence "when things were over, when the time had come" (Morrison 108) – white people "don't know when to stop" (104). Greg Thomas therefore attests to Baby Suggs first "spiritual" and then "supernatural maroonage" as she "divests herself from this order of being, psychically, metaphysically, and then physically" ("'Neo-Slave Narratives'" 226), with her death representing "her ultimate act of plane-changing" (227). Denver on the other hand grows up with a great fear of the outside world where, as Baby Suggs tells her, there "was no defense" (244). Her preferred place of retreat is an enclosure of "five boxwood bushes" in the woods beyond the house (28). After finding out about her mother's murder of her sister and the time Denver and Sethe spent in jail together, Denver further retreats from the

---

in a room for more than a year (119, 256). She let the child that was conceived of the repeated rapes die, committing a passive form of infanticide if you will (258–59, cf. Gillespie 38). The brief but repeated mention of sexual torments, such as Ella's and Vashti's, which Sethe and Stamp Paid assume Beloved may have also gone through (119, 235), of course, recall the dangers enslaved women, such as Jacobs, were exposed to. As Mitchell notes, the list of hints at such sexualized violence against Black women in the novel is particularly long (96–97).

world as she loses her hearing and stops speaking for some time until the baby ghost tries to crawl the stairs of the house and the spirit's presence becomes "full of spite" (103–05).

In contrast to Baby Suggs and Denver, Denver's older brothers Howard and Buglar refuse the stasis and confinement of the house and instead of retreat or death, they leave the family. As the text suggests, the ghost of their dead baby sister scares them off at age thirteen one at a time (3). Beaulieu criticizes Howard's and Buglar's escapes as "decidedly unmanly" because they run instead of "confronting their fears" and their escape could not be understood with "the usual male desire to roam, a recurrent trope in African American literature" (62). According to Thomas, however, Buglar and Howard – like Paul D – "continue the community's long narrative of running, in 'the North,' and they continue their unnamed maternal grandmother's narrative of movement-resistance" ("'Neo-Slave Narratives'" 214–15) that Sixo also practices. They become second- or third-generation fugitives, as it were, who "literally open this novel" by running out of the narrative frame (214).

Even though Sethe is no longer enslaved, on the run, or imprisoned, and she does not – like Paul D, Howard, and Buglar as well as Aminata in the following close reading – continue to move and travel after arriving in the US north, Sethe's physical settledness in the house is significantly unsettled, first, through the baby ghost's frequent "mischief" (Morrison 103) and, second, through Sethe's "rememories" of the past that Paul D's and Beloved's arrival in the house trigger. When at risk of recapture, Sethe chooses literal death (infanticide and suicide) over the reenslavement, a 'rememory' that haunts her mentally and physically. Asking about the "moral ground" on which Sethe stands, when she commits infanticide to save her children from re-enslavement – a motif that will reappear in Hill's neo-slave narrative as well –, Wilderson draws a "structural analogy which highlights how both Black Power activists and Toni Morrison's characters (Toni Morrison herself!) are void of relationality" and death becomes "a synonym for sanctuary" ("Black Liberation Army" 176). While Rushdy suggests that Sethe wants "life and safety [...] for her children and herself, not death, and escape, not oblivion" ("Slavery and Historical Memory" 243), Greg Thomas locates "[a]n African logic of 'supernatural maroonage' [...] at the center of Morrison's reinscription of Margaret Garner's history and family in fiction" ("'Neo-Slave Narratives'" 221). This logic is, however, "for many obscured by the Western conception of 'death,' 'suicide,' and 'infanticide'" (223). In other words, Thomas proposes to understand Sethe cutting her daughter's throat with a handsaw, which Stamp Paid describes as "outhurt[ing] the hurter" (234), as an attempt at a supernatural form of escape, a flight into a state of being that is 'elsewhere,' "[o]ver there. Outside this place, where they would be safe" (Morrison 162).

Sethe imagines death as a sanctuary for her children, but just like the forty-six prisoners' experience of captivity and fugitivity, it is caught up in what Childs aptly describes as the "Middle Passage carceral model." Beloved's cryptic recollections of where she was before she arrived on Bluestone Road suggest that death – or another form of being after death – was an ongoing Middle Passage for her. Beloved arrives on Bluestone Road as a young woman with skin that appears "new, lineless[,] and smooth" (50) after having entered Sethe's world by crossing a "bridge" (75) and walking "out of the water" of a stream "fully dressed" (50), the stream being a place "where spirits are traditionally thought to dwell," as Greg Thomas notes ("'Neo-Slave Narratives'" 223). Before that, Beloved was in a place reminiscent of the hold of a ship packed with bodies dead and alive. In a conversation with Denver that precedes their narrative duet of Sethe's escape/Denver's birth story, Beloved describes the place as "dark" and "hot," "with nothing to breath down there and no room to move in" but "[h]eaps" of people, some of them "dead" (75). Years after Sethe's escape, the daughter who returns from the dead and/or from the midst of captives of the Middle Passage gives Sethe a sense of restlessness without much physical movement that culminates in a circular exchange between the two about "the Misery," in which, confined to the house, Sethe tries to find ways to care for Beloved, while Beloved feeds on these attempts (243–44).

Rody suggests that Beloved's monologue (210–14) "carries a vast ancestral memory" from the Middle Passage (217), while Valerie Smith contends that it "represented the preconscious subjectivity of a victim of infanticide" that speaks to readers both "from the grave" and from the Middle Passage ("Neo-slave Narratives" 179–80). Describing the formal characteristics of Beloved's monologue, Smith observes that the layout of this monologue ("sentences, phrases, or individual words, [...] separated by spaces, not by marks of punctuation" and "[o]nly the first-person pronoun and the first letter of each paragraph [...] capitalized") "places all the moments of Beloved's sensation and recollection in a continuous and eternal present" (180). Broeck refers to Beloved's monologue as "the literary space of a telling narrative void," made up of "urgent language fragments, arranged more like a dazed and always already collapsing chant than the swift and artful narrative stridency characteristic for the novel" ("Enslavement as Regime" par. 3). With this "litany of stumbling passages the origins of which remain lost to the reader," "[a]s if in an echo of Adorno's dictum, *Beloved* seems to forbid itself any narrative after, and of the Middle Passage" (par. 3). Indeed, I would argue that the novel does not only forbid narrative after and of the Middle Passage, but that it questions the narratibility of the Middle Passage, slavery, and their afterlives in the post-Civil War era. As Smith contends, *Beloved* is the "speakable version" of enslavement and flight ("Neo-slave Narratives" 178). "*Beloved* thus points to a paradox central to any attempt to represent the body in pain" (178), or rather, as I would argue, a body in flight: its "unnarratability, indeed the inadequacy of language" (179), while at the same time "one can never escape narrative" (178).

Greg Thomas concludes that there is "a maroonage of the body, a maroonage of the spirit, and a maroonage of the mind, all on full narrative display" in *Beloved* ("'Neo-Slave Narratives'" 227). The novel both manages to give its characters the capacity to share what was "not a story to pass on" (Morrison 275) from the position of social death, and actual death for that matter, to flee, to "sound ordinary notes of care" (Sharpe, *In the Wake* 132), and to take momentary refuge. At the same time, however, *Beloved* complicates any linear, unambiguous escape plot and easy notions of arrival, freedom, care, and home with fantastic, gothic elements, non-linear, highly mediated narration. It also dissolves, first, the difference between the enslaving South and the non-enslaving North and the time before, during, and after the Civil War through its "forward-haunting" elements of confinement ("'You Ain't Seen Nothin'" 274), and, second, the narration and its syntax as such when it addresses death and the Middle Passage. Morrison has not only written the contradictions and fragmentations of chattel slavery and its prevailing afterlife into the texture of her novel, but also gestures toward the ways in which unconfined freedom, care, and healing of and for the enslaved and fugitive refuse plotting and (narrative) closure. The novel creates a fugitive capacity to witness and struggle against social death that is always already restricted and meant to remain ultimately unplottable.

Futile homemaking, impossible homecoming, and "perpetual migration" in Lawrence Hill's *Someone Knows My Name*

Archival material inspired not only Toni Morrison but also Lawrence Hill to write a neo-slave narrative, which – like *Beloved* twenty years earlier – became Hill's most well-known publication. Similar to *Beloved*, Hill's *Someone Knows My Name* grapples with issues of enslavement, flight, freedom, and kinship in the era of chattel slavery in North America through the eyes of a Black female fugitive protagonist. While the narrative unfolds, Aminata – like Sethe – loses her parents, children, and husband; she runs away from her legal owner to free territory in the North; and becomes ultimately reunited with one of her loved ones. First published as *The Book of Negroes* in Canada in 2007 and then re-published in the United States, Australia, and New Zealand under the title *Someone Knows*

My Name,²¹³ Hill's is "arguably, *the* most widely circulated novel by a black Canadian author" and possibly the most well-known Canadian neo-slave narrative (Medovarski 27–28).²¹⁴ The epic historical novel spans the autodiegetic narrator Aminata Diallo's life from 1745 to ca. 1805. In the frame narrative, Aminata writes her autobiography for the abolitionist cause in London in 1802. Like *Flight to Canada*, as a *mise en abyme*, Aminata's story is largely that of the novel. It begins with Aminata's capture during her childhood in what today may be Niger or Mali in 1756 and ends with her death of old age in London in the care of her daughter. In-between lie the Middle Passage, plantation slavery in South Carolina, escape from her second legal owner Solomon Lindo during a

---

213 The original title refers to the historical document that records the names and personal details of about 3,000 formerly enslaved who had served the loyalists and were therefore evacuated from New York City to Nova Scotia during the American Revolution (see Browne 547–58). Upon their arrival in what was then British North America, the 'Black loyalists' were promised a share of land as free settlers. Photocopies of pages of the book are included as a peritext to the novel that can be considered "an alternative imagining of the events surrounding the archive that could not be fully realized with the historical documents" (Browne 558). As Oluyomi Oduwobi, Harry Sewlall, and Femi Abodunrin note, the Canadian title also "echoes Langston Hughes's (1952) *The First Book of Negroes* […] as well as Langston Hughes and Arna Bontemps's (1958) *The Book of Negro Folklore*" (385). The American title is a direct quote from the side character Chekura in conversation with Aminata (Hill, *Someone* 81) and references James Baldwin's essay collection *Nobody Knows My Name* (1961, cf. Duff 252). His US American publisher made Hill change the title because of marketing concerns which Andrea Medovarski criticizes as a "profound act of erasure concealing the historical significance of the title in both Canadian and American contexts" that "illustrates the power dynamics at work in the literary marketplace" (24). I use the American title in its entirety and abbreviated form throughout, but cite the 2010 Black Swan edition that was published under the original Canadian title.

214 As Domínguez points out, *Someone* "was not Hill's first attempt at the form of the slave narrative" (104). His second novel *Any Known Blood* (1997) about five paternal generations of a fictional Black family in North America includes the narrative of an ancestor's escape into Canada in 1850 via the Underground Railroad. Domínguez observes a shift from *Any Known Blood*'s focus on "roots" to *Someone*'s preoccupation with "routes" (92).

Other well-known neo-slave narratives by Black Canadian writers are, for instance, Dionne Brand's *At the Full and Change of the Moon* (1999) and Esi Edugyan's *Washington Black* (2018). Afua Cooper's *The Hanging of Angélique: The Untold Story of Canadian Slavery and the Burning of Montréal* (2006) may be considered a mix between slave narrative, neo-slave narrative, and historiography. *The Hanging* tells the story of the historical figure Marie-Joseph Angélique, a Portuguese-born enslaved Black woman who was convicted, tortured, and killed for setting fire to large parts of Montréal in 1734 in order to escape. The book combines biographical narrative with historical analysis as it revisits the trial transcript (cf. Cooper, "Marie-Joseph Angélique"). As Greg Thomas notes, Cooper considers Angélique's story as 'the first known Black slave narrative in North America'" as the records predate known slave narratives in North America by at least forty years ("'Neo-Slave Narratives'" 218–19).

stay in New York City at the outset of the American Revolutionary War in 1775, and the challenges she faces as an emancipated settler in Nova Scotia and as a returnee in Freetown, Sierra Leone in 1792.

Literary critics have discussed Hill's novel from various perspectives, focusing, for instance, on the role of Africa and Africans (Domínguez); trauma (Krampe); currency and financial capitalism (Medovarski); the representation of the Black urban subject in London (Ball), and of the Jewish side character, the indigo inspector Lindo, as "intermediary" and accomplice of slavery (Casteel).[215] Scholars have also illuminated the novel's intertextual relations to various writing traditions, such as Black autobiography and slave narratives by Olaudah Equiano and Frederick Douglass (Duff, Sagawa and Robbins, Yorke), the early Black feminism of Sojourner Truth (Duff 250–51, Oduwobi, Sewlall, and Abodunrin 392),[216] or the (postcolonial) female bildungsroman (Oduwobi, Sewlall, and Abodunrin).[217] While the narrative spans the Atlantic hemisphere, including West Africa, the United States, and Canada, and thereby broadens Paul Gilroy's concept of the Black Atlantic (Domínguez 100, Duff 252, Yorke 141–42), scholars have also considered its intervention into dominant, specifical-

---

[215] As Sarah P. Casteel shows, *Someone* "introduces a wealthy Jewish broker figure" with Lindo "who at some point suffers the loss of both his wife and children; purchases a female slave [Aminata] from an abusive Christian master; treats the slave respectfully and facilitates her rise; and finally in a *prise de conscience* grants her freedom" (119). Yet, "Solomon's Pygmalion-style experiments with her education" and the fact that he arranged the selling of her son and eventually becomes a plantation owner himself draws "attention to less obvious stakeholders in the slavery economy who become compromised by their very presence within an inhumane system" (119).

[216] Sojourner Truth (1797–1883) was a formerly enslaved preacher, abolitionist, and women's rights activist. She published her slave narrative, *The Narrative of Sojourner Truth*, which she had dictated to Olive Gilbert, in 1850. Truth is most renowned for the speech "Ain't I a Woman?" she is said to have given at the Ohio Women's Rights Convention in Akron, Ohio from 1851, the title of which Aminata uses in modified form in a conversation with Lindo. As Duff notes, the "anachronistic nature" of this "brief yet defining intertextual reference […] makes Aminata a fictional foremother to the famed abolitionist and women's rights activist" (251).

[217] Oduwobi, Sewlall, and Abodunrin discuss *Someone* as a postcolonial neo-slave narrative with intertextual relations to classic white female Victorian bildungsromane, such as Charlotte Brontë's *Jane Eyre* (1847). They name, for instance, language similarities, narrative perspective, and their shared "confessional mode," as well as similarities in their gendered maturation processes and didacticism, while emphasizing that Hill adds an important "postcolonial" twist (390–91). In contrast, Berndt understands *Someone* primarily as a fictional autobiography and sees the development of the protagonist already "almost complete when Aminata is introduced as a child," as her parents teach her "the knowledge to live (bio-)," care for others (esp. midwifery), and "to tell the story of her life (-graphy)" (storytelling and a desire for literacy) ("In Search of" 167–68).

ly Canadian discourses and the Canadian literary scene. As Naomi Pabst notes, the longstanding African Canadian history and Canada's history of slavery is "constantly being re-discovered, then re-forgotten, forever re-subsumed by the dominant narrative of Canada as a haven for fugitive slaves" (27) and "the promise" of a multicultural society that foregrounds immigration and national or cultural origin before race and (anti-)Blackness (Krampe 74–76).[218] Scholars, such as Katrin Berndt, Christine Duff, and Christian Krampe argue that Hill – in perfect post-modern neo-slave narrative fashion – has Aminata write against this re-forgetting and re-subsuming of Black Canadian history and the history of slavery in Canada not least in order to critically comment on the present.[219]

Someone is divided into four "books," each split into several chapters of which the first of every book returns to the time and place of narration, before detailing every station of her long odyssey chronologically until the last chapter comes "full circle by returning to the same geographical and temporal spaces as the first" (Duff 242).[220] As Duff points out, by regularly returning to the time of narration in London, readers are reminded of Aminata's "narrative authority" over her story (241). The plot and the traditional realist style lean towards romance and sentimentalism and clearly stand in the tradition of slave narratives, many of which – including Jacobs's – Hill studied while writing *Someone*.[221]

---

[218] The history of the Black Loyalists and Black Canada plays an important role in *Someone*'s realist representation of captivity and fugitivity in North America. A detailed discussion of Black Canadian culture, history, and literature, and its relation to their African American counterparts exceeds the scope of this chapter, however. George Elliot Clarke has published several collections of Black Canadian literature and critical essays that offer introductions to the field (*Eyeing the North Star*, *Odysseys Home*, *Directions Home*; see also Bast; Siemerling, *Black Atlantic Reconsidered*; Walcott, *Black Like Who?*). For a recent collection of Black Canadian literature and art, see Antwi and Chariandy's 2017 special issue. For a recent critical discussion of "Black Canadas" and their relation to and conflicts with African America, see their conversation with Walcott in the issue (Walcott, Antwi, and Chariandy) as well as Pabst.
[219] Cf. also Medovarsky who reads *Someone* as a "powerful commentary on our contemporary hypermonetarized moment" (15).
[220] An exception is book four, which Aminata introduces as she settles in Freetown in 1792 (Hill, *Someone* 391, cf. Duff 241).
[221] Lawrence Hill explained in his keynote lecture "Blackness in Canada and the Accidents of Geography: Mining Fiction and Essays to Meditate on Black History, Identity, and Intersectionality" and in a private conversation at the 40th annual conference of the Association for Canadian Studies in German-Speaking Countries, 14–17 Feb. 2019, Grainau, Germany, that he read dozens of slave narratives while writing *Someone* and drew particular inspiration from Jacobs's narrative. In the peritexts "A Word About History," "Further Reading," and his "Acknowledgements," Hill further explains his historical research and lists various sources (*Someone* 493–98; see also Sagawa and Hill 315).

Hill's novel represents an important intertext to female-authored/narrated slave narratives, not least because no autobiographical text by an enslaved woman who has survived the Middle Passage exists (cf. Hartman, "Venus in Two Acts" 3). With respect to form, particularly the *in ultimas res* beginning represents an intriguing twist on the formulaic *ab ovo* openings of slave narratives (see ch. 2.1.) and the *in medias res* beginnings of the Black Power autobiographies (see ch. 2.2.). As Stephanie Yorke notes, Hill's neo-slave narrative has above all clear intertextual relations to slave narratives of the late eighteenth century, such as *The Interesting Narrative of the Life of Olaudah Equiano, Or Gustavus Vassa, the African, Written by Himself* (1789), which is mentioned repeatedly in the novel, not least as a potential model for Aminata (*Someone* 425, 460, 471; cf. Medovarski 21–22).[222] Both *Someone's* and *The Interesting Narrative*, Yorke observes, switch between the perspectives of the mature narrating I and the younger narrated I – that we also know from Black Power autobiographies –, while seeming less constraint in style and content than the typical slave narrative of the nineteenth century that appeared more formulaic due to publishing and editing pressures (100, 132).[223] As a neo-slave narrative, Duff and Yorke both further argue, *Someone* goes beyond the slave narrative tradition "by exposing the political machinations that dictated both its form and content" (Duff 242), using "a mode of literate orality," and "challeng[ing] racial mythologies that have been created through the mediated manufacture of the genre" (Yorke 129, 134).

In a remarkable intertextual reference to *Beloved*, Aminata calls her life "my own private ghost story" at the beginning of the narrative (19). Yet, counter to Morrison's novel, which claims to be "not a story to pass on" (275), Aminata asks unequivocally "what purpose would there be to this life I have lived, if I could not take this opportunity to relate it?" (19). Aminata uses witnessing and storytelling as strategies of survival and to work through the traumas of slavery (cf. Berndt, "In Search of" 166–67; Krampe 67–74), while claiming the abil-

---

[222] Olaudah Equiano's slave narrative was a publishing success in the late eighteenth century and can be considered the "originator" of the writing tradition (Encyclopedia Britannica, "Olaudah Equiano"). Whether Equiano (1745–1797) was born and captured in today's Nigeria and then shipped to North America aged 11, as the first chapter of his narrative suggests, is contested among historians (see Lovejoy, "Autobiography and Memory"; Carretta).

[223] While I agree with Yorke's basic argument and historical contextualization, her understanding of the slave narrative tradition appears overly generalized and reductionist when she claims, for example, that "Aminata is permitted to tell a compelling story" in contrast to the slave narrative which merely functioned as "a piece of evidence for use in a white-authored philosophical discussion" (132). In my analyses of Douglass's and Jacobs's slave narratives in chapter 2.1., I show how despite serious discursive constraints Douglass and Jacobs strategically manipulated the writing tradition for their purposes.

ity, authority, and right to tell her own story ("'My life. My words. My pen'" [*Someone* 472]). "While Aminata's reaction to the trauma of the Middle Passage is to survive in order to testify" (70), Krampe argues, other characters, such as Fanta and Fomba, who are also abducted from Bayo and shipped across the Atlantic with Aminata, demonstrate less articulate and more violent reactions to trauma. Recalling Sethe's and Ella's infanticides, Fanta kills two newborn babies to save them from slavery, whereas Fomba – not unlike Sixo and Paul D on the chain gang – retreats from the world and loses his ability or willingness to communicate and interact with his surrounding (Krampe 69–70). *Someone*'s and *Beloved*'s attitudes toward storytelling and writing diverge in important ways, however. For Aminata, storytelling, literacy, and writing – while no paths to unconfined freedom – certainly play a central role in her coping with the trauma of slavery and social death. In contrast, in *Beloved* writing "becomes a feature of the machinery of slavery" that schoolteacher uses against the enslaved (Rushdy, "Slavery and Historical Memory" 241), and, as I have shown in my close reading of the novel, *Beloved* indirectly casts the plottability of enslavement and flight into question.

These different attitudes find reflection in the narrative styles. In contrast to Morrison's and Reed's earlier neo-slave narratives, which experimented with language, satire, and non-realist elements in their rewritings of the slave narrative tradition, Hill's neo-slave narrative uses a more traditional, linear narration that tries to tell a story about the transatlantic history of slavery truthfully (cf. Sagawa and Hill 316). Critics, such as Nehl, have expressed discomfort with the linearity, coherence, and happy ending of the novel, criticizing what he deems an "ultimately triumphant account of an enslaved woman's life" for using "unconvincing melodramatic plot devices" and for "offering narrative closure and naively celebrating the healing power of literature" that "trivializes the horrors of slavery" (16, 23). He criticizes *Someone* for following "the commercially successful tradition of female-authored neo-slave narratives," such as Morrison's, while neglecting "the aesthetic and ethical challenge" involved in writing a literary archive of enslaved women's suffering and resistance as well as "the theoretical intricacies involved in 'the practice of speaking for others' (Linda Alcoff)" (23). In his attempt to recover voices and complex experiences of enslaved women in North America and to highlight "the liberating power of the act of writing," Nehl argues that Hill employs a "'fairy-tale'" like plot (23) that works towards narrative coherence and "closure" where, as Morrison and Hartman would have it, there is "none" (Hartman, "Venus in Two Acts" 8).[224]

---

[224] Nehl also contends that Hill "offers an unconvincing teleological conception of history and

Indeed, Hill seems to go to great pains to achieve coherence and closure by including elements of melodrama, persevering hope, and a somewhat happy ending in a world that denies African and African-descended people freedom and the status of 'the Human.' He reflects on the effort to create a readable narrative out of the horrific history of slavery:

> It's [...] hardest of all to write it [a novel about slavery] with enough lightness that a reader will want to ingest it. If you wrote it in a certain way, nobody would even want to turn past page one. So how do you write a story of horror – of an Atlantic holocaust – without killing a reader's interest? (qtd. in Pierre)

That was years after Reed and Morrison had already proven that "readers' interests" are not necessarily that easily "killed" by experimental narratives about the atrocities of slavery. Put differently, *Someone* seems to ultimately strive for accessibility and readability through a reliable and authoritative narrator, whose perspective other more experimental narratives, such as *Beloved* and *Flight to Canada*, call into question.

While I share Nehl's critique of the overly coherent and closed narrative structure and the (partially) happy ending, I do not understand those narrative moves necessarily as signs of a failure to narrate persuasively the atrocities of chattel slavery. Instead, I suggest seeing them as part of *Someone*'s passionate but futile effort to create subjectivity, agency, and narratibility in an environment that denies African-descended people communal and familial relation, home, refuge, and return. As I will show, the perpetuity of community building, homemaking, and migration complicate the predominantly coherent, linear, and closed narrative structure and make the endlessness of the "horrors" of slavery manifest. In this way, I argue, *Someone*, too, sheds light on the ways in which fugitive African and African-descended people continue to be on the run and socially dead after escaping from territories of legal slavery. Like slave narrative writers/narrators, they have no safe place to flee to in the transatlantic world that continues to be structured by the transnational reach of the time of slavery and its anti-black 'afterlives.' In other words, *Someone* can be understood as an attempt to write a home for the captives and fugitives of North American chattel slavery that already implicates its failure. Nehl argues with Maria Diedrich's words that "Aminata comes to realize that 'home is where freedom is'" (152). But what if there is no freedom and therefore no home for the captives and fu-

---

a reductive reconciliatory interpretation of eighteenth-century black life" (22) while rightfully questioning the possibility of returning to an ancestral home and the notion of Canada as a safe haven for fugitives (136, 144).

gitives in this anti-black world other than as perpetual flight? Or as Wilderson and Sharpe might ask with the metaphor of the slave ship: What if – even though she moves from one place to another – Aminata ultimately remains a captive "in the hold of the ship" (Wilderson, *Red, White, and Black* xi; Sharpe, *In the Wake*)? If read closely, Aminata's "perpetual migration" (Hill, *Someone* 401) is exposed as constant flight during which she has to constantly build and rebuild makeshift places of refuge and improvised communities of support. Ultimately, hope is repeatedly projected onto distant places (Hartman's 'elsewhere') and into the future until death is the only place left.

Futile homemaking

As Nehl notes, the many obstacles Aminata survives with comparably little harm to her mental and physical health and the many successes she accomplishes against all odds throughout the narrative may be considered hyperbolical – yet in a very different sense than Reed's ironic representation of slavery and flight. After having crossed the Atlantic several times and before dying of old age, Aminata manages to meet the English King; testify in court against the transatlantic slave trade and slavery; write her autobiography without "guidance" from eager London abolitionists (Hill, *Someone* 472); *and* becomes miraculously reunited with her daughter May who had been taken from her fifteen years earlier in Nova Scotia. Aminata appears like a "larger-than-life protagonist" who "leads an extraordinary life involved in a number of extraordinary events" (Berndt, "In Search of" 166). Considering the death toll of the coffle in West Africa, the Middle Passage, and the plantation system in the Americas, the "fact of her sheer physical survival" and "her ability to lead a purposeful life with a distinct possibility of retaining her capacity for love" appear not only "astonishing" (Krampe 69, 80), but almost unlikely. As the British official William Armstrong stationed on the coast of Sierra Leone explains to Aminata when she asks to arrange her return to the village she was abducted from as an eleven-year-old girl, her story is "[h]ard to believe really. That you were there [in Bayo], and taken overseas, and are back here now … You must understand, it's unusual" (Hill, *Someone* 434).

Just like *Incidents* tells an unbelievable story of a seven-year long self-confinement in an attic space and asks for its readers' trust in the truthfulness of the account, *Someone* also works hard to create an authoritative narrative voice of Black female agency against the stark backdrop of the realities of the actual and social deaths of the (formerly) enslaved. Moreover, like Jacobs and the Black Power autobiographers of the 1970s, Aminata shows that she is not that "unusual" by including in her own exceptional story "the many travels and endeavours, both forced and deliberately taken, of people of African descent in the eighteenth century" (Berndt, "In Search of" 166). While working for the British to document in the "Book of Negroes" the people of African descent who are evacuated from New York City in 1783 to Nova Scotia (Canada) in recognition of their services to the British Crown, she explains that she too "had imagined, somehow, that my life was unique in its unexpected migrations. I wasn't different at all, I learned. Each person who stood before me had a story every bit as unbelievable as mine" (Hill, *Someone* 306). While "the book" she writes for the Loyalists provides only limited space for documentation, her autobiography gives voice to a "number of ordeals and [...] countless hardships [that Aminata survives and tells about] in place of the millions of real-life African slaves," using the "generic attributes of oral storytelling" to represent the community rather than the individual (Berndt, "In Search of" 167). Like the autobiographical forbears in the historical era of slavery and in the Black Power movement, her story is a spectacular *pars pro toto* condensation of narratives of Black life in the diasporic communities Aminata becomes a part of during her lifetime (Krampe 63, 71; Yorke 135, 137).

The life Aminata can relate, however, is predominantly what she retrospectively calls a "life of losses" (Hill, *Someone* 469). Even though she is ultimately reunited with her daughter May, she loses all other people dear to her – especially her parents, her shipmate and later husband Chekura, and their son Mamadu. Lamentation of the loss of her family as a "series of tortures" (Yorke 142) are repeated throughout the narrative (cf., e.g., Hill, *Someone* 368, 402), calling into question whether the happy reunion at the end can counterbalance the loss. Chekura manages to return to Aminata twice for short periods of time, in which their two children are conceived, before he dies in a shipwreck on the way to Nova Scotia (385–86). Because he seems to reappear "every time she moves from one community to another," Yorke describes Aminata's husband "as a one-man social network and [...] co-witness" to her story (139). Yet, the time the two of them spend together at the same time and in the same place in North America is very brief, so that Chekura is mostly present in Aminata's story through his absence. The same seems true for Aminata's parents, who are killed trying to protect Aminata from capture in Bayo (Hill, *Someone* 39–41) and then only reappear occasionally in mental conversations "from the spirit lands" (54, see also 41–42, 150).[225] The traumatic loss of Aminata's parents

---

[225] Aminata's imaginary conversations with her parents through which she seeks support and

is only "rivalled" by the loss of her own children (Krampe 68). Her son Mamadu is still a baby when he is sold away and dies a year later of small pox (Hill, *Someone* 201, 240–41); May is a toddler when the Witherspoons, a white loyalist family, who Aminata had trusted, abduct her during an outbreak of anti-black violence against the residents of Birchtown (361). Aminata describes her lost children in physical terms "like phantom limbs, lost but still attached to me, gone but still painful" (366), a pain that "never really went away. The limbs had been severed, and they would forever after be missing" (37). The simile, which also provides a chapter title (347–67), indirectly recalls Chekura as well, who like lost limbs is not only primarily present through his absence, but who – when he manages to return to Aminata in Charles Town and in New York's "Canvas town" briefly – reappears with hair fallen out and an increasing number of fingers missing (236, 304). Like Sethe's chokecherry tree scar on her back, Aminata's metaphorical and Chekura's literal amputations illustrate the psychological and physical toll on enslaved people who try to create and uphold kinship ties, an effort that leads to mental and bodily dismemberment. While one of their children returns to Aminata and seems to promise future kin as pregnant May comforts Aminata at her deathbed, Aminata remains a metaphorical "amputee" (Krampe 68) with the majority of her family "forever after [...] missing" (Hill, *Someone* 37).[226]

Aminata reacts to the progressing loss of her family not only with grief, pain, and despair as expressed in the amputee-metaphor and her desire to bear witness. Out of persevering hope, she also repeatedly migrates to find a place of freedom and home where she could reunite with her family and feel free eventually. As Aminata tells Chekura in New York City, they are not "far from free," but they "aren't there yet," until they leave "the Thirteen Colonies" (305). Hill explains in an interview that "[t]he desire to find home" is haunting Aminata throughout the novel (Sagawa and Hill 311). In fact, her "quest for home" (Sagawa and Hill 311) and the notion of being close but not 'there yet' leads Aminata to undertake her long odyssey and to continue to project hope elsewhere. However, the shacks Aminata calls home on the indigo plantation on St. Helena Island in South Carolina, in New York City's "Canvas Town," in Birchtown (Nova Scotia), and Freetown (Sierra Leone) only offer refuge in makeshift communities

---

advice in difficult situations recall Angela's conjured imaginary communities of prisoners and enslaved ancestors while locked up in her jail cell (see ch. 2.2.).

226 As Portia Owusu notes, the disintegration of Beloved's body and Sethe's family both also reflect "the dismembering of the collective African body, comprising of communities and familial relationships, broken during slavery" (34). As my analysis suggests, this seems also true for Aminata, her family, and her West African roots.

under constant threat of attack, dispersal, and destruction, for example, from legal owners, "slave traders," and lynch mobs. Family and community for Aminata therefore remain far from what Yorke describes as "coherent and continuous" (139). Instead, communal and kinship relations have to be build and rebuild again and again not only when Aminata moves from one place to another but also in one and the same place, where family and community members continue to be sold away, abducted, or killed. Under these circumstances, as Hortense Spillers argues, "'kinship' loses meaning, since it can be invaded at any given and arbitrary moment by the property relations" ("Mama's Baby, Papa's Maybe" 74) and their spatial and temporal proliferations.

An example of the makeshift communities Aminata builds and rebuilds throughout the narrative is Birchtown, where Aminata experiences extreme poverty, a hostile climate, segregation, anti-black violence, and the sheer struggle for survival. While Andrea Medovarski argues that "Birchtown presents an idealized socialist vision" outside of "financial capitalism" (20, 21), the community forms out of what the British grant and/or deny the Black settlers, and what the surrounding white community tolerates, which leads Pilar C. Domínguez to describe Birchtown as "a location of black disempowerment" (101). At the same time, Nova Scotia also represents a turning point in Aminata's quest. When she receives confirmation that Chekura died in a shipwreck and the chances for May to return shrink, she decides to "take what was left of my body and spirit and join the exodus to Africa" (*Someone* 386). Aminata's decision to go on a voyage again and re-settle on the African continent marks not only the beginning of her attempt at returning to her "homeland" but also of her abandonment of hope for familial reunion and successful homemaking. Aminata again uses the metaphor of familial loss as bodily disintegration and adds the loss of hope ever to reunite with them: "Chekura was dead. Mamadu was dead. May had been gone for five years. If she was still alive, she probably didn't remember me, and most certainly wasn't coming back. I missed all three of my loved ones so terribly that my body, it seemed, was half missing" (386). From now on, Aminata's further movements are motivated less by hope for community, family, freedom, and home, and more out of despair and a shrinking number of places to turn to. Since the reach of anti-blackness and eighteenth-century slavery extends beyond the territories of legal slavery, she increasingly projects hope into a future 'yet to come' and onto distant territory. As Domínguez puts it, Aminata "is an African forced to make her home in the diaspora" (102), yet this 'diasporic home' turns out to be what Hartman describes as "the domain of the stranger [that] is always an elusive *elsewhere*" (*Lose Your Mother* 4).

Impossible homecoming

Upon her arrival in Freetown, the colony of the British Sierra Leone Company also "proves to be an illusory refuge" (Sagawa and Robbins 15). The legally free Black settlers are confined to the settlement that they need to build from scratch, living again in deep poverty, scarcity, and completely dependent on the Company. At the same time, they are exposed to the immediate risk of recapture and re-enslavement, as the trading in enslaved people continues outside of the settlement. While "many Nova Scotians seemed content to build their homes and work for the Company," Aminata "felt that being obliged to stay within town was like staying on an island off the coast: I wasn't yet free to reclaim my homeland" (Hill, *Someone* 400). The threat of being captured "next door at the slave-trading post" (401) and send along with a coffle and through the Middle Passage a second time causes Aminata to admit that "no place in the world was entirely safe for an African, and that for many of us, survival depended on perpetual migration" (401). Aminata's notion of "perpetual migration" is borne out of her futile attempts of sustainable community building and homemaking in North America and West Africa. Clearly, Hill positions Aminata in "a never-ending diasporic loop" of migration (Berndt, "In Search of" 166) and a "limbo of wanting to witness and having to suffer terribly throughout the process of witnessing" (Krampe 70). Like Sethe who wonders why there was "[n]o misery, no regret, no hateful picture" her brain refused to accept (Morrison 70), Aminata takes it all in and tells about it; for, like Linda, she is unable to find a "loophole of retreat" not at risk of being invaded by the enslavist forces that rule the late eighteenth-century Black Atlantic.

When a coffle passes right through Freetown, Aminata observes a young captive girl who, she imagines, "felt that we could save her if we truly wanted to do so" (Hill, *Someone* 414), just like Assata's daughter who believes that Assata could leave the prison if she only tried. Yet, as soon as some of the "Nova Scotians" attempt to free the captives, they are overpowered and several settlers killed (413–16). Picking up the notion of a Christian paradise and anachronistically echoing King's "I Have a Dream"-speech, Aminata writes on the tombstone of one of the killed, Thomas Peters "*leader of Nova Scotian settlers. Fought for freedom, and is free at last*" (417). Of course, readers may be reminded of another tombstone with the inscription of the single word *Beloved* here. But while *Someone* pushes home and freedom repeatedly into an uncertain future and place, and then – when the presumed "homeland" on the African continent remains out of reach – into freedom in death, *Beloved* complicates the question of freedom *in* death not least through the haunting ghost and its reappearance as a character.

Accepting the lack of refuge and the perpetuity of migration for African-descended people, Aminata still holds on to her quest for home, when she decides to move on further, now projecting her miraculously persevering hope for return onto her home village Bayo. Yet, she notices a change in her quest, for she "wasn't concerned any longer with the things I wanted to do, but rather, with the place I wanted to be. All I wanted to do was return to the place where my life began" (452) despite doubts whether the community still existed, whether the inhabitants would welcome her, and whether she would still be able to fit in after slavery and her odyssey had changed her (450–52, 459). At this point, Aminata considers Bayo not only as her place of origin but also as the only possible end point of her life (story), where she may find rest from her perpetual migrations and hopefully die in peace.

Yet, while making plans for her reverse-trek, Aminata comes close to questioning whether freedom really lies in death. When she asks Armstrong for assistance to return to Bayo, she looks out of the window of his chambers in a fort on Bance Island, in which she, too, had been held captive as a young girl, and spots "the captives" down below her:

> I hated myself for doing nothing to help the captives escape their wretched confines. I tried to tell myself I was powerless to free them, but in truth, the mere sight of them made me feel complicit and guilty. The only moral course of action was to lay down my life to stop the theft of men. But how exactly could I lay down my life, and what, in the end, would it stop at all? (433)

Aminata seems to question not only whether dying while trying to free the captives – like Thomas Peters had – would change anything in the course of history, the larger transnational system of enslavement of Black people, and its afterlives. She also seems to question indirectly whether she owned her life so that she could give it for the sake of others. What differentiated Aminata, the returnee, from the girl she saw in the coffle who seemed to belong to nobody "but her captors" (413)? After all, she had been that girl herself years ago, having become with Spillers's words "*being for* the captor" ("Mama's Baby, Papa's Maybe" 67) in the coffle to the coast. As Aminata notes a little later, it seemed "that the trading in men would continue for as long as some people were free to take others as their property" (Hill, *Someone* 440).

When she finally makes the trek to Bayo in 1800 with the assistance of Alassane, a "Fula trader" (446), Aminata falls repeatedly ill (453, 456) and eventually finds out that the trader intends to sell her back into slavery (455). Aminata decides to escape immediately for "I could not go on living if all my years of longing for liberty and homeland were to lead me back to the neck yokes and ankle chains of my childhood abduction" (455). Luckily, she finds shelter in a small local community where she can recover and pays back her hosts by telling stories from her life. While recovering, she reflects on her life-long quest for home and lets go of her "greatest desire" to return to Bayo:

> From the day I was stolen, thoughts of home had made it impossible for me to feel I belonged anywhere I lived. Perhaps if I had been able to keep my husband and to live for years with him and our children, I would have learned to feel settled in a new place. But my family never settled in its nest. We never had any nest at all. But after I heard Alassane's words, I felt no more longing for Bayo – only a determination to stay free. And now as I waited for my strength to return in a hut belonging to people I didn't even know, I let go of my greatest desire. I would never go back home.
>
> [...] I would sooner swallow poison than live twenty more years as the property of another man – African or toubab. Bayo, I could live without. But for freedom, I would die. (459)

Aminata not only connects freedom with death again, she also reveals several underlying assumptions relevant for her understanding of freedom and home. First, she suggests that because she had experienced a familial and communal home in Bayo before her capture, she was unable to belong anywhere else afterwards. As John C. Ball notes, *Someone* "affirms a local, rooted identity – one left behind in Bayo and endlessly longed for thereafter – over the violent exigencies of circum-Atlantic travelling routes and the transnational, relational identities they generate." Second, she assumes that had she been able to sustain a family in the diaspora, she might have found a new place to make and sustain a home and therefore give up her desire for return. Third, instead of giving up hope all together because of the seemingly endless reach of slavery, on the way to Bayo Aminata finally separates the concepts of homecoming and freedom, abandoning any hopes for the former, and professing to die for the latter rather than be re-enslaved. In other words, Aminata acknowledges the impossibility of home-making and the relation between enslavement and social as well as physical death only when she reaches both physical and structural limits to migrate further due to her health and the immediate danger of re-enslavement.

Whereas the narrated I can only acknowledge on the reverse-trek to Bayo that she will not be able to return to her home village, the narrating I already suggests this when Aminata makes the trek in the opposite direction. As a young abducted girl walking naked in the coffle from Bayo to the West African coast, the narrating I recalls the narrated I's thoughts, assuring herself mantra-like: "Surely I would get free. Surely this would end. Surely I would find a way to flee into the woods and make my way home" (Hill, *Someone* 45). I suggest that the repetitiveness and the conditional of those words call the possibility of freedom and homecoming into question long before Aminata would attempt to take the same route into the opposite direction as a grown woman.

The narrative also suggests that there is no safe haven for Black people but death, social and physical, in the eighteenth-century Atlantic world before the attempted return trek leads Aminata to identify freedom with death as both the only possible end of a 'life of losses' and as the last projection surface for hope of a better future. Having fled from Lindo, Aminata feels unsafe in New York City and fears recapture – just as Douglass did upon his arrival in the city as a fugitive. She hides for a few days in the woods outside of the city where African and African-descended people bury their dead. When she stumbles into a burial ceremony, she is welcomed by the mourners and joins their dance and "wailing" (271) because

> all they had to do was look at me and hear my own sobs in my maternal tongue and they knew that I was one of them. The dead infant was the child I had once been; it was my own lost Mamadu; it was every person who had been tossed into the unforgiving sea on the endless journey across the big river. (272)

Aminata and the mourners form a temporary, improvised community of shared feelings of loss of home, family, and community "in the wake" (Sharpe, *In the Wake*). After the ceremony, Aminata asks one of the women where they lived (i.e., whether they had a home) and whether they were free or enslaved (i.e., whether there is freedom in this part of the world). The woman replies, "None of us are truly free, until we go back to our land" (Hill, *Someone* 272). After they exchange information about their ethnic origin (Ashanti, Fula, Bamana), the woman continues to describe what freedom as homecoming may mean:

> 'I work in house for man from England who say he make me free one day. But there is no free in this land. There is only food in your belly and clothes on your back and roof to hold off the rain. Home is only place that's free. That baby we done buried is on her way home. You see the coloured glass?'
>
> 'The beads around her waist?'
>
> 'They bring her spirit clear across the water, and take her home where she belongs.' (272)[227]

For the mourners, death appears to be the only return to community and home available apart from the makeshift community of loss that they can offer each other in the diaspora. Yet, death does not automatically mean return to their an-

---

[227] This burial scene may also remind readers of the baby girl that Sethe buries and who returns from her grave as a grown woman with memories of the Middle Passage in *Beloved*.

cestors and their ancestral land. Homecoming seems to require either a death at "home" or ceremonial attempts to sending the deceased back. While Aminata's parents, who died in Bayo, are now in the ancestral world and a limited form of mental communication is possible between daughter and parents, Chekura and Mamadu die in the grips of slavery in North America, receive no form of burial, and appear completely lost to Aminata. Sustainable communities, belonging to anybody "but [their] captors" (413), and homecoming are structural impossibilities for the (formerly) captive, enslaved, and fugitive because of the expanding reach of the time and place of slavery – no matter whether they call a place of birth their home and other people their family or community.

While acting as village storyteller for a month in the community that offers her shelter after she abandons her trip to Bayo, Aminata quickly finds a new purpose that brings her to cross the Atlantic one more time. She wants to tell her story in support of the abolitionist cause (465), projecting her hope onto abolitionism instead of any form of individual freedom or homemaking. Unsurprisingly, upon her arrival in London, physical exhaustion, having "grown tired and old," rather than successful homemaking lead Aminata to have "no determination to go somewhere" else but to be "interred right here, in the soil of London" after having testified before "the King" (465, 20). As Ball points out, London functions as "last resort when other places and people have failed" Aminata; it seems to represent "the best place – the only place – to escape the perils and disappointments of the Atlantic world," while working towards the liberation of her people.

After Bayo is beyond reach, Aminata chooses London as the place where she wants to die: "Africa is my homeland. But I have weathered enough migrations for five lifetimes, thank you very much, and I don't care to be moved again" (Hill, *Someone* 20). The passive construction 'to be moved' can be interpreted in several ways. First, it draws attention to the fact that Aminata's "perpetual migration[s]" (Hill, *Someone* 401) should not be considered primarily active and voluntary migrations but forced flights out of lack of alternatives. Second, her desire not "to be moved again" also seems to suggest that Aminata does not assume to return to her "homeland" after death. Third, the quote also recalls a similar statement by Sethe in the first chapter of *Beloved* discussed above, when Sethe tells Paul D that she will not leave the possessed house on Bluestone Road 124 ("No more running – from nothing. I will never run from another thing on this earth. I took one journey and I paid for the ticket, but let me tell you something, Paul D Garner: it cost too much! Do you hear me? It cost too much" [Morrison 15]). Moreover, Aminata's refusal to be moved physically, perhaps also emotionally, recalls Sethe's bedridden state in the drawing room in the last chapter of *Beloved* after her embodied daughter-ghost had left the house for good. In contrast to Sethe, however, it takes Aminata many more forced migrations to conclude eventually that she will die rather than move on in search of the "elusive elsewhere" (*Lose Your Mother* 4). Yorke describes Aminata as a "refugee without a potential refuge, migrating away from, and toward, suffering" who in the end gains a "supply of 'food, and quills and ink and paper' and a skeleton of family" (142, 138). Thus, even though the novel has "the texture of a happy ending, all Aminata's hard work really secures is a quiet deathbed" (138).

Remarkably and surely in line with its overt drive for closure, *Someone* nonetheless ends its narrative with the word 'home.' The last few sentences of the novel read:

> May kisses me on the forehead and is gone. The girl has young legs and moves like a cyclone. I, with bones afire, have no more tolerance for walking. I will cross no bridges and board no ships, but stay here on solid land [...] and lie back on this bed of straw. [...] They [the pregnant May and her husband] can wake me with the news, when they come home. (Hill, *Someone* 487–88)

While Aminata seems to suggest that 'home' refers here to the actual home she was ultimately able to make for herself, her daughter, and her daughter's growing family in London, I suggest that it should rather be understood as the end to perpetual migration and the repose death heralds, after Aminata had "trouble dying" (1) – as the first sentence of the novel goes – for so long.

"Perpetual migration"

While her first escape in New York City is a conscious flight from slavery, like Raven in *Flight to Canada*, Aminata's later migrations are marked by perpetuity. Expecting home and freedom constantly at the next station of her journey, Aminata believes in the reunion with family members and, after losing her son, husband, and daughter, in the possibility of return to her village community against all odds. In contrast to Raven who gives up his quest for freedom elsewhere upon his arrival in Canada, when homemaking and homecoming prove impossible for Aminata in New York City, Canada, and West Africa, death provides the last projection surface for futurity. While *Someone* ultimately ends with survival and reunion, futile homemaking, community building, and homecoming call into question the extent to which the ending of the novel can be considered 'happy' at all.

As Hill explains about his driving questions in writing the novel: "I said to myself, Wow, what kind of life might a woman have had if she were on one of these vessels [that headed to Freetown]? [...] Where was she born? Where was she abducted? Where was she raised in slavery and how on earth did she get to New York and then to Nova Scotia? And why would she leave after all of this?" (qtd. in Pierre). One possible answer to Hill's last question that I have pursued here by reading in and between the lines of *Someone*, *Beloved*, and *Flight to Canada* is that there is no home on the run and no end to fugitivity for people racialized as Black in the Atlantic world of the historical era of slavery. *Beloved* reflects this by creating the ghost protagonist turned flesh, through Sethe and Paul D's 'rememories,' the infanticide, the chain gang, the claustrophobic house on Bluestone Road, non-linear, highly mediated narration, and the momentary collapse of plot and syntax when addressing death and 'the hold' of the Middle Passage. It also blurs any clear lines between enslaving South and non-enslaving North and between the time before, during, and after the Civil War, ultimately refusing the desire for a closed narrative, and remaining ambiguous about the possibility of its re-telling. Reed addresses the complexity and perpetuity of African American flight in the nineteenth-century through his broad cast of archetypical Black male fugitives whose highly mediated flights ironically end not in Canada but on a plantation in the gothic American South. In contrast, *Someone* goes into great pains to create a speaking subject from the abyss of social death and to narrate a story of survival and reunion. In search of kin and home in the diaspora and her "homeland," Aminata is forced into constant fugitivity and futile homemaking. Focusing on the historical time of slavery and its immediate aftermath in the nineteenth century, these neo-slave narratives fundamentally question the narratibility of the traumas of slavery and the possibility of its overcoming as well as the recovery of lost voices. They show in different and differently explicit ways that the characters' makeshift homes, communities, and "freedom dreams" remain in the clutches of anti-blackness, while at the same time testifying to their care for each other, their endurance and struggle to overcome the Black border to unconfined forms of freedom, which are not within actual reach but sought after in flight. Teju Cole's novel *Open City* that will be discussed in the last section of this chapter catapults these questions into the twenty-first century.

The "fugitive notes" of Teju Cole's *Open City*

The following close reading that examines forms of flight in Teju Cole's 2011 novel *Open City* will close this chapter on fugitive fictions of slavery and flight by broadening its perspective beyond neo-slave narratives set in the eighteenth and nineteenth century. In contrast to the African American novels of slavery discussed so far, *Open City* is set in the twenty-first century and narrated from the perspective of a Nigerian German New Yorker whose narration subtly juxtaposes fragments of a neo-slave narrative with narratives of other forms of violence that have affected Black people and people of color through the centuries on both sides of the Atlantic in the context of migration, colonialism, genocide, imprisonment, and war. *Open City* is in many ways a border and genre crossing narrative of a 'new' Black diaspora in the United States that has arisen out of and speaks to recent Black African migration across the Atlantic from Africa south of the Sahara to the United States in the late twentieth and early twenty-first centuries. It joins the ranks with prize-winning "Afropolitan" novels, such as *Americanah* by Chimamanda Ngozi Adichie (2013), *Virgin of Flames* (2007) by Chris Abani, *We Need New Names* (2013) by NoViolet Bulawayo, *The Beautiful Things That Heaven Bears* (2007) by Dinaw Mengestu, and *Ghana Must Go* by Taiye Selasi (2013) (cf. Ede; Gehrmann; Goyal, *Runaway Genres* ch. 5; Varvogli).[228] Due to their complex transnational and diasporic, at times cosmopolitan disposition, these realist novels negotiate, among many other things, contemporary concepts of (West) African as well as US American Blackness and anti-blackness.[229] They

---

[228] The term "Afropolitanism" was coined by Achille Mbembe and gained popularity in the wake of Selasi's well-received essay "Bye-Bye, Babar" from 2005. Afropolitanism addresses an increased global mobility of African middle and upper classes and attempts to dissociate Africanness and African descent from dominant stereotypes of poverty, famine, environmental catastrophe, and political upheavals. Instead, it emphasizes global mobility, flexibility, hybridity, and hipness. According to Selasi, Afropolitans are a generation of well-educated, mobile, middle- to upper-class Black people with close ties to places and communities on the African continent as well as the metropolises of the Western world today. The above-mentioned writers and their novels' protagonists fit Selasi's description in many ways, while also striving for "more philosophical depth and ethical weight" as propagated by Mbembe (Gehrmann 64; cf., e. g., Fongang, Varvogli). Susanne Gehrmann defines Afropolitanism as "[c]osmopolitanism with African roots" (61) because it is "[c]osmopolitan in scope, anti-essentialist, open to cultural and intellectual hybridization, but endowed with a particular consciousness for Africa's historical wounds" (64). In this way, Afropolitan writing also looks back on a substantial history of African literature and storytelling. For an introduction to African literature, see Gunner and Scheub. For a brief overview of West African literatures in English, which Cole and the majority of the novels mentioned above also belong to, see Berndt, "West Africa." For a study of the role of the slave trade in Anglophone West African literature of the twentieth and early twenty-first century, see Murphy, *Metaphor*.

[229] As Jared Sexton points out, "[t]hough they are distinguished often enough, and often enough distinguish themselves, from the native-born black population in the United States, black immigrants cannot avail themselves of the racial capital of nonblack immigrants of color. Instead, they find themselves consistently folded back into the category of homegrown blackness, as it were, and subject to the same protocols of social, political, and economic violence, especially in subsequent generations" ("People-of-Color-Blindness" 53).

depart from the primary focus of the neo-slave narratives analyzed so far as they trace Black presences in North America not only to the history of the transatlantic slave trade and slavery but also to more recent movements between North America and the African continent and resulting demographic shifts in urban US America,[230] without losing sight, however, of the transatlantic history that includes both. As Yogita Goyal notes, Afropolitan novels bring together discourses around "immigration and slavery" that until recently "US cultural histories have usually treated [...] as two distinct stories, and their collision [...] presages a number of conflicts and challenges to expected ways of narrating both America and Africa" (*Runaway Genres* ch. 5). Showcasing the many facets of the Black Atlantic (Gilroy) as both Black and African as well as contemporary and centuries old, these novels not only "herald[...] an African literary renaissance," but also call for rethinking concepts, such as diaspora (ch. 5), "at the critical and theoretical crossroads where African-American, immigrant, and postcolonial literary studies meet" (Varvogli 236). Thus, focusing on *Open City* as the last step in my analysis of fugitive fictions of slavery and flight as well as captivity and fugitivity in Black American literature at large returns us afresh to both the border concepts *and* the theories of anti-blackness this study started out with (see ch. 1.2. and 1.3.).

Scholars have grappled with grouping Afropolitan novels, such as *Open City*, within established writing traditions. Caren Irr, for instance, describes *Open City* together with Mengestu's and Abani's aforementioned novels as "African migrant fiction" in a chapter that also discusses Caribbean, South and Central European, as well as South Asian American fiction as "digital migrant" novels that belong to an emergent genre that she terms the "geopolitical novel" (see Irr, ch. 1, esp. 37–42). *Open City* has also been read as part of Nigerian, African, transnational, "Pan-African American," and world literature (Gamso; Goyal, "The Transnational Turn"; Dalley; Li; Saint; Suhr-Sytsma). Furthermore, scholars have examined Cole's second work of fiction for its representation of post-9/11 New York City (Golimowska 55–69, Miller 198–204) and the memorialization of the terrorist attacks and its racialized aftermaths (Edwards, Freedman

---

[230] Between 2000 and 2016, the Black African population in the United States rose from about 574.000 to 1.6 million (Anderson and López). As the *New York Times* reported in 2014, in the first decade of the new century "more black Africans arrived in this country on their own than were imported directly to North America during the more than three centuries of the slave trade" (S. Roberts). According to estimates, while about 12 million people were captured on the African continent and about 10.7 million survived the Middle Passage and arrived in the Americas, only about 388.000 of them arrived directly in North America. The majority landed in the Caribbean and South America (Trans-Atlantic Slave Trade Database, "Estimates Only Embarked"; "Estimates Only Disembarked").

177–86, Mihăilă 286–94, O'Gorman 40–75). Thus, while Hamish Dalley, for example, examines *Open City* as "third generation Nigerian literature" that questions the concepts of generation and nation in terms of their historical and territorial (non-)affiliation (26–32), Erica R. Edwards reads *Open City* together with Bulawayo's and Adichie's work as part of the "New Black Novel," "crafted in and among the ruins of the long war on terror" in the United States (679). Yet, the primary analytic frame for literary analyses of *Open City* has been the novel's negotiation of dominant and subversive concepts of cosmopolitanism (Ba and Soto, Breger, Elze, Epstein, Fongang, Johansen, Hallemeier, Krishnan, Oniwe, Sollors, Vermeulen). Many of these studies identify its protagonist Julius as a cosmopolitan flâneur only on the surface of the narrative, whereas the novel as a whole is understood to question or complicate this existence, revealing cosmopolitanism as a "ruse" that "begs its own undoing" (Krishnan 689, 683). While Alexander Greer Hartwiger, for instance, uses Julius to develop his concept of a postcolonial flâneur (2–5), Pieter Vermeulen identifies the protagonist as a *fuguer*, a psychological concept that described "'mad travelers'" in nineteenth-century France (42). Claudia Breger on the other hand argues that *Open City* confronts "cosmopolitanism with the legacies of historical violence in uneven powerscapes to the effect of opening up the concept toward more transnational perspectives" (117). Due to the diversity of the topics addressed in the novel, scholars have also examined, for example, the role of photography and music as intertextual reference points and aesthetic approaches in the novel (Epstein, K. Jacobs, Neumann and Kappel, Reese and Kingston-Reese) as well as migrancy, mobility, and the relation of pedestrian mapping and African epistemologies (Gamso, Varvogli, Cumpsty).

Thus, while scholars have looked at more and more aspects of the novel's broad reference frame over the last few years, the role of slavery and flight has been surprisingly neglected. Admittedly, the term *fugitive* appears literally only once in *Open City*, when the autodiegetic narrator Julius recounts a spontaneous visit to Notre Dame de la Chappelle during his winter vacation in Brussels, Belgium, and describes the idiosyncratic music he hears:

> I noticed, just then, a dissonance in the sound of the organ music. There were distinct fugitive notes that shot through the musical texture, like shafts of light refracted through stained glass. I was sure it was a Baroque piece, not one I had heard before, but with all the ornamentation typical of the period, yet it had taken on the spirit of something else – what came to mind was Peter Maxwell Davies's "O God Abufe" – a fractured, scattered feeling. (Cole, *Open City* 138)

A little later Julius realizes that the music is in fact being played from a recording and that the sound of a vacuum cleaner is causing the "dissonance." He starts to

2.3 Fugitive Fictions of Slavery and Flight — 225

wonder where the woman, who is cleaning the aisle, might be from and why she is in this Brussels church:

> I thought that she, too, might be here in Belgium as an act of forgetting. Her presence in the church might doubly be a means of escape: a refuge from the demands of family life and a hiding place from what she might have seen in the Cameroons or in the Congo, or maybe even Rwanda. And perhaps her escape was not from anything she had done, but from what she had seen. It was a speculation. I would never find out, for she possessed her secrets fully as did those women that Vermeer painted in this same gray, lowland light; her silence seemed absolute. (140)

The term "fugitive" is followed by the closely related terms "escape," "refuge," and "hiding place" which become associated with "silence" and the "forgetting" of violence either witnessed or perpetrated. This familiar concept of fugitivity, I argue, runs as a common thread through Cole's novel. Because Julius loosely identifies the woman as Cameroonian, Congolese, or Rwandan, he considers Belgium a sanctuary for her even though she goes about the physically straining work of cleaning, precarious work primarily assigned to Black women, poor people, people of color as well as migrants and refugees throughout the global North. The juxtaposition of Julius's existence as an affluent US American tourist of German and Nigerian descent with a taste for Baroque music and seventeenth-century Dutch art with the assumed realities of a Black female migrant cleaner is telling for the larger tensions *Open City* grapples with in terms of race, ethnicity, sexuality, class, origin, and gender, and the ways in which they bear on people's lives and movements. In speculating about the woman, however, Julius might actually be speaking more accurately about himself, projecting forgetting and flight onto her as a metaphorical "two-dimensional canvas, safely backdated to the seventeenth century" (Clark 193) and suppressing their ramifications in his own (hi)story in order not to undermine his cosmopolitan demeanor.

Julius, a psychiatry resident at Presbyterian Hospital in New York City, takes long evening walks through Manhattan as well as his intermittent travel destination Brussels between fall 2006 and fall 2007. As readers find out during the course of the novel, which unfolds at Julius's "walking pace" (Cole, *Open City* 3), he, too, 'escaped' from his Nigerian past and seems to seek "refuge" in his flâneur-like existence in New York City and Brussels. While wandering, Julius muses about the cities' architecture and their checkered histories, recites conversations and individual stories of the people he meets, and contemplates art, music, literature, and philosophy. Sometimes the walks and musings evoke memories of Julius's childhood and youth in Nigeria; these, however, irrupt only as fragments of otherwise incomplete or obscure narrative strands. Taken together those fragments convey that his Nigerian father had died of tuberculosis when Julius was fourteen years old and that Julius had become estranged from his white German mother after his graduation when he left his home in Lagos in order to receive higher education in the United States, cutting most family ties. Towards the end of the novel, it also transpires that as a teenager in Lagos Julius in all likelihood raped Moji Kasali, the older sister of a Nigerian high school friend, an incident he seems to have completely forgotten by the time he reencounters her by chance many years later in Manhattan.

Given the apparent amnesia of this traumatic incident and the ways in which Julius's narrative seems to circle persistently around addressing the presumed rape through physical walking and mental wandering, Werner Sollors asks whether "[i]n his aimless walks in the city, is he perhaps in reality always running away from something? Does he wish to forget what his life was like before he started walking?" (245). While showing the breadth and diversity of digressing references in the novel, Sollors observes a strong emphasis on the history of the Holocaust and the Second World War and therefore argues that "whether there can be cosmopolitanism after the Holocaust and the many other twentieth-century atrocities that it overshadows is the question at the heart of the novel" (242). Indeed, Karen Jacobs argues that the history of the Jewish diaspora and the legacy of the Holocaust suppress the narrative's ability to engage the African diaspora, gesturing towards "the scope of the African diaspora's representational impossibility" (90, 92). While I share Sollors's observation about the novel's digressions and Jacobs's observation about the eclipsing of the Black diaspora, I argue that a concept of fugitivity as developed in ch. 1.4. that emerges from the history of the transatlantic slave trade and slavery and acknowledges their racialized and gendered afterlives sheds further light on Julius's constant compulsion to walk, overshare, and evade. In what follows, I examine the narrative strategies that foreground certain textual strands, such as the walks and other people's stories, to claim readers' primary attention while evading other more fugitive aspects of the story and its narrator that remain, to paraphrase Julius, "secret" for the most part and only sometimes "shoot through" and "fracture" the narrative "texture" (Cole, *Open City* 138). My reading of *Open City* therefore foregrounds flight as a form of physical and mental movement performed by both the text and its narrator, who flees from an association with his past and the past of the Atlantic world, without, however, suggesting this privileged middle-class protagonist of German and Nigerian descent be mistaken for a refugee. Although the sheer abundance of anecdotes and encyclopedic excursuses seem to bury traumatic memories of Nigeria and the Black Atlantic, they nevertheless haunt the narrator, disrupting the narrative structure at certain points, and gesture toward what Julius wants to evade. His cosmopolitan demeanor, obsessive walking and musings about art and high culture can offer only temporary, improvised refuge (Sollors 246). Ultimately, the text finds itself caught in the gendered history and anti-black legacy of the transatlantic slave trade and chattel slavery as well as the multiple ways in which Julius is implicated in these legacies: as a "brother [...] from the Motherland" (Cole, *Open City* 186), an achieving "young black man [...] in that white coat" of a doctor (210), and a man of "mixed race" (195).[231] As Breger notes, the narrator is also "both victim and apparent perpetrator" (120) of anti-black gender violence, on the run from being identified and self-identifying with this violence and the ways in which it becomes part of the Hartmanian afterlife of slavery. Anti-black violence, I argue, forms the mostly obscured ground on which Julius's narration is built, while *Open City*'s fugitive narrative technique ostensibly refrains from addressing anti-blackness, even as it subtly negotiates the narratibility of the racialized and gendered "scenes of subjection" (cf. Hartman, *Scenes of Subjection*) that Julius refuses to associate with personally.

In order to make this subtle negotiation visible, I suggest an attentive reading strategy that – similar to my analysis of *Someone* – pits the novel against its protagonist and looks for breaches in the narrative flow, eruptions of the transatlantic history and the personal past of the protagonist. Trying to resist the allure of the narrative voice that creates what Rebecca Clark calls an "inescapable narrative intimacy" and "parasitically and inescapably embeds its readers deep in the labyrinthine folds of its narrator" (195), I examine in a first step the strategies that divert the attention away from the narrator and his past. Julius not only appears to hide behind his wanderings, his encyclopedic knowledge, and the stories of minor characters – many of which are in fact literal stories of migration, flight, and imprisonment. The text also offers reasons to approach his narrative with skepticism as it subtly but repeatedly casts doubt on the trustworthiness of stories and the reliability of narration, as well as the ability of the mind to remember. Dalley suggests a "reading method," not unlike my own that recognizes "not only [...] what is visible, but also [...] what has been erased" in Julius's story. While Dalley argues that Nigeria is "the object that is most insistently *absent* to Julius" (30), my analysis focusses on the narrative strategies of flight that refuse identification with the history of the transatlantic slave trade and slavery which – whenever it finds its way into the text – also triggers involvement with Julius's personal past in Nigeria. After examining the narrative strategies of oversharing and evasion, my analysis focusses, secondly, on textual in-

---

[231] While characters in *Open City* try to claim Julius in these ways, Cole refuses, as Goyal notes, to make Julius "an allegorical figure representing the migrant, or the cultural hybrid, or the mixed-race figure (the tragic mulatto), or the been-to, or what Adichie calls the 'Americanah'" (*Runaway Genres* ch. 5).

stances where diversion and deception seem to fail and the suppressed transatlantic past as well as Julius's personal past in the form of his father's death and the rape of Moji interrupt the narrative flow.

Oversharing walks, musings, and fugitive stories to mistrust

In more than half of *Open City*'s twenty-one chapters, Julius's solitary walks through Manhattan and Brussels structure the narrative and provide frames into which intellectual digression and smaller anecdotes are embedded. Upon the novel's *in medias res* opening, the protagonist describes his compulsion to walk as a counterpoise to his workdays at the hospital, as a "therapy" that met a "need" and soon became "the normal thing" to do at least twice a week and on one of the weekend days (Cole, *Open City* 7, 6). He often feels compelled to go on his "aimless wandering[s]" (3) driven by an unconscious force that makes him suddenly launch on those walks without a specific destination in mind (44–45, 188). Recounting them usually involves detailed descriptions of the long route Julius takes that suggest primarily a "map-like familiarity with the topography of the place" (Elze 89) as well as reports on the immediate surrounding he passes through, (over)sharing his observations about people, architecture, and the natural environment (see, e.g., 46–47). Details about weather conditions, changing seasons, or occasional time designations (see, e.g., 8, 43, and 174) roughly define the time frame and chronology of the two parts of the novel.[232] The narrative sequences of walks serve as inspiration and starting points for Julius's musings and more often than not trigger associative digression into art, history, and music.[233] Such chains of association framed by walks not only take over much of the narrative attention and contribute to the slow progression of the plot; they also substantially fragment the novel as it falls apart

---

[232] For a detailed discussion of how space is created in the narrative primarily through the topographical order of the map, while time is characterized through anachronisms, see Elze.
[233] Salient examples of these digressions are, for instance, Julius's detailed recollections of a visit to a music store (16–17), the American Folk Art Museum (56–40), the Cloister Museum (236–38), and a classical concert in Carnegie Hall (249–54). For a detailed summary of Julius's broad and diverse cultural reference frame, see Sollors (234–37, 240–45).

Julius's digressions may also resemble those of the protagonist John in David Bradley's neo-slave "novel of remembered generations" *The Chaneysville Incident* (1981; Rushdy, "Neo-Slave Narrative" 95). As Valerie Smith notes, "in the early sections of the novel," the historian John "tends to fall into extended, pedantic stories about sociological phenomena and historical events" behind which he hides his feelings ("Neo-slave Narratives" 173–74).

into short threads – not unlike those of *Beloved* – that are picked up at different points in the text so that a plot is hard to identify at times.

This narrative pattern is already noticeable by chapter 4. In the first section, the fragmentation develops from the juxtaposition of Julius's walk through the Financial District that frames the chapter with his reproduction of a psychiatry session with his patient M. from earlier that day and digression inspired by the places Julius passes through during his walk (Cole, *Open City* 43–51). When Julius recalls how he stopped by Trinity Church, for example, "the unpremeditated idea that I might go inside and pray for M." introduces a paragraph that reproduces what M. "had said to me earlier" from the patient's first-person perspective and with almost no interjections from the narrator (48). Afterwards Julius continues to observe the surroundings of the church and its graveyard, followed by an excursus on the church's history and its relation to Dutch settlement, whale hunting, and the novel *Moby Dick* (49–51). The digression is taken a step further through an indented paragraph composed of a quotation from "the story recounted by the Dutch settler Antony de Hooges in his memorandum book" reproduced in smaller print reminiscent of citations in academic and non-fictional works (50) that may also remind readers of Reed's intertextual citation strategy.

Apart from the ritualistic walks (cf. Cumpsty) and the intellectual digression, plot and narrator are concealed further, as Julius repeatedly shifts narrative focus away from himself toward other people's stories. The vignettes of the many minor characters emerge from Julius's observations of and conversations with people he meets on his walks, travels, and at work. He narrates their stories from a seemingly omniscient perspective (Clark 186, Edwards 676) in indirect speech with occasional interjections until – more often than not – the mediation of the narrator seems to disappear and blends into an embedded first or third-person narrative of the minor characters themselves. In this way, *Open City* offers mediated insight into a large cast of mostly secondary characters, many of which are migrants or fugitives, just like Julius, that draw attention away from the protagonist.[234] Birgit Neumann and Yvonne Kappel contend that Julius loses control of his narrative when other character's narratives take center stage and his voice can no longer be differentiated from those of the minor characters, "pervasively

---

[234] Julius's old literature Professor Dr. Saito, for example, had moved from the United Kingdom to the United States where he and his parents were interned during the Second World War because of their Japanese descent (Cole, *Open City* 13–14). Farouq and Khalil, whose acquaintance Julius makes in Brussels, migrated from Morocco to Belgium to seek a better life. Dr. Maillotte, who Julius meets on his flight to Brussels, moved from Belgium to the United States and continues to travel between the two places on a regular basis.

displac[ing] notions of a single voice and unified self" (44). In contrast, I argue that Julius tries to keep narrative control by using other people's stories as self- and reader distraction, i.e., as a means to escape from associating with the legacy of slavery and gendered anti-black violence. As Clark points out in her analysis of Julius as a parasitic narrator-protagonist, "[t]here is something uncomfortably invasive about the way he uses others for story-making [...], feeding off of their unacknowledged personal trauma" (192). "He collapses all of the stories he hears and retells them in his own affectively flat, unhurried, unidirectional univocality. At the same time, he freely and invasively diagnoses, deconstructs, and symptomatizes any and all narratives that are not his own" (Clark 189). Whereas readers discover plenty about minor characters like patient M., the white Belgian physician Maillotte, or the Moroccan migrant Farouq, minor characters and readers learn very little about Julius and his familial background, with the narrator only gradually and reluctantly adding small, ambiguous pieces to the puzzle that is his past in Nigeria. In other words, while Julius is a storyteller (Varvogli 250), in contrast to autobiographical storytellers such as Aminata, his idiosyncratic autodiegetic voice strives for a 'parasitic' omniscient homo- or even hetero-diegesis, while attempting to evade autonarration.

Importantly, Julius often distrusts the stories of others he puts into the center of his narration. As a psychiatrist in training, Julius is well aware of the unreliability of peoples' accounts, their own judgment about themselves, and the ability of the mind "to deceive itself" through the mind's many "blind spots" (Cole, *Open City* 239, 238). Moreover, while looking at a young crowd of Rwandans in a nightclub in Brussels, Julius confesses that he

> felt some of that mental constriction – imperceptible sometimes, but always there – that came whenever I was introduced to young men from Serbia or Croatia, from Sierra Leone or Liberia. That doubt that said: These, too, could have killed and killed and only later learned how to look innocent. (139)

Here Julius continues his deliberations about reasons for flight and the perpetration, witnessing, and silencing of violence that he began when thinking about the woman in the Brussels church in the same chapter, interpreting, as Clark notes, the clubbers' faces as flat, readable masks (193). Certainly, Julius's mistrust of other people, especially men, and his doubts about their innocence should make readers wary of the reliability of his judgment and fugitive voice as it might mask a potential complicity with and perpetration of violence on his part, foreshadowing what will later emerge as Moji's rape accusation. When Maillotte, who dominates their conversations with anecdotes about her life and only occasionally asks a few polite questions about Julius's background

(Cole, *Open City* 88–93, 141–42), inquires about his first language, he considers telling her that he spoke German to his mother as a child but ultimately decides against it because he "didn't want to get into the intricacies of the story" (142).[235] Unpacking the "intricacies" of Julius's fugitive "story" buried under the weight of other, more cohesive narrative fragments about walks, excursuses, and other peoples' stories represents one of the major challenges *Open City* poses.

Memories of slavery and Nigeria

Towards the end of the long walk from Wall Street Station in chapter 4, Julius arrives at the Upper New York Bay from where he can make out the Statue of Liberty and Ellis Island in the distance. The sight triggers the first reference to the transatlantic slave trade in the novel, as Julius notes that Ellis Island "had been built too late for those early Africans – who weren't immigrants in any case – and it had been closed too soon to mean anything to the later Africans like Kenneth or the cabdriver, or me" (54–55). Kenneth is a Barboudian museum guard who Julius first meets during an earlier visit to the American Folk Art Museum and reencounters in a restaurant he enters after finding Trinity Church closed (53–54, 51); the "cabdriver" picks Julius up after he leaves the American Folk Art Museum to go home (40–41). Julius feels that both Kenneth and the cabdriver lay claims on him as a 'fellow African,' claims, he explains to the reader, he is not "in the mood for" (40).[236] Julius is reluctant to identify with his African ancestry, the African diaspora, and the legacy of slavery and the slave trade attached to both – quite contrary to his German heritage with which he actively wants to reconnect on his trip to Brussels (31–32).[237] As Nicholas Gamso notes,

---

[235] As Sollors notes, the narrator repeatedly reports on "conversations in which Julius tells the reader what he did not say" to his counterpart, gesturing again towards his unwillingness to relate information about himself to others (234).

[236] In fact, during their second encounter in the restaurant Julius thinks that Kenneth was not only claiming him as a 'fellow African.' He also feels that Kenneth's eyes were "asking a question. A sexual question." that causes Julius to excuse himself hastily (Cole, *Open City* 54). Thus, apart from racial affiliation, sexual orientation and class affiliation also seem to play a role in Julius's refusal to relate with people (Elze 92).

[237] Julius first expresses the urge to reconnect with his "oma" right after two white children from "a family of out-of-towners" label and racialize him as a "gangster" (32, 31). Breger reads this sudden urge as "an impulse to affirm his (white) European roots" against the racist remark. Germane to my argument, she proposes that "this reaction could code his sophisticated cosmopolitanism as a gesture of flight and attempted detachment from those other parts of his

Julius "feels confined by the mandate to identify – to give meaning to itinerancy, to migrancy, to blackness" (61). However, on being confronted now with two of the most well-known symbols of European immigration to New York in the nineteenth century and the immigrants' 'freedom dreams,' Ellis Island and the Statue of Liberty, Julius cannot help but acknowledge the confining Black border and see traces of the Middle Passage in their shadows.[238]

> Ellis Island was a symbol mostly for European refugees. Blacks, "we blacks," had known rougher ports of entry: this, I could admit to myself now that my mood was less impatient, was what the cabdriver had meant. This was the acknowledgement he wanted, in his brusque fashion, from every 'brother' he met. (Cole, *Open City* 54–55)

Julius suddenly understands that the cabdriver and Kenneth seek "acknowledgement" of shared experiences of anti-blackness stemming from an inextricably connected transatlantic history of the slave trade and slavery that Julius cannot escape from no matter how hard he tries.

This idea of diasporic 'brotherhood' appears a few times in the novel and remains rather problematic for Julius (40–41, 86–89). On one occasion, Julius exchanges the greeting he associates with the concept with two young Black men in Harlem on his way home; the two reappear in the company of another teenager shortly afterwards to rob, beat, and injure Julius. Recounting the incident, he explains in more detail that what he had felt were

> only the most tenuous of connections between us, looks on a street corner by strangers, a gesture of mutual respect based on our being young, black, male; based, in other words, on our being 'brothers.' These glances were exchanged between black men all over the city every minute of the day, a quick solidarity worked into the weave of each man's mundane pursuits, a nod or smile or quick greeting. It was a little way of saying, I know something of what life is like for you out here. (212)

In *Open City* Julius's ambiguous experiences of solidarity among Black men and its limits complicate the legacy of the transatlantic slave trade for Black communities today, what Hartman so pointedly describes as the "fugitive legacy" of "those who stayed behind" and of "the children of the captives dragged across

---

background that are constantly projected (back) upon him in New York as well as Brussels" (120).

238 The new permanent exhibition of the Statue of Liberty Museum, which opened to the public on Liberty Island in 2019, actually rebuts the idea that Lady Liberty was only ever dedicated to white European immigration. Instead, in the 1860s the statue was "originally designed to celebrate the end of slavery," but by the time the statue was officially presented to the public in 1886 this original dedication had been abandoned (Brockell).

the sea" (*Lose Your Mother* 234, 232). The novel indirectly addresses fundamental differences in terms of history, class, and origin that may divide Black communities in New York City without, however, losing sight of the anti-black violence directed gratuitously and indiscriminately against those communities. It divides and binds together Black people in *Open City* through their shared "grammar of suffering" (Wilderson, *Red, White, and Black* 11, 37) just as they may become united in their "anagrammatical" refusal of, survival despite, resistance against, and flight from past, present, and ongoing forms of violence (Sharpe, *In the Wake* 75, 77).

This first direct reference to the Middle Passage in the novel is followed by a brief allusion to the slave trade at the end of chapter 4 that contextualizes it within other violent histories that have left their marks on US society in general and marginalized communities in particular. Julius's long walk through Downtown Manhattan takes him eventually to the site of the terrorist attacks of 11 September 2001, which was "a massive construction site" at the time (Cole, *Open City* 57). He contemplates the "disaster" but immediately connects it with other earlier "atrocities" (58), coming to the conclusion that

> The site was a palimpsest, as was all the city, written, erased, written again. There had been communities here before Columbus ever set sail, before Verrazano anchored his ship in the narrows, or the black Portuguese slave trader Esteban Gómez sailed up the Hudson; human beings had lived here, built homes, and quarreled with their neighbors long before the Dutch ever saw a business opportunity in the rich furs and timber of the island and its calm bay. Generations rushed through the eye of the needle, and I, one of the still legible crowd, entered the subway. I wanted to find the line that connected me to my own part in these stories. (59)

As Jens Elze notes, "[b]y [...] directing his gaze deeper and deeper into the construction site he retrogradingly tells the spatial history of the place to pre-Columbian times" (98). Thus, Julius links the more recent atrocity of Islamist terrorism not only to the history of the slave trade but also to that of settler colonialism and the genocide of Native American people in the region, while also noting the "Christian Syrian" and "the Lebanese" communities that had been displaced when the Twin towers were built (59). More importantly however, after mentioning the participation not only of white Italian Christopher Columbus and Giovanni da Verrazano but also of Black Portuguese Esteban Gómez in settler colonialism and the slave trade, for once Julius wonders about his own part "in these stories," i.e., about the ways in which he might be implicated in these complex transatlantic histories.

This insight is followed in the next chapter by more stories of minor characters that address more atrocities and again de-center Julius's role. His conversation with the Liberian Saidu at a detention center in Queens quickly focusses on Saidu's story, told from a third-person perspective, about his long flight from war-torn Liberia via Mali, Morocco, Spain, and Portugal to the United States. Julius mediates the story "in the manner of an extradiegetic narrator" (Haensell 341), apparently recounting the unfiltered "details of [Saidu's] life" (Cole, *Open City* 70) and providing "a sympathetic ear to a story that, for too long, [Saidu] had been forced to keep to himself" (64). In the "carceral ambience" of the detention center (Gamso 71), Julius positions himself as a disinterested reporter and "the listener, the compassionate African who paid attention to the details of someone else's life and struggle," not least in order to impress his girlfriend (Cole, *Open City* 70). While doing so, Julius hardly mentions the fact that he too migrated from a country on the West African coast to the United States, admittedly under different circumstances and for different reasons. Visiting Saidu as part of "the Welcomers," a group of his girlfriend's parish (60), Julius ironically narrates Saidu's story without "sympathy [...]or understanding" (Goyal, *Runaway Genres* ch. 5), distancing himself from any possible connection to the Liberian detainee. Here, Saidu's story indirectly points to the different forms of privileges stemming from class, origin, and a proximity or distance to whiteness and Western, more specifically US American cultures, that define the possibilities and circumstances of migration and travel to the United States today.

After two thirds of the "stylized and formulaic" story (Goyal, *Runaway Genres* ch. 5), Saidu's narration becomes increasingly vague and Julius eventually comments on what has been shared so far. He expresses serious doubt about the story's truthfulness since Saidu "had, after all, had months to embellish the details, to perfect his claim of being an innocent refugee" (Cole, *Open City* 67). Julius wonders, "naturally, [...] whether it wasn't more likely that he had been a soldier" (67), a perpetrator of violence as opposed to its victim, whereas Saidu seems to anticipate doubt and explicitly maintains his innocence by explaining that pickpocketing a few times in Spain "was the only crime he ever committed" (68). Julius recounts Saidu's story, one of the longest about a minor character uninterrupted by other narrative fragments in the novel, only to eventually question its validity. He seems to want to avoid feeling associated with or implicated in this story of flight and incarceration, and the related questions of victimhood and perpetration. The complex transatlantic history of the country Saidu hails from also remains unmentioned, even though this information could have been excellent digression material for Julius. Yet, the fact that, in the nineteenth century, free and formerly enslaved African Americans founded Liberia after having fled North America to start a new life on the continent of their ancestors further complicates the transatlantic history Julius prefers not to address. As Goyal notes, this vignette shows "how Julius neither fears nor honors any 'rememory' that Saidu might evoke"; as "an Afropolitan figure [he] fails to connect with an equally recognizable figure of the detainee, refugee, and illegal immigrant" (Goyal, *Runaway Genres* ch. 5), even though – or because – Saidu could easily be perceived as Julius's "alter ego" (Sollors 240).

The narrative unit about the Haitian "bootblack" Pierre, who shines Julius's shoes in Penn station in Midtown Manhattan, immediately follows Saidu's story, while they both precede Julius's more explicit reflections on truthfulness, witnessing, and perpetration that he makes in reference to his stay in Brussels. Pierre's story is told primarily as an embedded first-person narrative without any commentary from Julius (Cole, *Open City* 71–74). In the cases of Saidu, the church cleaner, and the Rwandan clubbers, Julius expresses doubt about their trustworthiness and their role in recent histories of war and genocide at their places of origin. In Pierre's case, however, Julius does not even mention the fact that the elderly Haitian's life story covers historical events in Haiti and New York City from the eighteenth and nineteenth century that span more than a lifetime. Apparently, Pierre fled Haiti with the family he "was in the service" of during the Haitian Revolution in 1791 and came to New York City enslaved (72). Working as a hairdresser, he became eventually able to "purchase freedom for" his sister, his future wife, and himself (73). That Julius does not comment on the story that reaches him "with that peculiar sense of metamorphosis one experiences on waking up from an afternoon nap" (71), by pointing for example to its temporal inconsistencies or insertion of an excursus on Haitian history, is striking. Even more so when we note that towards the end of the novel's first part, Julius finds strange satisfaction in identifying "minor lapses" in Farouq's stories that, as Julius admits, cannot even be considered mistakes (114).

Why does Julius not mistrust a story that evidently bends historical time such as Pierre's but doubts that of Saidu and Farouq, and also questions the blamelessness of the Rwandan clubbers? Pierre's language betrays his origin in earlier times and the style of his story is reminiscent of slave narratives.[239] Even though he remains restrained in his literal reference to the history of slavery (he uses the term "service") and only briefly mentions "the terror of [Napoleon] Bonaparte and the terror of [Dutty] Boukman" (72), Pierre's story of self-emancipation without flight from his legal owner represents a more direct reference to chattel slavery and its transnational dimension in the novel. Told by a Haitian who cleans the shoes of business people in one of the largest under-

---

[239] Haensell also suggests reading Pierre's story as a slave narrative and points to stylistic similarities, "complete with the register and rhetoric of the slave narrative, but at the same time part of Julius's intradiegetic world-making" (342), whereas Isabel Soto convincingly argues that the sequence is actually a "re-writing" of the life of the historical figure Pierre Toussaint (378).

ground transport hubs in the Western hemisphere, Pierre's neo-slave narrative literally locates the transnational, ongoing influence of slavery in the twenty-first-century New York "underground catacombs" (70), lurking beneath the surface of the US American 'city of dreams.' By inserting a neo-slave narrative into the material underground landscape of today's Manhattan, *Open City* draws a map of the afterlife of slavery in New York that is buried underneath Julius's digression but resurfaces at certain instances. Following Saidu's narrative of war, flight, and incarceration, Pierre's neo-slave narrative shows how, with Sharpe's words, "[t]he ongoing state-sanctioned legal and extralegal murders of Black people" as well as their captivity form "the ground we walk on," which if acknowledged, may also provide "the ground from which we [...] attempt to speak, for instance, an 'I' or a 'we' who know, an 'I' or a 'we' who care" (*In the Wake* 7).

Yet, instead of directly acknowledging that 'ground' when confronted with Pierre, the ghostly neo-slave narrative leads directly to yet another reference to slavery and Jim Crow. When Julius leaves Pierre, he stumbles into the aftermath of a demonstration he describes in terms that recall the Draft Riots that took place in this area of Midtown Manhattan in 1863 (cf. Miller 199). Musing about the relation of past and present, a haunting sight strikes him:

> That afternoon, during which I flitted in and out of myself, when time became elastic and voices cut out of the past into the present, the heart of the city was gripped by what seemed to be a commotion from earlier time. [...] What I saw next gave me a fright: in the farther distance, beyond the listless crowd, the body of a lynched man dangling from a tree. The figure was slender, dressed from head to toe in black, reflecting no light. It soon resolved itself, however, into a less ominous thing: dark canvas sheeting on a construction scaffold, twirling in the wind. (74–75)

This trick of the eye, such that Julius's mind literally deceives him, is not only a nightmarish allusion to the anti-black violence of the Draft Riots but also more generally to slavery and the Jim Crow era with an instant and intense effect on Julius and his narration.[240] In fact, this momentary lapse into the time of slavery and Jim Crow, together with the scenes discussed above, forms a small but noticeable cluster of traces and recurrences of chattel slavery and the transatlantic slave trade during Julius's walks in the first part of the novel. While Saidu is

---

240 This is, in fact, already the second reference to the US American history of lynching in the novel. On a walking "detour" through Harlem, Julius mentions in passing "enlarged photographs of early-twentieth-century lynchings of African-Americans" that were displayed on a table next to stands of "the Senegalese cloth merchants, the young men selling bootleg DVDs, [and] the Nation of Islam stalls" (18).

locked up in a detention center and Pierre is confined to the underground, Julius seems to be caught in his narrative of evasion, haunted by the Hartmanian "ghost of slavery," that may recall the 'ghost stories' of Aminata and Sethe discussed above, who are also "still looking for an exit from the prison" (*Lose Your Mother* 133).

In the next chapter, the cluster culminates into recollections from Julius's time as a teenager in Lagos and at a military boarding school in Northern Nigeria, presumably around the time the rape would have happened (Cole, *Open City* 76–85). After evoking the Middle Passage, recounting Saidu's narrative of flight and incarceration and Pierre's neo-slave narrative, as well as observing the specter of the Draft Riots against the backdrop of today's New York, "voices" from Julius's family history now "cut out of the past into the present" (74). Memories of death, sexualized violence, militarized education, and migration in his family find their way into the narrative. Among other things, Julius tells of the early death of his father from tuberculosis, of his mother's memories of her early childhood in Germany right after the end of the Second World War, and their increasing estrangement in the immediate aftermath of his father's death (78–81). Rape is also mentioned in this fragmentary family history. Retrospectively thinking about his conversations with his mother after his father's death, Julius "surmised that [his mother's mother,] my oma, heavily pregnant, had likely been one of the countless women raped by the men of the Red Army that year in Berlin, that so extensive and thorough was that particular atrocity, she could hardly have escaped it" (80). The detailed description of harsh physical punishment at his school – Julius is "caned" and shamed publicly for alleged stealing – follows, giving him "callous self-confidence" paired with a positive reputation with "the girls" (84). Julius remains mostly incapable or unwilling to explain the impact of those memories and experiences that involve an assumed rape and allusions to the awakening of teenage sexuality at this point or later in the novel. The chapter ends with Julius explaining how he migrated to the United States after his graduation without telling his mother (84–85). Clearly, the juxtaposition of the fragments of memory encourage a reading of Julius's move from Nigeria to the United States as a flight from traumatic memories of death and sexualized violence, all of which foreshadow Moji's rape accusation, an episode that remains unmentioned for most of the novel. Nonetheless, the clustered references to slavery and captivity that have entered the text in the two preceding chapters disrupt the otherwise seemingly impersonal flow of digressive and evasive walks, musings, and conversations and enable Julius to start sharing – at least partially – some of what hitherto had been left unsaid or seemed utterly unsayable.

In the second part of the novel, chattel slavery and its relation to New York City are mentioned only twice, but much more explicitly than in the first part.

Slavery and the slave trade become more conspicuous reference points and increasingly entangled with the plot as the novel progresses towards finally addressing the suppressed rape. At the beginning of part 2, Moji appears for the first time as what seems to be yet another migrant side character of *Open City*. When she approaches Julius in a supermarket in Manhattan shortly after his return from Brussels, Julius does not recognize her as the "older sister" of Julius's "school friend Dayo" until she introduces herself and he recalls having met her "two or three times in Lagos" at Dayo's home (156). Explaining how he and Dayo had lost touch, Julius briefly mentions in passing the event that Moji would later refer to as the night of the rape: "Then there had been a party at his house, a wild one, with lots of drinking" (157). Strikingly, Julius introduces the "reencounter" by musing about two senses of the past. On the one hand he sees "a secure version of the past that I had been constructing since 1992" through "reiteration" and, on the other, an "irruptive, sense of things past" in which – like Sethe's 'rememories' – "something or someone long forgotten, some part of myself I had relegated to childhood and to Africa" suddenly reappears (156). Moji is such a "someone long forgotten," but her reappearance does not ensure, as Julius claims, "that what seemed to have vanished entirely existed once again" (156). Readers learn in the penultimate chapter of the novel that Julius's 'rememory' of Moji and his account of the teenage party are incomplete when she confronts him with her traumatic memory of the rape. While this memory has accompanied Moji every day of her life as she explains to Julius during her disclosure (244), the absence of that memory in Julius's narration seems to start haunting him in his present life in New York City.

Even though Julius seems initially to have no recollection of Moji at all and appears not to remember nor care to narrate the events of that night from his perspective – whether at this point or later in the novel – in the chapter immediately following their chance meeting, Moji's "apparition" unsettles Julius profoundly (156, cf. Haensell 335). He forgets to bring his checkbook to a meeting with his accountant and cannot remember his bank account pin number in order to withdraw the amount in cash he owes him (Cole, *Open City* 160 – 62). He hastily interprets his sudden forgetfulness as a sign of age, but his memory lapse continues to worry him; he is troubled and "awed by this unsuspected area of fragility in myself" (162). Fleeing from an in-depth engagement with this emotional agitation, Julius ventures on another walk through the Financial District of Lower Manhattan "down to Battery Park" from where he could again "see [...] the glimmering green figurine of the Statue of Liberty" (162) he had been looking at earlier when pondering the Middle Passage and 'brotherhood.' Passing a playground, he suddenly recalls that "this had been a busy mercantile part of the city in the middle of the nineteenth century" and describes how – even though "[t]rading in slaves had become a capital offense in the United States in 1820" – New York bankers had continued to profit from slavery in various ways (163). In comparison to earlier mentions of slavery and the slave trade, this digression on the historical connections between the slave trade and the New York banking business in the so-called free North is much more detailed, explicating, for example, the role Moses Taylor (1806–1882) played first as a "sugar merchant" and then "board" and president of the City Bank (163). Julius even cites the *New York Times* from 1852, which criticized the bankers' guilty involvement in the trade as "equivalent to the slave traders themselves" (163). In the first part of the novel, a culmination of indirect references to slavery and the slave trade triggers an involvement with Julius's family history, whereas now it seems to serve Julius's effort to evade telling Moji's story.

It does not seem coincidental, then, that Julius also stumbles over physical evidence of slavery in Downtown Manhattan before readers finally learn of Moji's accusation through Julius's indirect reproduction of her account. Leaving a diner two weeks after the mugging in Harlem, Julius comes across the African Burial Ground Monument, a symbol for the communal history of chattel slavery in New York City buried under today's streets of Lower Manhattan. As Julius explains, "in the seventeenth and eighteenth centuries" what now appears as a "small patch of grass" closed "for renovation" measured "some six acres" (220). "Into this earth had been interred the bodies of some fifteen to twenty thousand blacks, most of them slaves, but then the land had been built over and the people of the city had forgotten it was a burial ground" (220). Most of the burial ground, just like Pierre's underground neo-slave narrative, "was now under office buildings, shops, streets, diners, pharmacies, all the endless hum of quotidian commerce and government" (220).

After glimpsing the economic entanglement of the banking industry with slavery at a playground in the Financial District, standing at a spot that memorializes the burial ground with a monument, Julius feels "steeped in [...] the echo across centuries, of slavery in New York" (221). He ponders the "evidence of suffering" of the "excavated bodies" in form of "blunt trauma" and "broken bones" as well as the "hints of African religions" scholars had found in the graves in form of "shells, beads, and polished stones" (221). In his typical digressive fashion, Julius also elaborates on the problem of "cadaver" theft at the time, quoting "a petition by free blacks in defense of their dead" that eventually ensured that "the buried bodies of innocent blacks" were "left in peace and neglect" (221).[241]

---

[241] For a brief discussion of the "cadaver trade" beyond the slave trade from a historical perspective, see Berry.

Concerned about the difficulty of conceiving of the buried as "truly people" (222), Julius thus addresses not only physical traces of anti-black violence but also signs that attest to the ways in which Black people refused to be stripped of their communal histories, cultural identities, and a sense of themselves as humans by caring, in Sharpe's words, for the "dead and dying" (*In the Wake* 10, 12–13).[242] The "beads" in the excavated graves (Cole, *Open City* 221), of course, form an intertextual reference to the afore-cited vivid description of a burial ceremony that Hill's protagonist Aminata stumbles across when arriving as a fugitive at the outskirts of New York City at the outset of the American Revolutionary War (272). Rebekah Cumpsty argues that Julius "reinvest[s] the city with its dead with whom he identifies as part of the Black Atlantic" (310) in her analysis of walking as a "ritual" that conjures the dead and "Yoruba cosmology" in an otherwise "secular" environment in the novel. Walking and narrating other people's stories are Julius's attempt at caring when he stumbles over remnants of atrocities, like the African Burial Ground.[243]

Significantly, the scene in which the teenagers violently mug Julius after greeting him in seeming solidarity closely precedes this last reference to slavery in the novel. Indeed, Julius had been pondering the attack and its bodily and psychological repercussions in the diner from where he had left to visit the site. Cole connects the two scenes even more directly when Julius ignores the "cordon" to the monument, steps onto the ground and lifts a stone from the grass. At that exact moment, an injury that persisted from the beating causes "pain [to] sho[o]t through the back of my left hand" (Cole, *Open City* 222). Put into a larger historical context, the pain functions as a physical reminder of Julius's own bodily implication in the complex history of the transatlantic slave

---

[242] Sharpe asks, "What does it mean to defend the dead? To tend to the Black dead and dying: to tend to the Black person, to Black people, always living in the push toward our death?" She suggests "a method of encountering a past that is not past [...] along the lines of a sitting with, a gathering, and a tracking of phenomena that disproportionately and devastatingly affect Black peoples any and everywhere we are" (*In the Wake* 10). I propose to understand the historical remnants of African burial ceremonies in eighteenth-century New York, the activism by Black New Yorkers in the 1990s for the African Burial Ground National Monument (Willoughby) as well as Cole's and Hill's literary fashioning of these histories as signs of care 'in the wake.'
[243] Concern for inadequate care in the wake of death is also expressed in George Jackson's prison letter collection *Blood in My Eye*, whose title and respective epigraph cites Jackson's parents exclaiming "blood in my eye" when they could not find their second son Jonathan's gravestone (cf. G. Thomas, "'Neo-Slave Narratives'" 227). (Joy James attributes the quote to Jackson's mother and explains that the first issue of *Blood in My Eye* cites his father, while the second does not give any source ["Introduction: Violations" 11, 16]).

trade, chattel slavery, and their afterlives as victim, witness, and perpetrator (cf. Li 79–80).[244]

Afterwards, another of Julius's memories from Nigeria follows this textual strand about slavery in New York City. Memories around his "father's burial in May 1989" (223) in the next chapter not only appear as a continuation of the memories of Julius's teenage years that were recounted in chapter 6; they also seem to evolve naturally from the visit to the African Burial Ground Monument. Sitting in the "movable catacombs" (7) of the New York subway on his way to work, Julius remembers that it is the anniversary of his father's burial, the memory of which had been "complicated" with famous "depictions of burials" and death over the years (228). The associative connection between Julius's chancing upon the historical burial ground and his father's burial manifests itself when Julius imagines his "father with coins on his eyes, and a solemn boatman collecting them from him, and granting him passage" (228). This quote echoes not only canonized representations of burials and death but also the "bodies […] found with coins over their eyes" at the burial ground Julius pondered earlier (221). It may also recall fugitives' passage on boat from the enslaving South to the presumed free North and ultimately also the Middle Passage as addressed, for example, in *Beloved*. Fugitive memories from Julius's personal past and the (hi)stories of the transatlantic slave trade and chattel slavery follow each other so closely to the point where they intermingle.[245]

Withheld from most of the narrative, and foreshadowed through various strategies of evasion, Julius finally narrates the rape accusation in the following penultimate chapter of the novel. He begins, however, by recounting his long stroll to the party Moji had invited him to at her boyfriend's apartment (231) and then departs from the chronology of the evening to describe how he leaves the party in the early hours of the morning for his long walk home (241–43). Withholding to the limit Moji's disclosure, as a transition between the walk

---

244 For an alternative reading of this scene, see Haensell, who interprets Julius feeling the pain in his hand as "rooting him firmly, instantly, in the present" (339–40).

245 In her discussion of the representation of Brussels in *Open City* and *Heart of Darkness* by Joseph Conrad (1899), Elisabeth Bekers notes how Cole's description of the subway cars as cramped "movable catacombs" in which people travel underground "also evoke the conditions of the slave-packed ships that supported Europe's trans-Atlantic trade triangle" (359). Yet, practicing what Michael Rothberg describes as "multidirectional memory," Cole also locates slavery alongside the Holocaust in the underground and in close connection to Julius's family history. Riding the subway on the day of his father's burial anniversary, the subway car's ventilation system reminds Julius of the "final terrible moments in the camps, moments that no one has survived to give first-hand testimony of, when the Zyklon B was switched on and all the human captives breathed in their deaths" (Cole, *Open City* 229; cf. Sollors 242).

and the accusation scene, Julius muses again extensively about the mind and the role of male perpetrators of violence:

> Each person must, on some level, take himself as the calibration point for normalcy, must assume that the room of his own mind is not, cannot be, entirely opaque to him. Perhaps this is what we mean by sanity: that, whatever our self-admitted eccentricities might be, we are not the villains of our own stories. (243)

Here Julius gives a rather different account of the workings of the mind than that posited just a few pages earlier when he had expressed a critical awareness of deceiving minds and blind spots, ophthalmic and mental (238–39). Julius seems to prepare himself and the readers for what follows.

After these prologue-like remarks, Julius retells how Moji confronts him with the rape accusation on the balcony of her boyfriend's apartment overlooking the Hudson River at the end of the party. In this "non-account of Moji's rape" (Clark 183), Julius describes how Moji said that he "had forced [himself] on her" while she was almost unconscious from drinking during the party that "her brother Dayo had hosted" and explains that she "still carried this hurt" today (Cole, *Open City* 244). The confrontation is mostly mediated through Julius's narrative voice who refers to Moji in third-person and to himself in first person and only briefly slips into a first-person monologue from Moji's perspective. Both before and after mediating Moji's accusations, Julius does not share his own point of view on the matter. Thus, the degree to which Moji truly "reverses Julius's penetrating, analyzing gaze," "pierces the psychiatrist's presumed omniscience" (Edwards 677), and thereby "displaces the primacy of Julius's voice" (Neumann and Kappel 51), as several critics have suggested, remains unclear, not least due to the complex mediation that disallows readers to identify with either Moji or Julius (Goyal, *Runaway Genres* ch. 5). Nonetheless, Moji anticipates that Julius will not say anything in response and seems to understand well this strategy of forgetting or suppressing painful memories of his past in Nigeria. She directly points to the fault at the core of this strategy, which the novel at large also exposes: "Things don't go away just because you choose to forget them" (Cole, *Open City* 245). For both the rape and slavery remain present absences or, rather, fugitive presences in the novel, forming the ground on which Julius walks obsessively and narrates evasively. Even though the narrator tries to suppress traumatic memories of his past in Nigeria and the larger past of the Black Atlantic, their irruptive re-appearances ultimately unsettle the deceptively self-possessed surface of Julius's narrative and reveal the flight from those memories as the narrative's driving force.

The unnamed narrator of Cole's first work of fiction *Every Day Is for the Thief* – who shares many characteristics with *Open City*'s Julius such that they could actually be one and the same protagonist (Suhr-Sytsma 345) – describes his personal past and that of Nigeria as "perhaps [...] connected, the way the small segment of a coastline is formed with the same logic that makes the shape of the continental shelf" (*Every Day* 145). This simile also seems to describe the ways in which Julius, his relationship to Moji, and his narration are all caught up in the gendered history and anti-black legacy of the transatlantic slave trade and chattel slavery. Examining *Open City* for its enslavist absences and hauntings as well as its narrative strategies that aim to divert and evade, the novel clearly enacts a constant flight from Julius's personal past and from the anti-black world he is implicated in. The narrator's endless walks and musings fragment the plot and thrust Julius's intellectual narrative voice to the forefront, pushing the many loose textual strands and his obscure history out of the narrative focus. The walks and musings distract from Julius's conflicted past and frame his German Nigerian migrant identity as an intellectual cosmopolitan, or Afropolitan in the United States.

However, being middle-class, well educated, and well-travelled does not allow Julius to wander the metropolis peacefully. Recollections of slavery, the slave trade, (forced) migration, and incarceration as well as sexualized violence, as in the stories of Pierre, Saidu, and Moji, keep haunting Julius. They temporarily interrupt the elusive flow of Julius's erudite, seemingly detached, narrative voice. What links those fragments is the anti-black gender violence of the Black border and Julius's attempt to avoid addressing or acknowledging this common ground. Framing Julius as a fugitive reveals his excessive cosmopolitan musings as narrative strategies designed to distract from his own past and the diaspora he is inevitably a part of. He struggles to flee from the legacy of slavery and from his own implication in the violence attached to these legacies. While Stephen Miller might be right to claim that Julius "is never sure what story is his story: the slave story or the immigrant story" (200), the novel shows that both are simultaneously inhabited by Julius and that they are inextricably related with each other and the gendered history of anti-black violence, captivity, and flight.

Reading the text in the context of an anti-black world in the gendered afterlife of slavery also shows how the novel questions the very concepts it invokes. As Goyal notes, "*Open City* models the insufficiency of older languages – those of psychotherapy for trauma, the arts as generating empathy, cosmopolitan exchange as a response to difference, racial and diasporic solidarity, the enigmas of immigrant arrival" (*Runaway Genres* ch. 5). Through its nuanced negotiation of the diversity of contemporary Black communities and their histories in New York

City, the novel questions any monolithic notions of the African diaspora, solidarity, and racialized and gendered identities, if not the concepts as such. *Open City* does not address the surveillance and policing of the movement of Black people, specifically Black men in public spheres as explicitly as Garnette Cadogan does by illustrating the impossibility of the "Romantic experience of walking alone [while black]" in US cities from a Black Caribbean immigrant perspective (142). However, the narrative technique of flight nonetheless questions the usefulness of concepts such as flânerie and cosmopolitanism in the face of the Black border.

Ultimately, *Open City* meditates on the narratibility of the legacy of gendered anti-black violence in the long-standing tradition of neo-slave narratives from the perspective of the supposed post-black era (Edwards 670–73) through the mediation of some and silencing of other gendered and racialized narrative voices that address the abjection they are subjected to and impose on others. Cole thus also raises questions about fugitives' inescapable complicity in what they flee from and the awareness that there might be no real end to their "perpetual migration" (Hill, *Someone* 401). While Julius ultimately performs his own twisted form of care 'for the dead and dying' by 'rememorying' fragments of his family history and the history of the Black diaspora in *Open City*, Moji is left painfully alone with her testimony of rape. She remains literally uncared for, not to mention the absolute silence about African American women in this novel (cf. Li 79). As Gamso concludes, Moji "is the site par excellence of storytelling, the terra nullius for the enactment, the becoming of an idea: chauvinism, development, exception" (74). *Open City* thus also illustrates how anti-black violence specifically affects Black women such as Moji or the seemingly nameless and origin-less cleaner from Brussels and the ways in which their struggles to survive and refuse violence is structurally silenced, while "disallowing empathy as the primary response to suffering" (Goyal, *Runaway Genres* ch. 5). In fact, their stories of flight, refusal, and survival prove to be even more absent from *Open City* than Nigeria, Julius's personal history, and the history of slavery and the slave trade. Theirs, Cole seems to suggest, are the ultimate fugitive stories of *Open City*, the traces of which may be too fragmented to be reconstructed but deserve to be carefully uncovered all the same.

Fugitive insights of narratives of slavery and flight

In her analysis of "fictions of the new African diaspora," Goyal argues that measuring *Open City* "against the elements of the slave narrative helps read its generic singularity" which "is not to propose that *Open City* has anything in common with a neo-slave narrative" (*Runaway Genres* ch. 5). Goyal supports this claim by pointing out that

> [...] Julius would prefer not to acknowledge any call to share on the basis of race or nationality, displays no sentimentalism, offers no jeremiad or ethnography. His constant travel is neither difficult nor especially interesting. Rather, the novel's treatment of slavery and allied forms of historical violence helps clarify its distinctiveness and its peculiar negotiation of exceptionalism and exemplarity, its status as an African book, an American novel, a black Atlantic travel narrative, a quintessentially global novel. Reading its formal turns – through the genres of the imperial romance, the black Atlantic travel narrative, the modernist bildungsroman or kunstlerroman, a reverse heart-of-darkness impressionist tale, an anticolonial antiromance, and modernist flânerie – further illuminates key aspects of contemporary revisitings of slavery. (*Runaway Genres* ch. 5)

Thus according to Goyal, *Open City*'s "generic singularity" may set the novel apart not only from its Afropolitan counterparts but also from the neo-slave narratives, Black Power autobiographies, and slave narratives discussed in this study. After all, the novel's narrative mode is "more akin to that of a diary or internal monologue than of memoir or autobiography" (Haensell 333). While I agree with Goyal's observations about the novel's broad genre mixing and Julius's attempts to dissociate from others and disallow sentimentalism, this study's focus on fugitivity and captivity reveals how *Open City* has in fact much in common with neo-slave narratives, slave narratives, and Black Power autobiographies and is transversely related and indebted to those writing traditions.

As the preceding close reading of *Open City* showed, with Pierre's story Cole inserts a neo-slave narrative fragment into the novel that draws on the tradition of autodiegetic neo-slave narratives, such as *Someone*. While overtly a more traditional realist novel like Hill's, not least with the anachronisms of Pierre's story and the novel's narrator, *Open City* also experiments with narrative fragmentation and mediation and questions the narratability of slavery, flight, and trauma, like *Beloved*. Whereas Pierre's ghostly appearance in the underground of twenty-first century New York City may recall Beloved's fleshly return in Morrison's novel, *Open City*'s unreliable narrator Julius uses stubborn oversharing and evasion to cope with the history of slavery and his own personal story in contrast to Beloved's collapse of narrative when confronted with her 'rememories' of the Middle Passage. However, Julius also refuses to adopt a self-referential, consciously autobiographical voice, such as the one we encountered in the close reading of *Someone* with its enthusiastic fictional autobiographer, witness, and storyteller Aminata. All in all, *Open City* uses and refuses generic fragments and narrative styles that speak to Rushdy's neo-slave narrative types of the historical novel, the "pseudo-autobiographical slave narrative," and "the novel of remembered generations" ("Neo-Slave Narrative" 90). *Open City* forms a "palimpsest narrative" that broadens Rushdy's notion of familial remembered generations (*Remembering Generations* 8–9) to a twenty-first century perspective on remembered and suppressed transatlantic familial and communal histories across the Atlantic from the colonial era until today.

Oversharing other people's stories in highly mediated form, while avoiding detailed introspection with respect to his family and Moji, Julius's narrative voice gives a disturbing twist to African American political autobiographies' and slave narratives' emphasis on the community over the individual that often goes hand in hand with a lack of introspection (see ch. 2.1. and 2.2.; cf. Perkins, ch. 1). Julius's narrative restraint with respect to important incidents in his life, such as the fallout with his mother and his rape of Moji, also recalls in an unsettling way slave narrative writers/narrators' and Black Power autobiographers' strategic silences, yet without the clear political message of the autobiographical texts. Middle-class male twenty-first century Black diasporic subjects, Julius seems to suggest, have reached a position of privileged mobility and seeming narrative power from which they can venture on limitless travels and walks and speak of, to, and about whatever and whoever they choose. As my analysis of the novel showed, however, *Open City* insists that Julius's assumed freedom of movement and narration remains a fallacy in the twenty-first century – not only if we look at the less privileged position of Black African migrant women, such as Moji and the cleaner, as well as Black African refugees, such as Saidu, but also when we closely consider Julius's fugitive narrative strategies. The sexualized anti-black violence against Black women pervasive in slave narratives, political autobiographies of the Black Power movement, and neo-slave narratives re-appears in *Open City* as that which is evaded and needs to be exposed all over again.[246] Not figuring as transgressions in an anti-black world, the violence of slavery is connected, conflated, and buried together with the violence committed against Black female bodies in *Open City*'s narrative underground. Julius's narration mostly conceals this violent base and only partially excavates its remnants as a form of care "in the wake" (Sharpe, *In the Wake*). Thus, *Open City*'s fugitive narration subtly negotiates – similar to Black Power autobiographies, such as Shakur's, and neo-slave narratives, such as Morrison's – the (im)possibility of autonarration and the mediation, appropriation, and violation of Black

---

[246] See Douglass, *Narrative*; H. Jacobs; Hill, *Someone*; Morrison; Shakur, *Assata* as discussed in this study, but also, e.g., Butler; Cleaver; Gyasi; G. Jones, *Corregidora*.

gendered voices and bodies, reflecting on the narratibility of the Black gendered self.

After novels that focus more exclusively on the historical time of slavery and its immediate aftermath have already fundamentally questioned the plottability of slavery and its anti-black afterlives, Cole's novel hurls these long-standing interrogations of 'perpetual migration' across Europe, Africa, and North America into our twenty-first century setting. As a post-9/11 novel, *Open City* shows how metropolitan cities, such as New York, have been and continue to be 'open' to the 'permutations' of enslavist, anti-black violence that may be hidden behind male-centered notions of cosmopolitanism. While *Open City* was published during the first term of the Obama presidency, frequently celebrated as the inauguration of a post-racial era, the novel seems to follow the argument – against Julius's efforts to suggest otherwise – that structural and interpersonal violence against those not considered fully 'Human' since the transatlantic slave trade "do[es] not go away just because [we] choose to forget them." Read in this way, *Open City* indirectly reflects the emergence of a counter-discourse to post-racialism and a refocusing on the pervasiveness of anti-black violence during Obama's second term, when the Black Lives Matter movement and Afro-pessimism also emerged. Oscillating between migrant novel and neo-slave narrative that also uses narrative strategies of slave narratives and Black Power autobiographies, among others, *Open City* thus draws our attention again to the borders of migration in US Latinx studies, the Black border I proposed as a concept in Afro-pessimism, *and* the theories of Black feminist fugitivity examined in the first part of this study. It reiterates in a restrained form what the other narratives examined in this study have suggested more openly for centuries – that the Black border between social death and the 'Human' world established in the historical era of slavery remains erect in the twenty-first century, fencing off the anti-black, white supremacist, settler colonialist, ableist, and patriarchal civil society that slavery made.

Fugitive Conclusions or the Inescapability of Captivity, Flight, and Fugitive Narration

At the beginning of this study, I asked 'Afro-pessimistically,' what if the social death that the transatlantic slave trade and chattel slavery brought over people racialized as Black has been never-ending in North America and beyond. And if so, in what ways have literature, theory, and cultural analysis been equipped to account for Black social life that endured and continues to endure social death? Similar to Crawford's notion of "post-neo-slave narratives," this study uncovered how *Open City* and the other narratives analyzed in its three analytic chapters dwell on the anti-black "architecture of the unknown," test the limits of different literary forms to narrate stories of Black enslavement, captivity, and flight, and thereby ask readers to adopt a "counter-literacy" that may critically unpack the narratives and the world they speak to ("Post-Neo-Slave Narrative" 72, 71). With the help of state-of-the-art concepts of captivity and fugitivity in Afro-pessimist and Black feminist theory and Bal's cultural analysis as theoretical framework and methodological approach elaborated in part 1, this study followed the conceptual 'travels' of confinement and flight through three major Black writing traditions in North America from the 1840s to the early twenty-first century. It revealed how Afro-pessimist and Black feminist theories together significantly illuminate the narratives' conceptualizations of captivity and fugitivity towards an elsewhere and 'else when.' The study also showed how the autobiographical and fictional archive's centuries-old intertextual engagement with the seeming paradox of social life of the socially dead adds further complexity to what I have called the Black border in Afro-pessimist theory between those positioned as fully 'Human' and those racialized as Black outside of the 'Human' realm.

Summarizing the various forms of captivity and fugitivity that were identified in the narratives helps illuminate their contributions to a theorization of captivity and fugitivity as long-standing and ongoing conditions and practices of Black social life in social death. It also serves to evaluate the results of this study and its relevance for (German) North American studies, African American studies, and Black studies more broadly. As this study showed, all narratives examined in this study describe captivity and fugitivity not as singular events but as reiterative or ongoing experiences and practices, questioning the possibility of ultimate arrival and homemaking at a place of unconfined freedom. As will be summarized in the following, the forms of captivity and flight that create the notion of the proliferation and perpetuity of confinement and flight in the narratives can be described as (1) literal or physical, (2) ambiguous or unconscious, and (3) mental or spiritual, narrativized through specific literary strategies.

Firstly, all narratives that were examined involve confinement and flight in the most literal sense – not least, because they were selected as part of this study's corpus specifically for their captivity and flight plots. On the one hand, legal slavery as a form of captivity in enslaving territory in the historical era of slavery plays a central role in all of the examined slave and neo-slave narratives, just as incarceration in prisons and jails appears in all Black Power autobiographies. However, imprisonment can also be found on the margins of some slave and neo-slave narrative plots, pointing to the close relation between these forms of captivity. For example, in Harriet Jacobs's *Incidents* Linda's enslaved family is imprisoned to force her out of her hiding (*Incidents* 81–82). In Toni Morrison's *Beloved*, Sethe and Denver are jailed after the infanticide, whereas Paul D is held in a box in the ground as part of a chain gang after his escape attempt; and in *Open City*, Saidu is imprisoned in a detention center for illegalized immigration after fleeing from war in Liberia.

On the other hand, literal forms of flight refer to conscious, physical escape from captivity in the narratives. They involve movement across borders of estates, states, and nations, away from a place of captivity to a place assumed less confining. In almost all slave and neo-slave narratives under scrutiny literal flight takes place from enslaving to non-enslaving territories in the historical era of slavery, i.e., from enslaving southern states to non-enslaving northern states as well as Canada (see, e.g., the literal flights of Raven, 40s, Stray, Aminata, Sethe, and Paul D).[247] War also appears as a reason for escape for characters, such as Paul D and Saidu, but less frequently and only at the margins of the main plots of neo-slave narratives. In Black Power autobiographies, fleeing from the FBI, the police, and from prison into the underground or exile represents a central part of the plot whose details the narratives tend to avoid – just as the slave narratives.

Secondly, the close readings revealed that all of the narratives involve forms of escape and confinement that may be physical or literal but the characters and/or narrators do not explicitly describe or conceive of them as such – at least not initially. Ambiguous confinement includes places of refuge after literal escape from legal owners, prisons, and the police, that frequently take on confining characteristics, such as isolation or confinement in hideouts. Examples in the narratives are the underground and exile in Davis's and Shakur's autobiographies, the house on Bluestone Road for Sethe and Denver in Morrison's

---

[247] Cole's novel is an exception here because in its neo-slave fragment the enslaved Pierre flees from Haiti to New York City with his legal owner and buys his loved ones and himself free before leaving his owner's family.

novel, and Emancipation City and Canada for Reed's protagonists. Moreover, Black Power autobiographers, such as Jackson and Shakur, also describe growing up in poor Black urban communities in the 1950s and 60s as confining. They emphasize, as Davis does, that the reach of physical imprisonment extends into the prisoners' non-imprisoned communities, making the case for understanding life in poor Black urban communities in the twentieth century as a form of neo-slavery, not least with the tropes of the prison as a plantation and society as a prison. Thus, ambiguous forms of confinement in the narratives refer to the proliferation of what Christina Sharpe and Frank Wilderson have described as "the hold of the Middle Passage" during the historical era of slavery, which mutates after emancipation into the holding cell of prisons and chain gangs, as well as the constraints of poverty (Wilderson, *Red, White, and Black* 315; see also Sharpe, *In the Wake*).

Apart from ambiguous forms of captivity, the analyses also demonstrated the narratives' conceptualization of ambiguous forms of flight. Like its literal counterpart, ambiguous flight describes movement away from a place of literal or ambiguous confinement but without the awareness or clear articulation of that movement as escape. Onwards movement after the literal escape from captivity occurs throughout the writing traditions and can be understood, as I have argued, as an unconscious form of flight. The many travels, migrations, and resettlements of characters, such as Josiah, Frederick, Linda, Aminata, and Raven, between enslaving and non-enslaving territory, Canada, the UK, and West Africa after their initial escapes are driven by an ongoing search for true safety, a home, family, community, livability, and unconfined freedom projected onto different places and into an unknown future. In this way, especially slave narratives and neo-narratives emphasize the perpetuity of escape, what Aminata calls euphemistically "perpetual migration" (Hill, *Someone* 401). They also expose the places available to fugitives outside of enslaving territory during the historical era of slavery as an "improvement" in their "condition" or "comparative freedom" at best (H. Jacobs 156; Douglass, *My Bondage*; cf. N. Roberts 71–73) and as places of refuge vulnerable to anti-black captivity at worst, such as in Reed's construction of nineteenth-century Canada. Even frequent, seemingly banal movement through cities, such as New York and Chicago, in the twentieth and twenty-first centuries proved in the narrative analyses as escapes from the dangers of classed and gendered anti-blackness and their memories. Examples can be found not only in Julius's move from Nigeria to the United States, his cosmopolitan travel to Brussels, and his aimless walks in New York City, but also in the ways in which Assata and George in their childhood and teenage years run away from the confinements of their family homes onto policed streets. As the close readings made clear, narratives of captivity and fugitivity particularly profit from a critical reading lens that focusses on both anti-black structures and performances of refusal and escape because it can unmask confining structures even in legal freedom as forms of captivity, and movements away from confined forms of freedom as flight.

Thirdly, alongside literal and ambiguous or unconscious forms of captivity and flight, the analyses in part 2 also identified experiences of confinement and performances of flight without or with only very little physical enclosure or movement. Mental or spiritual forms of flight are often performed in the context of the literal captivity of slavery or the prison in the examined narratives. In the Black Power autobiographies, Eva's "astro-space projection" as explained to fellow prisoner Assata (Shakur, *Assata* 59), Jackson's revolutionary and fugitive letter writing praxis in his prison cell as "thought on the run" (Koerner 166), and Angela's mental conjuring of communities of fugitives and political prisoners to fight the isolation of her jail cell belong to this category. Mental and spiritual escape in the narratives is also central to life after literal escape that may take on confining characteristics. In the neo-slave narrative *Beloved*, the preacher Baby Suggs provides "Black care" as a spiritual form of escape to the local community in the clearing (C. Warren, "Black Care" 44–46; G. Thomas, "'Neo-Slave Narratives'" 227). In *Flight to Canada*, the different strategies of Raven, Stray, and 40s after their successful literal escapes from the plantation – i.e., activism and art, commercialization of the sexualized Black body, and militant eremitism – together with Robin's coup to take over the plantation from within represent memorable examples of escape as a mindset that struggles with confining features as well. Similarly, after the infanticide and the appearance of the baby ghost, Denver's temporary loss of hearing and speech as well as her frequent retreats into the close-by woods while growing up isolated in the house on Bluestone Road also oscillate between inner forms of escape and confinement. Thus, inner fugitivity describes mental or spiritual strategies of flight that help characters endure ongoing forms of confinement (literal, ambiguous, and mental), not least through retreat or self- and intramural care "in the wake" (Sharpe, *In the Wake*).

The close readings also demonstrated that most forms of literal escape first begin as an inner process that may later be transformed into action. In Douglass's slave narrative, Frederick learns to read and write and then physically fights with the plantation owner Covey in order to reach the mindset he needs to attempt to escape physically as well. In Shakur's autobiography, her grandmother's dream of Assata's escape appears to not only announce but also miraculously bring about her later prison escape/liberation, whereas in *Beloved* Sixo frequently steals himself away from the plantation to meet his girlfriend, before learning about the North Star and the Underground Railroad and plotting their escape. In Morrison's novel, the complex relation between different forms of captivity and flight also becomes clear in Hi Man's work as the leader of the chain gang. As Christina Sharpe has suggested and I discussed in the novel's close reading in some detail, Hi Man's calls redirect anti-black violence – from the morning rapes to the day's forced prison labor, from forced prison labor to the nighttime imprisonment in a small box in the ground, and from imprisonment to rape in the morning again. His "ordinary note[s] of care" both allow the prisoners to escape one type of violence temporarily in exchange for another and thereby help them endure and communally survive the violence until an opportunity for joint literal escape occurs (Sharpe, *In the Wake* 132).

As these selected examples from the analytic chapters already illustrate, the different forms of captivity and flight (literal, ambiguous/unconscious, mental/spiritual) are necessarily interconnected which is why the suggested categorizations remain necessarily tentative and provisional. Moreover, these examples also point to the fugitives' reoccurring struggle with the (im)possibility of care for themselves and others in the examined narratives. This can also be seen in Linda's escapes and confinements in Jacobs's slave narrative. Linda initially tries to escape from her legal owner's enslavist reach into a relationship with her white neighbor. When this does not ensure her protection, she hides in an attic space located at the center of the enslaving system at risk of invasion, but hidden from her owner's view for seven years before fleeing to the northern states, where she still feels confined and at risk. Her escape on a boat to the 'free North' clearly represents a literal form of escape. Her relationship with Sands and her hiding in the garret, however, move between literal and mental forms of flight as well as ambiguous and literal, life-threatening forms of confinement in the name of self-protection and familial care, epitomized in Jacobs's "loophole of retreat."

Together with the (im)possibility of care, death and death-boundedness thus represents another related trope around which different forms of captivity and flight revolve in several of the narratives I analyzed in this study. Linda risks her life self-confining in an attic in order not to abandon her enslaved children, while George Jackson and his brother Jonathan lose their lives in open revolt, refusing the literal, ambiguous, and mental captivity of late-twentieth-century US neo-slavery. In Hill's neo-slave narrative, death represents familial and communal loss but also appears to be the only place left for Aminata where unconfined freedom may still lie or where she may at least find her final rest after "a life of losses" and perpetual escapes (Hill, *Someone* 469). Sethe, too, envisions the killing of her children and herself as a form of care and a last resort, with death figuring again as a potential place of refuge that may save her children from the ongoing social death of reenslavement, but which might – as *Beloved*'s fragmented 'rememories' seem to imply – also hold the ongoing captivity of the Middle Passage for them.

Last but not least, my analyses of the different writing traditions also focused on the ways in which the texts exhibit specific narrative strategies that address the various forms of captivity and fugitivity and their interrelations with death, death-boundedness, and the (im)possibility of care not only on the level of characters and plot but also in their narrative form. These strategies involve, for example, purposeful ellipses around literal escapes or time underground in the plots of all of the autobiographical texts due to extratextual reasons of safety. In contrast to fiction, in autobiographical narratives strategic ellipses serve to protect people involved in the escape, including the narrators/writers themselves, or to keep escape routes safe for other fugitives to use. They also recall the complex relation between writer/narrator, editor, and intended as well as potentially hostile readerships and question unconfined forms of freedom through extratextual events, such as Shakur's ongoing isolation in exile and Jackson's violent death in prison. In neo-slave narratives, such as *Beloved*, *Flight to Canada*, and *Open City*, ellipses together with evasion, allusion, mediation, and fragmentation frequently replace direct, singular narration of experiences of captivity and flight. Indirectly responding to the autobiographical writing traditions and their clear political messages, such neo-slave narratives challenge the narratibility of Black captivity, flight, and freedom.

The close readings also underscored the relevance of deemphasizing the narrating I in the majority of the narratives. Especially female-authored slave narratives and Black Power autobiographies frequently focus on the pervasiveness of experiences of captivity in the African American community at large and put the narrators' individual gendered experiences of enslavement and imprisonment into the service of the Black liberation movements of the time, the metaphorical lowercasing of the autobiographical I that *Assata* takes literally. Slave narratives, such as *Incidents*, also use narrative camouflaging as fugitive narrative strategies to bring their political messages across disguised in familiar narrative forms, such as sentimentalism and the gothic. Questioning the position of the narrating I and experimenting with genre conventions also plays an important role in novels, such as *Beloved*, *Flight to Canada*, and *Open City*. Their non-linear, circular, and fragmented narrations together with unreliable, changing narrative perspectives and focalizations meditate on the possibilities of remembering and narrating experiences of slavery from the position of the formerly enslaved and fugitive – a position which Afro-pessimist and Black feminist fugitive theory also interrogates.

Like Hartman's notion of the "afterlife of slavery" and Warren's "Black Time," all of the narratives also systematically destabilize linear concepts of time and place. Even though literal escape in slave narratives included in this study occurs only once, Douglass and Henson were compelled to rewrite and revise that initial escape throughout their lives. Henson's and Douglass's reiterations in several autobiographies but also Reed's intertextual approach to different writing traditions, histories, and discourses point to the ways in which narrating experiences of slavery, escape, and arrival elude a linear, single-stranded story of closed events. Similarly, Paul D's and Sethe's initial escape stories are told and retold in Morrison's neo-slave narrative in a fragmented way, suspending linear and coherent concepts of time and place which is further reinforced by Paul D oscillating for many years between literal captivity and literal as well as ambiguous flight during and after the Civil War. Episodic, fragmented, circular, and highly mediated narration, such as can be found in *Open City*, *Someone*, and *Flight to Canada*, seems to lend itself to narratives of captivity and flight, with the narrative macrostructure reflecting and reenacting the characters' experiences of the proliferation of confinement and flight across geographies and through time. What is more, Black Power autobiographers' tropes of neo-slavery, the prison as plantation, and society as prison also address the historical connection between captivity in the historical time of slavery and imprisonment and impoverishment of Black communities in the twentieth century that blurs the lines between the inside and outside of prisons and the time before, during, and after captivity. Similarly, repetitions, anachronisms, and satire in slave narratives and neo-slave narratives also illuminate the relation and overlap of past and present narratives of slavery and other forms of confinement, as well as today's forms of racialized imprisonment and migration.

As this brief summary of the results of the analytic chapters grouped around different forms of captivity and flight and their narrative strategies illustrates, a focus on anti-blackness and the workings of the Black border and fugitivity in literary analyses of African American and Black diasporic literature proves fruitful. After all, the close readings illuminated the many ways in which narratives of literal flight and confinement are infused with many more interrelated forms of captivity and fugitivity that refuse clear definition as well as linear and coherent temporal and spatial placement in the narratives. The analyses of anti-black structures of captivity in the narratives allowed for a reading against the narratives' typical drive for plot closure and coherence in their construction of time and place that revealed the pervasiveness and proliferation of confinement and death. Combined with a focus on flight, this study's focus also allowed for an appreciation of the momentum of enduring social life in social death, reflected not least in forms of care for the fugitive self and others. As part 2 showed, to Afro-pessimistically observe the structural make-up of US civil society and the modern world at large reflected in Black North American literature does not diminish the importance of past and present liberation movements, such as abolitionism, Black Power, and Black Lives Matter. On the contrary – when considered through an Afro-pessimist *and* Black feminist fugitive lens, narratives of captivity and fugitivity also highlight the dire need for and the long history of Black intramural care and performances of refusal, not least found in and enacted through literature. In this way, the study demonstrated how fugitivity as a concept offers perspectives that underline the structural impossibility of Black life across the Black border, while adding further nuance to the rigid equation of Blackness as only and always already captive "Slaveness" (Wilderson, *Red, White, and Black* 52). It does so without prematurely assuming unambiguous agency, resistance, and subjectivity, neither in literal captivity nor in legal freedom. Instead, the close readings of this study recognized performances of Black fugitive life "in the wake" (Sharpe, *In the Wake*) that have occurred and been reflected upon outside of civil society for hundreds of years, dreaming beyond the current "afterlife of slavery" (Hartman, *Lose Your Mother*) towards a time and place that may offer the seeds of "otherwise worlds" (King, Navarro, and Smith; cf. Crawley).

With its analyses of a selection of Black North American narratives of captivity and flight, this study has pointed to several areas for further research. Of course, there is a rich archive of North American slave narratives, neo-slave narratives, and Black political autobiographies that exceeded the scope of this study, but whose engagement with captivity and flight on the level of content and form deserves further attention. Some examples were mentioned in this study in passing (see, e.g., Abu-Jamal; Butler, W. W. Brown, *Narrative*; Cleaver; E. Jones; G. Jones; C. Johnson; X). Analyzing forms of captivity and flight in late nineteenth- and early twentieth-century African American autobiographies and fiction also seems promising (see, e.g., Bontemps, Ellison, Herndon, Petry, R. Wright). Captivity and flight in other Black diasporic genres, such as poetry and drama deserve attention as well, just as a more detailed examination of recent Black North American contributions to the writing traditions discussed here (see, e.g., Edugyan, Gyasi, Whitehead, Woodfox). Moreover, casting a more detailed look at specifically Black Canadian, Nigerian American, or Afropolitan narratives of captivity and flight which this study touched upon with the close readings of Henson's, Hill's, and Cole's writing seems also more than worthwhile, especially with respect to their localized histories of Blackness, anti-blackness, confinement, and flight (see, e.g., Adichie, Edugyan, Gyasi).

The study's literary analyses also pointed to research gaps in the area of autobiography studies and narratology. Chapters 1.1. and 1.5. together with the close readings in part 2 raised questions that narratologists and autobiography scholars should investigate if they want to address Black North American writing traditions in their full complexity. As this study showed, apart from Border studies concepts, the terminology of autobiography studies and narratology are also stretched to their limits with respect to fictional or autobiographical narratives that tell of the social life of a socially dead narrating and narrated I, or a protagonist from a third-person perspective, who is no subject Afro-pessimistically speaking. In what ways can autobiography scholars and narratologists examine such narratives without adopting the assumptive logics of agency, comparability, and subjectivity that their terminology and disciplines are frequently based on and which Afro-pessimism calls into question? What alternative generic terms other than 'autobiography' and 'narrative' may better describe texts whose narrative strategies actually question the basic definitions of these text types, such as the existence of a narrating or narrated subject, a plot structure of development, transformation, and change, as well as linear and coherent constructions of time and place? As I have shown, combining Afro-pessimist with Black feminist fugitive theoretical concepts, such as Tina Campt's fugitive refusal and Christina Sharpe's 'wake work,' can direct literary analysis into the right direction, just as detailed analyses of the texts' narrative strategies, such as Shakur's lowercasing of the autobiographical I, did. Nevertheless, an adequate investigation of the methodological consequences of Afro-pessimist and Black feminist fugitive theory in autobiography studies and narratology remains a desideratum.

The same seems to be true for in-depth interdisciplinary interrogations of how Black captivity and fugitivity may relate to, or rather exists alongside other racialized forms of confinement and flight, such as in non-Black Jewish American, Asian American, or Indigenous literatures and theory, without assuming a basic common ground for comparability by default. Apart from this study's contrasting of Border studies concepts with Afro-pessimist theory, what more can we learn from a juxtaposition of literary representations and theoretical interrogations, for instance, of captivity and death during slavery and in the concentration camps of the Holocaust, Japanese American internment during the Second World War, Native American reserves, or immigrant detention? What could such questions illuminate, what would they obliterate, and at what and whose costs? Not least, a novel, such as *Open City* that fragments its narrative with references to all of these forms of racialized captivity calls for literary studies research to address these questions. Recent work, such as the essay collection edited by Tiffany Lethabo King, Jenell Navarro, and Andrea Smith on anti-blackness and settler colonialism, might point in the direction of a nuanced theoretical and methodological approach that may acknowledge and at the same time also critique and exceed the Afro-pessimist "incommensurability" of the position of Blackness in and between Black studies and other disciplines, such as Indigenous studies. In line with such pathbreaking scholarship, I suggest German American studies and North American studies more generally must not connive at the far-reaching insights of recent Afro-pessimist and Black feminist theory as well as long-standing Black North American writing traditions. Instead, particularly white American studies scholars should grapple with the fundamental questions the theory and literature examined in this study raise with respect to the Black border, captivity, and fugitivity, and how they relate to our research practices and their underlying assumptions. We must ask ourselves, where the Black border runs through our own work, where it is reinforced, what its implications are, and what our work may contribute towards its dismantling.

Works Cited

Abani, Chris. *The Virgin of Flames*. Vintage, 2008.
Abu-Jamal, Mumia. *Death Blossoms: Reflections from a Prisoner of Conscience*. South End Press, 1996.
Abu-Jamal, Mumia. *Have Black Lives Ever Mattered?*. City Lights Books, 2017.
Abu-Jamal, Mumia. *Live from Death Row*. Addison-Wesley, 1995.
Adams, Rachel. *Continental Divides: Remapping the Cultures of North America*. U of Chicago P, 2009.
Adichie, Chimamanda Ngozi. *Americanah*. Fourth Estate, 2013.
Ahmed, Sara. "Declarations of Whiteness: The Non-Performativity of Anti-Racism." *borderlands*, vol. 3, no. 2, 2004. http://www.borderlands.net.au/vol3no2_2004/ahmed_declarations.htm. Accessed 11 Nov. 2015.
Alexander, Michelle. *The New Jim Crow: Mass Incarceration in the Age of Colorblindness*. New Press, 2012.
Aljoe, Nicole N. "Caribbean Slave Narratives." *The Oxford Handbook of the African American Slave Narrative*, edited by John Ernest. Oxford UP, 2014, pp. 362–70.
Allatson, Paul. *Key Terms in Latino/a Cultural and Literary Studies*. Blackwell, 2007.
Anderson, Monica, and Gustavo López. "Key Facts about Black Immigrants in the US." Pew Research Center, 24 Jan. 2018. http://pewrsr.ch/2E2rH4N. Accessed 31 Mar. 2020.
Andrews, William L., and Nellie Y. McKay, editors. *Toni Morrison's Beloved: A Casebook*. Oxford UP, 1999.
Anim-Addo, Joan, and Maria Helena Lima, editors. "The Neo-Slave Narrative Genre." Special Issue Part 1, *Callaloo*, vol. 40, no. 4, 2017, pp. 1–222.
Anim-Addo, Joan, and Maria Helena Lima. "The Power of the Neo-Slave Narrative Genre." *Callaloo*, vol. 40, no. 4, 2017, pp. 3–13.
Antwi, Phanuel, and David Chariandy, editors. "Writing Black Canadas." Special Issue, *Transition*, vol. 124, 2017, pp. 31–211.
Anzaldúa, Gloria E. *Borderlands/La Frontera: The New Mestiza*. Aunt Lute Books, 1987.
Applebaum, Barbara. "Critical Whiteness Studies." *Oxford Research Encyclopedia of Education*. Editor in chief George W. Noblit. Oxford UP, 2016, pp. 1–25. https://oxfordre.com/education/view/10.1093/acrefore/9780190264093.001.0001/acrefore-9780190264093-e-5. Accessed 27 May 2020.
Armstrong, Tim. "Slavery and American Literature 1900–1945." *The Cambridge Companion to Slavery in American Literature*, edited by Ezra F. Tawil. Cambridge UP, 2016, pp. 204–18.
Ba, Souleymane, and Isabel Soto. "The Problematics of Openness: Cosmopolitanism and Race in Teju Cole's *Open City*." *Atlantic Studies: Global Currents*, vol. 18, no. 3, 2021, pp. 298–315.
Bachmann-Medick, Doris. "From Hybridity to Translation: Reflections on Travelling Concepts." *The Trans/National Study of Culture*, edited by Doris Bachmann-Medick. De Gruyter, 2014, pp. 119–36.
Balagoon, Kuwasi. *A Soldier's Story: Revolutionary Writings by a New Afrikan Anarchist*, edited by Matt Meyer and Karl Kersplebedeb. Kersplebedeb and PM Press, 2019.
Balagoon, Kuwasi. *Look for Me in the Whirlwind: The Collective Autobiography of the New York 21*. Vintage Books, 1971.
Bal, Mieke. *Travelling Concepts in the Humanities: A Rough Guide*. U of Toronto P, 2002.

Ball, John C. "Infinite Worlds: Eighteenth-Century London, the Atlantic Ocean, and Post-Slavery in S.I. Martin's *Incomparable World*, Lawrence Hill's *The Book of Negroes*, David Dabydeen's *A Harlot's Progress*, and Thomas Wharton's *Salamander*." *Transnational Literature*, vol. 5, no. 2, 2013. http://dspace.flinders.edu.au/jspui/bitstream/2328/26707/1/Infinite_Worlds.pdf. Accessed 10 Jan. 2020.
Baraka, Amiri. *Home: Social Essays*. Akashi books, 2009.
Baraka, Amiri. *Autobiography of LeRoi Jones*. Lawrence Hill Books, 1997
Barnwell, Cherron A. "A Prison Abolitionist and Her Literature: Angela Davis." *CLA Journal*, vol. 48, no. 3, 2005, pp. 308–35.
Baroni, Raphaël. "Tellability." *The Living Handbook of Narratology*, edited by Peter Hühn et al. Hamburg University, 2014. http://www.lhn.uni-hamburg.de/article/tellability. Accessed 10 Aug. 2020.
Bast, Heike. "African-Canadian Literature and (Post-)National Imaginary: Tracing a Black Literary Tradition in Canada." *Canada in Grainau: A Multidisciplinary Survey of Canadian Studies After 30 Years / Le Canada à Grainau : Tour d'horizon multidisciplinaire d'études canadiennes, 30 ans après*, edited by Klaus-Dieter Ertler and Hartmut Lutz. Peter Lang, 2009, pp. 257–81.
Baumbach, Sibylle, Beatrice Michaelis, and Ansgar Nünning, editors. *Travelling Concepts, Metaphors, and Narratives: Literary and Cultural Studies in an Age of Interdisciplinary Research*. WVT, 2012.
Beaulieu, Elizabeth A. *Black Women Writers and the American Neo-Slave Narrative: Femininity Unfettered*. Greenwood Press, 1999.
Beck, Janet Kemper. "I'll Fly Away: Ishmael Reed Refashions the Slave Narrative and Takes it on a *Flight to Canada*." *The Critical Response to Ishmael Reed*, edited by Bruce Allen Dick with assistance of Pavel Zemliansky. Critical Responses in Arts and Letters. Greenwood Press, 1999, pp. 133–40.
Bekers, Elisabeth. "From 'Sepulchral City' to 'Open City:' Hetero-Images of Brussels in Joseph Conrad and Teju Cole." *Atlantic Studies: Global Currents*, vol. 18, no. 3, 2021, pp. 351–67.
Belfrage, Sally. *Freedom Summer*. Viking Press, 1968.
Bell, Bernard W. *The Afro-American Novel and its Tradition*. U of Massachusetts P, 1987.
Berndt, Katrin. "In Search of Transcultural Canadian History: Lawrence Hill's *Any Known Blood* (1997) and *The Book of Negroes* (2007)." *Beyond "Other Cultures": Transcultural Perspectives on Teaching the New Literatures in English*, edited by Sabine Doff. WVT, 2011, pp. 157–73.
Berndt, Katrin. "West Africa." *English Literatures Across the Globe: A Companion*, edited by Lars Eckstein. Wilhelm Fink, 2007, pp. 61–85.
Berry, Daina Ramey. "Beyond the Slave Trade, the Cadaver Trade." *New York Times*, sec. Sunday Review, Opinion, 3 Feb. 2018. https://nyti.ms/2GG65Mx. Accessed 3 Apr. 2020.
Best, Stephen M. *The Fugitive's Properties: Law and the Poetics of Possession*. U of Chicago P, 2004.
Best, Stephen M. "On Failing to Make the Past Present." *Modern Language Quarterly*, vol. 73, no. 3, 2012, pp. 453–74.
Best, Stephen M. *None Like Us: Blackness, Belonging, Aesthetic Life*. Duke UP, 2018.
Beverly, William. *On the Lam: Narratives of Flight in J. Edgar Hoover's America*. UP of Mississippi, 2003.

Bey, Marquis. *Them Goon Rules: Fugitive Essays on Radical Black Feminism*. U of Arizona P, 2019.
Bey, Marquis. "Black Fugitivity Un/Gendered." *The Black Scholar*, vol. 49, no. 1, 2019, pp. 55–62.
Birk, Hanne, and Birgit Neumann. "Go-between: Postkoloniale Erzähltheorie." *Neue Ansätze in der Erzähltheorie*, edited by Ansgar Nünning and Vera Nünning. WVT, 2002, pp. 115–52.
Black Lives Matter Global Network. https://blacklivesmatter.com/. Accessed 19 Apr. 2019.
Blackmon, Douglas A. *Slavery by Another Name: The Re-Enslavement of Black Americans from the Civil War to World War II*. Anchor Books, 2008.
Blassingame, John W. *The Slave Community: Plantation Life in the Antebellum South*. Oxford UP, 1979.
Bloch, Ernst. "Reiseform des Wissens." *Merkur*, vol. 16, no. 169, 1962, pp. 201–10.
Bloom, Harold. *Toni Morrison's Beloved*. Chelsea House, 1999.
Bok, Francis, and Edward Tivnan. *Escape from Slavery: The True Story of My Ten Years in Captivity and My Journey to Freedom in America*. St. Martin's Griffin, 2004.
Bontemps, Arna Wendell, editor. *Great Slave Narratives*. Beacon Press, 1972.
Bontemps, Arna Wendell. *Free at Last: The Life of Frederick Douglass*. Dodd Mead, 1971.
Bontemps, Arna Wendell. *Black Thunder*. Beacon Press, 1992
Bradley, David. *The Chaneysville Incident: A Novel*. Open Road Media, 2013. http://gbv.eblib.com/patron/FullRecord.aspx?p=1807383. Accessed 9 June 2020.
Brand, Dionne. *At the Full and Change of the Moon: A Novel*. Grove Press, 1999.
Brand, Dionne. "Toni Morrison Transformed the Texture of English." *Globe and Mail*, 6 Aug. 2019. https://www.theglobeandmail.com/arts/books/article-toni-morrison-transformed-the-texture-of-english/. Accessed 7 Aug. 2019.
Braxton, Joanne M. "Autobiography and African American Women's Literature." *The Cambridge Companion to African American Women's Literature*, edited by Angelyn Mitchell and Danille K. Taylor. Cambridge UP, 2009, pp. 128–49.
Breger, Claudia. "Transnationalism, Colonial Loops, and the Vicissitudes of Cosmopolitan Affect: Christian Kracht's *Imperium* and Teju Cole's *Open City*." *Transnationalism in Contemporary German-language Literature*, edited by Elisabeth Herrmann, Carrie Smith-Prei, and Stuart Taberner. Camden House, 2015, pp. 106–24.
Brock, Jared A. *The Road to Dawn: Josiah Henson and the Story that Sparked the Civil War*. PublicAffairs, 2018.
Brockell, Gillian. "The Statue of Liberty Was Created to Celebrate Freed Slaves, not Immigrants, its New Museum Recounts." *Washington Post*, sec. Retropolis, 23 May 2019. https://www.washingtonpost.com/history/2019/05/23/statue-liberty-was-created-celebrate-freed-slaves-not-immigrants/. Accessed 30 Mar. 2020.
Broeck, Sabine. "Enslavement as Regime of Western Modernity: Re-reading Gender Studies Epistemology through Black Feminist Critique." "Black Women's Writing Revisited." Edited by Beate Neumeier. *Gender Forum: An Internet Journal for Gender Studies*, vol. 22, 2008. http://genderforum.org/black-womens-writing-revisited-issue-22-2008/. Accessed 3 Mar. 2015.
Broeck, Sabine. *Gender and the Abjection of Blackness*. SUNY Press, 2018.

Broeck, Sabine. "Lessons for A-Disciplinarity: Some Notes on What Happens to an Americanist When She Takes Slavery Seriously." *Postcolonial Studies Across the Disciplines*, edited by Jana Gohrisch and Ellen Grünkemeier. Rodopi, 2013, pp. 351–59.

Broeck, Sabine. "The Challenge of Black Feminist Desire: Abolish Property." *Black Intersectionalities: A Critique for the 21st Century*, edited by Monica Michlin and Jean-Paul Rocchi. Liverpool UP, 2013, pp. 211–24.

Broeck, Sabine. "Trauma, Agency, Kitsch, and the Excesses of the Real: *Beloved* within the Field of Critical Response." *America in the Course of Events: Presentations and Interpretations*, edited by Josef Arab, Marcel Arbeit, and Jenel Virden. VU UP, 2006, pp. 201–16.

Brown, Elaine. *A Taste of Power: A Black Woman's Story*. Anchor Books, 1994.

Brown, Vincent. "Social Death and Political Life in the Study of Slavery." *American Historical Review*, vol. 114, no. 5, 2009, pp. 1231–49.

Brown, William Wells. *Clotel, Or, The President's Daughter*. The Floating Press, 2009. *EBSCOhost*, http://search.ebscohost.com/login.aspx?direct=true&db=nlebk&AN=313980&site=ehost-live. Accessed 9 June 2020.

Brown, William Wells. *Narrative of William W. Brown, an American Slave: Written by Himself*. U of North Carolina P, 2011. *JSTOR*, www.jstor.org/stable/10.5149/9780807869604_brown. Accessed 8 June 2020.

Browne, Simone. "EVERYBODY'S GOT A LITTLE LIGHT UNDER THE SUN." *Cultural Studies*, vol. 26, no. 4, 2012, pp. 542–64.

Bruce, Dickson D. J. "Slave Narratives and Historical Understanding." *The Oxford Handbook of the African American Slave Narrative*, edited by John Ernest. Oxford UP, 2014, pp. 54–66.

Bukhari-Alston, Safiya. "Coming of Age: A Black Revolutionary." *Imprisoned Intellectuals: America's Political Prisoners Write on Life, Liberation, and Rebellion*, edited by Joy James. Rowman & Littlefield, 2003, pp. 125–34. *EBSCOhost*, http://search.ebscohost.com/login.aspx?direct=true&db=nlebk&AN=83397&site=ehost-live. Accessed 15 Sept. 2018.

Bukhari-Alston, Safiya. *The War Before: The True Life Story of Becoming a Black Panther, Keeping the Faith in Prison, and Fighting for Those Left Behind*, edited by Laura Whitehorn. Feminist Press, 2010.

Bulawayo, NoViolet. *We Need New Names: A Novel*. Little, Brown and Co., 2013.

Burnham, Michelle. "Loopholes of Resistance: Harriet Jacobs' Slave Narrative and the Critique of Agency in Foucault." *Incidents in the Life of a Slave Girl: Written by Herself*, edited by Nellie Y. McKay and Frances Smith Foster. Norton, 2001, pp. 278–94.

Butler, Octavia. *Kindred*. Beacon Press, 2004.

Cadogan, Garnette. "Black and Blue." *The Fire This Time: A New Generation Speaks About Race*, edited by Jesmyn Ward. Scribner, 2016, pp. 129–44.

Campa, Marta Fernández. "Counter-Memory and the Archival Turn in Dorothea Smartt's *Ship Shape*." *Callaloo*, vol. 40, no. 4, 2017, pp. 94–112.

Campt, Tina M. "Black Feminist Futures and the Practice of Fugitivity." Helen Pond McIntyre '48 Lecture. Barnard Center for Research on Women, Barnard College/Columbia University in the City of New York, 7 Oct. 2014. http://bcrw.barnard.edu/videos/tina-campt-black-feminist-futures-and-the-practice-of-fugitivity/. Accessed 10 Jan. 2015.

Campt, Tina M. *Image Matters: Archive, Photography, and the African Diaspora in Europe*. Duke UP, 2012.

Campt, Tina M. *Listening to Images*. Duke UP, 2017.

Campt, Tina M. "The Sounds of Stillness: Dwelling in the Visual Archive of Diaspora." Remapping the Black Atlantic: (Re)Writings of Race and Space conference. Center for Black Diaspora, DePaul University, 14 Apr. 2013. https://www.wbez.org/shows/wbez-news/the-sounds-of-stillness-dwelling-in-the-visual-archive-of-diaspora/887a9499-872e-471e-82cc-81ce5a34f455. Accessed 15 Mar. 2015.

Campt, Tina M., and Saidiya V. Hartman. "Conference Statement." Loophole of Retreat: A Conference, Guggenheim Museum, New York City, 27 Apr. 2019.

Carpio, Glenda R. "Conjuring the Mysteries of Slavery: Voodoo, Fetishism, and Stereotype in Ishmael Reed's *Flight to Canada*." *American Literature*, vol. 77, no. 3, 2005, pp. 563–89.

Carretta, Vincent. "Response to Paul Lovejoy's 'Autobiography and Memory: Gustavus Vassa, Alias Olaudah Equiano, the African.'" *Slavery & Abolition*, vol. 28, no. 1, 2007, pp. 115–19.

Casteel, Sarah P. "Port and Plantation Jews in Contemporary Slavery Fiction of the Americas." *Callaloo*, vol. 37, no. 1, 2014, pp. 112–29.

Castillo, Debra A., and María Socorro Tabuenca Córdoba. *Border Women: Writing from La Frontera*. U of Minnesota P, 2002.

Chernekoff, Janice. "Resistance Literature at Home: Rereading Women's Autobiographies from the Civil Rights and Black Power Movements." *A/b: Auto/Biography Studies*, vol. 20, no. 1, 2005, pp. 38–58.

Chevigny, Bell G., editor. *Doing Time: 25 Years of Prison Writing*. Skyhorse, 2011.

Child, Lydia Maria. "Lydia Maria Child: Letters to Harriet Jacobs." *Incidents in the Life of a Slave Girl: Contexts, Criticism*, edited by Nellie Y. McKay and Frances Smith Foster. Norton, 2001, pp. 193–95.

Childs, Dennis. "'An Insinuating Voice': Angelo Herndon and the Invisible Genesis of the Radical Prison Slave's Neo-Slave Narrative." *Callaloo*, vol. 40, no. 4, 2017, pp. 30–56.

Childs, Dennis. *Slaves of the State: Black Incarceration from the Chain Gang to the Penitentiary*. U of Minnesota P, 2015.

Childs, Dennis. "'You Ain't Seen Nothin': *Beloved*, the American Chain Gang, and the Middle Passage Remix." *American Quarterly*, vol. 61, no. 2, 2009, pp. 271–97.

Chinosole, editor. *Schooling the Generations in the Politics of Prison*. New Earth Publications, 1996.

Chinosole. *The African Diaspora and Autobiographics: Skeins of Self and Skin*. Peter Lang, 2001.

Cho, Sumi. "Post-Racialism." *Iowa Law Review*, vol. 94, no. 5, 2009, pp. 1589–649.

Christian, Barbara. "The Race for Theory." *Feminist Studies*, vol. 14, no. 1, 1988, pp. 67–79.

Churchill, Ward, and J. J. Vander Wall, editors. *Cages of Steel: The Politics of Imprisonment in the United States*. Maisonneuve Press, 1992.

Clark, Rebecca. "'Visible only in speech': Peripatetic Parasitism, or, Becoming Bedbugs in *Open City*." *Narrative*, vol. 26, no. 2, 2018, pp. 181–200.

Clarke, George Elliot, editor. *Eyeing the North Star: Directions in African-Canadian Literature*. McClelland & Stewart, 1997.

Clarke, George Elliot. *Directions Home: Approaches to African-Canadian Literature*. U of Toronto P, 2017.

Clarke, George Elliot. *Odysseys Home: Mapping African-Canadian Literature*. U of Toronto P, 2017.
Cleaver, Eldridge. *Soul on Ice*. Random House, 1992.
Clifford, James. "Notes on Travel and Theory." *Inscriptions*, vol. 5, no. 29, 1989, pp. 177–88.
Clifford, James. *Routes: Travel and Translation in the Late Twentieth Century*. Harvard UP, 1999.
Colbert, Soyica D., Robert J. Patterson, and Aida Levy-Hussen, editors. *The Psychic Hold of Slavery: Legacies in American Expressive Culture*. Rutgers UP, 2016.
Cole, Teju. *Every Day Is for the Thief*. Faber and Faber, 2014.
Cole, Teju. *Open City: A Novel*. Random House, 2011.
~~Coleman~~, Nathaniel Adam Tobias. "The Atlantic Ocean is Not the Only Black Sea." Futures of Black Studies: Historicity, Objectives and Methodologies, Ethics conference, University of Bremen, 25 Apr. 2014.
Collier-Thomas, Bettye, and V. P. Franklin. *Sisters in the Struggle: African American Women in the Civil Rights-Black Power Movement*. New York UP, 2001.
Columbia Global Centers I Paris. "The Sojourner Project: A Convening Curated by the Practicing Refusal Collective." https://www.facebook.com/events/CGCParis/the-sojourner-project/274090433438076/. Accessed 15 Mar. 2019.
Commander, Michelle D. "Poetics and Care in the Wake." *ASAP/Journal*, vol. 3, no. 2, 2018, pp. 310–14.
Commander, Michelle D. *Afro-Atlantic Flight: Speculative Returns and the Black Fantastic*. Duke UP, 2017.
Commander, Michelle D., editor. "Editor's Forum: Protest and/as Care." *ASAP/Journal*, vol. 3, no. 2, 2018, pp. 299–314.
Cooper, Afua. "Marie-Joseph Angélique." *The Canadian Encyclopedia*, 18 Feb. 2014, Historica Canada. https://thecanadianencyclopedia.ca/en/article/marie-joseph-angelique. Accessed 15 Mar. 2018.
Cooper, Afua. *The Hanging of Angélique: The Untold Story of Canadian Slavery and the Burning of Old Montréal*. HaperCollins, 2011.
Craft, William, and Ellen Craft. *Running a Thousand Miles for Freedom: Or, The Escape of William and Ellen Craft from Slavery*. Cambridge UP, 2013.
Crawford, Margo Natalie. "The Inside-Turned-Out Architecture of the Post-Neo-Slave Narrative." *The Psychic Hold of Slavery: Legacies in American Expressive Culture*, edited by Soyica D. Colbert, Robert J. Patterson, and Aida Levy-Hussen. Rutgers UP, 2016, pp. 69–85.
Crawford, Margo Natalie. "The Twenty-First-Century Black Studies Turn to Melancholy." *American Literary History*, vol. 29, no. 4, 2017, pp. 799–807.
Crawley, Ashon T. *Blackpentecostal Breath: The Aesthetics of Possibility*. Fordham UP, 2017.
Crisu, Corina A. "Stowe and Brown Revisited: Fiction-made Characters in Ishmael Reed's *Flight to Canada*." *Zeitschrift für Anglistik und Amerikanistik*, vol. 58, no. 2, 2014, pp. 103–18.
Cumpsty, Rebekah. "Sacralizing the Streets: Pedestrian Mapping and Urban Imaginaries in Teju Cole's *Open City* and Phaswane Mpe's *Welcome to Our Hillbrow*." *The Journal of Commonwealth Literature*, vol. 54, no. 3, 2019, pp. 305–18.

Dalley, Hamish. "The Idea of 'Third Generation Nigerian Literature': Conceptualizing Historical Change and Territorial Affiliation in the Contemporary Nigerian Novel." *Research in African Literatures*, vol. 44, no. 4, 2013, pp. 15–34.
Davis, Adrienne. "'Don't Let Nobody Bother Yo' Principle': The Sexual Economy of American Slavery." *Sister Circle: Black Women and Work*, edited by Sharon Harley and the Black Women and Work Collective. Rutgers UP, 2002, pp. 103–27.
Davis, Angela Y. *Angela Davis: An Autobiography*. Women's Press, 1990.
Davis, Angela Y. *Are Prisons Obsolete?*. Seven Stories Press, 2003.
Davis, Angela Y. "From the Prison of Slavery to the Slavery of Prison: Frederick Douglass and the Convict Lease System." *The Angela Y. Davis Reader*, edited by Joy James. Blackwell, 1998, pp. 74–94.
Davis, Angela Y. "Political Prisoners, Prisons and Black Liberation." *The Angela Y. Davis Reader*, edited by Joy James. Blackwell, 1998, pp. 39–52.
Davis, Angela Y. "Reflections on the Black Woman's Role in the Community of Slaves." *The Angela Y. Davis Reader*, edited by Joy James. Blackwell, 1998, pp. 111–28.
Davis, Angela Y. "The Past, Present, and Future of Assata's Message." *Women's Studies Quarterly*, vol. 46, no. 3, 2018, pp. 232–34.
Davis, Angela Y. *Women, Race, and Class*. Vintage ebooks, 1983.
Davis, Angela Y., and Bettina Aptheker, et al., editors. *If They Come in the Morning: Voices of Resistance*. Orbach and Chambers, 1971.
Davis, Angela Y., and Farah Jasmine Griffin. "Toni Morrison, Revolutionary Political Thinker." *New York Times*, sec. Opinion, 7 Aug. 2019. https://nyti.ms/2ZF3xaP. Accessed 8 Aug. 2019.
Davis, Thulani. "Recovering Fugitive Freedoms." *Social Text*, vol. 125, 2015, pp. 61–67.
Day, Iyko. "Being Or Nothingness: Indigeneity, Antiblackness, and Settler Colonial Critique." *Critical Ethnic Studies*, vol. 1, no. 2, 2015, pp. 102–21.
Delgado, Elena L., Ronaldo J. Romero, and Walter Mignolo. "Local Histories and Global Designs: An Interview with Walter Mignolo." *Discourse*, vol. 22, no. 3, 2001, pp. 7–33.
Dick, Bruce Allen, editor. *The Critical Response to Ishmael Reed*. With assistance of Pavel Zemliansky. Critical Responses in Arts and Letters. Greenwood Press, 1999.
Dillon, Stephen. *Fugitive Life: The Queer Politics of the Prison State*. Duke UP, 2018.
Domínguez, Pilar C. "In Search of a 'Grammar for Black': Africa and Africans in Lawrence Hill's Works." *Research in African Literatures*, vol. 46, no. 4, 2015, pp. 90–106.
Donahue, James J., Jennifer Ann Ho, and Shaun Morgan, editors. *Narrative, Race, and Ethnicity in the United States*. Ohio State UP, 2017.
Donnan, Hastings, and Thomas M. Wilson, editors. *A Companion to Border Studies*. Wiley-Blackwell, 2012.
Douglass, Frederick. "Douglass on his Escape from Slavery." *Narrative of the Life of Frederick Douglass, an American Slave, Written by Himself: Authoritative Text, Contexts, Criticism*, edited by William L. Andrews and William S. McFeely. Norton, 1997, pp. 104–11.
Douglass, Frederick. *My Bondage and My Freedom*, edited by Lerone Bennett Jr. Johnson, 1970.
Douglass, Frederick. *Narrative of the Life of Frederick Douglass, an American Slave, Written by Himself: Authoritative Text, Contexts, Criticism*, edited by William L. Andrews and William S. McFeely. Norton, 1997, pp. 1–80.

Douglass, Frederick. *The Life and Times of Frederick Douglass*, edited by George L. Ruffin. Digital Scanning, 2001, *EBSCOhost*, http://search.ebscohost.com/login.aspx?direct=true&scope=site&db=nlebk&AN=1424206. Accessed 25 Apr. 2020.

Douglass, Patrice D. "Assata is Here: (Dis)Locating Gender in Black Studies." *Souls*, vol. 22, no. 1, 2020, pp. 89–103.

Douglass, Patrice D., Selamawit D. Terrefe, and Frank B. Wilderson III. "Afro-Pessimism." *Oxford Bibliographies*, section African American Studies. Last modified 28 Aug. 2018. https://www.oxfordbibliographies.com/view/document/obo-9780190280024/obo-9780190280024-0056.xml. Accessed 4 Aug. 2021.

Douglass, Patrice D., and Frank B. Wilderson III. "The Violence of Presence: Metaphysics in a Blackened World." *The Black Scholar*, vol. 43, no. 4, 2013, pp. 117–23.

Drake, Kimberly. "Doing Time in/as 'The Monster': Abject Identity in African-American Prison Literature." *From the Plantation to the Prison: African-American Confinement Literature*, edited by Tara T. Green. Series Voices of the African Diaspora, edited by Chester J. Fontenot. Mercer UP, 2008, pp. 118–53.

Du Bois, W. E. B. *The Souls of Black Folk*. The Avalon Project: Documents in Law, History and Diplomacy, Lillian Goldman Law Library, 2008. http://avalon.law.yale.edu/20th_century/dubois_01.asp. Accessed 4 Apr. 2019.

Dubey, Madhu. "Neo-slave Narratives." *A Companion to African American Literature*, edited by Gene Andrew Garrett. Wiley-Blackwell, 2010, pp. 332–46.

Duff, Christine. "Where Literature Fills the Gaps: *The Book of Negroes* as a Canadian Work of Rememory." *Studies in Canadian Literature / Études En Littérature Canadienne*, vol. 36, no. 2, 2011, pp. 237–54.

Durán, Isabel. "The Brown/Mestiza Metaphor, Or the Impertinence against Borders." *Border Transits: Literature and Culture across the Line*, edited by Ana María Manzanas Calvo, Rodopi, 2007, pp. 119–45.

Ede, Amatoritsero. "The Politics of Afropolitanism." *Journal of African Cultural Studies*, vol. 28, no. 1, 2016, pp. 88–100.

Edugyan, Esi. *Washington Black: A Novel*. Knopf, 2018.

Edwards, Erica R. "The New Black Novel and the Long War on Terror." *American Literary History*, vol. 29, no. 4, 2017, pp. 664–81.

Ellison, Ralph. *Invisible Man*. Random House, 1995.

Eltis, David. "Europeans and the Rise and Fall of African Slavery in the Americas: An Interpretation." *American Historical Review*, vol. 98, no. 5, 1993, pp. 1399–423.

Elze, Jens. "Cosmopolitan Place, Postcolonial Time, and the Politics of Modernism in Teju Cole's *Open City*." *Zeitschrift für Anglistik und Amerikanistik*, vol. 65, no. 1, 2017, pp. 85–104.

Encyclopedia Britannica editors. "Ann Petry." *Encyclopædia Britannica*, Encyclopædia Britannica, inc, 24 Apr. 2020. https://www.britannica.com/biography/Ann-Petry. Accessed 8 June 2020.

Encyclopedia Britannica editors. "Ernest J. Gaines." *Encyclopædia Britannica*, Encyclopædia Britannica, inc, 11 Apr. 2019. https://www.britannica.com/biography/Ernest-J-Gaines. Accessed 15 Nov. 2019.

Encyclopedia Britannica editors. "Felipe Guáman Poma de Ayala." *Encyclopædia Britannica*, Encyclopædia Britannica, inc. https://www.britannica.com/biography/Felipe-Guaman-Poma-de-Ayala. Accessed 21 Apr. 2020.

Encyclopedia Britannica editors. "Margaret Walker." *Encyclopædia Britannica*, Encyclopædia Britannica, inc, 3 July 2019. https://www.britannica.com/biography/Margaret-Walker. Accessed 15 Nov. 2019.

Encyclopedia Britannica editors. "Nat Turner." *Encyclopædia Britannica*, Encyclopædia Britannica, inc, 5 Feb. 2020. https://www.britannica.com/biography/Nat-Turner. Accessed 29 Feb. 2020.

Encyclopedia Britannica editors. "Olaudah Equiano." *Encyclopædia Britannica*, Encyclopædia Britannica, inc, 29 July 2019. https://www.britannica.com/biography/Olaudah-Equiano. Accessed 11 Jan. 2020.

Epstein, Josh. "*Open City*'s 'Abschied': Teju Cole, Gustav Mahler, and Elliptical Cosmopolitanism." *Studies in the Novel*, vol. 51, no. 3, 2019, pp. 412–32.

Equiano, Olaudah. *The Interesting Narrative of the Life of Olaudah Equiano, Or Gustavus Vassa, the African, Written by Himself: An Authoritative Text, Contexts, Criticism*, edited by Werner Sollors. Norton, 2001.

Ernest, John. "Beyond Douglass and Jacobs." *The Cambridge Companion to the African American Slave Narrative*, edited by Audrey A. Fisch. Cambridge UP, 2007, pp. 218–31.

Ernest, John. "Introduction." *The Oxford Handbook of the African American Slave Narrative*, edited by John Ernest. Oxford UP, 2014, pp. 1–18.

Ernest, John, editor. *The Oxford Handbook of the African American Slave Narrative*. Oxford UP, 2014.

Essi, Cedric, Samira Spatzek, Paula von Gleich, Gesine Wegner, and Stephen Koetzing. "*COPAS* at Twenty: Interrogating White Supremacy in the United States and Beyond." *Current Objectives of Postgraduate American Studies*, vol. 20, no. 2, 2019, pp. 1–17. https://copas.uni-regensburg.de/article/view/336. Accessed 20 Mar. 2020.

Fanon, Frantz. *Black Skin, White Masks*. Translated by Charles Lam Markmann. Pluto Press, 2008.

Fisch, Audrey A. "Introduction." *The Cambridge Companion to the African American Slave Narrative*, edited by Audrey A. Fisch. Cambridge UP, 2007, pp. 1–7.

Fisch, Audrey A., editor. *The Cambridge Companion to the African American Slave Narrative*. Cambridge UP, 2007.

Fludernik, Monika, and Caroline Pirlet. "Narratology." *English and American Studies: Theory and Practice*, edited by Martin Middeke, Timo Müller, Christina Wald, and Hubert Zapf. J.B. Metzler, 2012, pp. 225–30.

Fongang, Delphine. "Cosmopolitan Dilemma: Diasporic Subjectivity and Postcolonial Liminality in Teju Cole's *Open City*." *Research in African Literatures*, vol. 48, no. 4, 2017, pp. 138–54.

Fontenot, Chester J. "Series Preface." *From the Plantation to the Prison: African-American Confinement Literature*, edited by Tara T. Green. Series Voices of the African Diaspora, edited by Chester J. Fontenot. Mercer UP, 2008.

Foucault, Michel. *Discipline and Punish: The Birth of the Prison*. Vintage Books, 1995.

Franklin, Benjamin. *Autobiography and Other Writings*, edited by Ormond Seavey. Oxford UP, 1998.

Franklin, Howard Bruce, editor. *Prison Writing in 20th-Century America*. Random House, 1998.

Franklin, Howard Bruce. *The Victim as Criminal and Artist: Literature from the American Prison*. Oxford UP, 1978.

Freedman, Ariela. "How to Get to 9/11: Teju Cole's Melancholic Fiction." *Representing 9/11: Trauma, Ideology, Nationalism in Fiction, Film, and Television*, edited by Paul Petrovic. Rowman & Littlefield, 2015, pp. 177–86.
Freeman, Elizabeth. *Beside You in Time: Sense Methods and Queer Sociabilities in the American Nineteenth Century*. Duke UP, 2019.
"Fugitive." *Merriam-Webster.com Dictionary*, Merriam-Webster, https://www.merriam-webster.com/dictionary/fugitive. Accessed 12 Jan. 2015.
Gaines, Ernest J. *The Autobiography of Miss Jane Pittman: A Novel*. Dial Press, 1971.
Gamso, Nicholas. "Exposure and Black Migrancy in Teju Cole." "Editors' Forum: Thinking the Global with Literature." Guest-edited by Heather H. Yeung. *New Global Studies*, vol. 13, no. 1, 2019, pp. 60–79.
Ganser, Alexandra, Katharina Gerund, and Heike Paul. "Introduction." *Pirates, Drifters, Fugitives: Figures of Mobility in the US and Beyond*, edited by Alexandra Ganser, Katharina Gerund, and Heike Paul. Winter, 2012. pp. 11–27.
Gates, Henry Louis Jr. *The Signifying Monkey: A Theory of African-American Literary Criticism*. Oxford UP, 1989.
Gates, Henry Louis Jr. "The Blackness of Blackness: A Critique on the Sign and the Signifying Monkey." *Literary Theory: An Anthology*. 2nd ed., edited by Julie Rivkin and Michael Ryan. Blackwell, 2004, pp. 987–1004.
Gates, Henry Louis Jr., and Charles T. Davis, editors. *The Slave's Narrative*. Oxford UP, 1990.
Gehrmann, Susanne. "Cosmopolitanism with African Roots: Afropolitanism's Ambivalent Mobilities." *Journal of African Cultural Studies*, vol. 28, no. 1, 2016, pp. 61–72.
Genet, Jean. "Introduction by Jean Genet." *Soledad Brother: The Prison Letters of George Jackson*. Coward-McCann, 1970.
Gerund, Katharina. "Angela Davis: The (Un)Making of a Political Fugitive in the Black Power Era." *Pirates, Drifters, Fugitives: Figures of Mobility in the US and Beyond*, edited by Alexandra Ganser, Katharina Gerund, and Heike Paul. Winter, 2012, pp. 279–301.
Gillespie, Carmen. *Critical Companion to Toni Morrison: A Literary Reference to Her Life and Work*. Facts on File, 2008.
Gilroy, Paul. *The Black Atlantic: Modernity and Double Consciousness*. Verso, 2002.
von Gleich, Paula. "Markus Nehl, *Transnational Black Dialogues: Re-Imagining Slavery in the Twenty-First Century* (Bielefeld: transcript, 2016), 212 pp." *Amerikastudien / American Studies*, vol. 62, no. 4, 2018. https://dgfa.de/wp-content/uploads/von-Gleich_62.4.pdf. Accessed 22 Dec. 2018.
von Gleich, Paula, Samira Spatzek, and Frank B. Wilderson III. "'The Inside-Outside of Civil Society': An Interview with Frank B. Wilderson, III." *Black Studies Papers*, vol. 2, no. 1, 2016, pp. 4–22. http://nbn-resolving.de/urn:nbn:de:gbv:46-00105247-16. Accessed 12 Oct. 2016.
Goffman, Alice. *On the Run: Fugitive Life in an American City*. U of Chicago P, 2014.
Goldberg, Jesse A. "The Restored Literary Behaviors of Neo-Slave Narratives: Troubling the Ethics of Witnessing in the Excessive Present." *Callaloo*, vol. 40, no. 4, 2017, pp. 57–77.
Golimowska, Karolina. *The Post-9/11 City in Novels: Literary Remappings of New York and London*. McFarland, 2016.
Gordon, Lewis R. *Bad Faith and Antiblack Racism*. Humanities Press, 1995.
Gordon, Lewis R. "Shifting the Geography of Reason in Black and Africana Studies." *The Black Scholar*, vol. 50, no. 3, 2020, 42–47.

Gordon, Lewis R. "Some Thoughts on Juliet Hooker's *Theorizing Race in the Americas*." *Blog of the APA*, sec. Diversity and Inclusiveness, 12 June 2018. https://blog.apaonline.org/2018/06/12/some-thoughts-on-juliet-hookers-theorizing-race-in-the-americas/#. Accessed 5 May 2020.

Gordon, Lewis R., Annie Menzel, George Shulman, and Jasmine Syedullah. "Critical Exchange: Afro Pessimism." *Contemporary Political Theory*, vol. 17, no. 1, 2018, pp. 105–37.

Gore, Dayo F., Jeanne Theoharis, and Komozi Woodard, editors. *Want to Start a Revolution? Radical Women in the Black Freedom Struggle*. New York UP, 2009.

Gould, Philip. "The Rise, Development, and Circulation of the Slave Narrative." *The Cambridge Companion to the African American Slave Narrative*, edited by Audrey A. Fisch. Cambridge UP, 2007, pp. 11–27.

Goyal, Yogita. *Runaway Genres: The Global Afterlives of Slavery*. New York UP, 2019. *ProQuest Ebook Central*, http://ebookcentral.proquest.com/lib/suub-shib/detail.action?docID=5844700. Accessed 21 Nov. 2019.

Goyal, Yogita. "The Transnational Turn and Postcolonial Studies." *The Cambridge Companion to Transnational American Literature*, edited by Yogita Goyal. Cambridge UP, 2017, pp. 53–71.

Graaff, Kristina. *Street Literature: Black Popular Fiction in the Era of US Mass Incarceration*. Winter, 2015.

Green, Tara T. "Introduction." *From the Plantation to the Prison: African-American Confinement Literature*, edited by Tara T. Green. Series Voices of the African Diaspora, edited by Chester J. Fontenot. Mercer UP, 2008, pp. 1–7.

Green, Tara T., editor. *From the Plantation to the Prison: African-American Confinement Literature*. Series Voices of the African Diaspora, edited by Chester J. Fontenot. Mercer UP, 2008.

Greenfield-Sanders, Timothy, dir. "Trailer." *Toni Morrison: The Pieces I Am*. Magnolia Pictures, 21 June 2019. https://www.tonimorrisonfilm.com/videos/. Accessed 8 Aug. 2019.

Gumbs, Alexis P. *Spill: Scenes of Black Feminist Fugitivity*. Duke UP, 2016.

Gunner, Elizabeth Ann Wynne, and Harold Scheub. "African Literature." *Encyclopædia Britannica*, Encyclopædia Britannica, inc, 23 May 2018. https://www.britannica.com/art/African-literature. Accessed 30 Mar. 2020.

Gunning, Sandra. "Reading and Redemption in *Incidents in the Life of a Slave Girl*." *Incidents in the Life of a Slave Girl: Contexts, Criticism*, edited by Nellie Y. McKay and Frances Smith Foster. Norton, 2001, pp. 330–53.

Gyasi, Yaa. *Homegoing*. Penguin Books, 2017.

Haensell, Dominique. "Going Through the Motions – Movement and Metahistory in Teju Cole's *Open City*." *Atlantic Studies: Global Currents*, vol. 18, no. 3, 2021, pp. 331–50.

Haley, Sarah, Shoniqua Roach, Emily Owens, and Keeanga-Yamhatta Taylor. "Confinement, Interiority, Black Feminist Study: A Forum on Davis's 'Reflections' at 50." *The Black Scholar*, vol. 51, no. 1, 2021, pp. 3–19.

Hallemeier, Katherine. "Literary Cosmopolitanisms in Teju Cole's *Every Day Is for the Thief* and *Open City*." *ariel: A Review of International English Literature*, vol. 44, no. 2–3, 2013, pp. 239–50.

Hames-García, Michael. *Fugitive Thought: Prison Movements, Race, and the Meaning of Justice*. U of Minnesota P, 2004.

Harney, Stefano, and Fred Moten. *The Undercommons: Fugitive Planning and Black Study*. Autonomedia, 2013.

Harris, Glen A. "Ishmael Reed and the Postmodern Slave Narrative." *Comparative American Studies an International Journal*, vol. 5, no. 4, 2007, pp. 459–79.

Harris, M. A., Morris Levitt, Roger Furman, and Ernest Smith. *The Black Book*. Random House, 2009.

Hartman, Saidiya V. *Lose Your Mother: A Journey Along the Atlantic Slave Route*. Farrar, Straus, and Giroux, 2007.

Hartman, Saidiya V. "Venus in Two Acts." *Small Axe*, vol. 12, no. 2, 2008, pp. 1–14.

Hartman, Saidiya V. *Scenes of Subjection: Terror, Slavery, and Self-Making in Nineteenth-Century America*. Oxford UP, 1997.

Hartman, Saidiya V. "The Belly of the World: A Note on Black Women's Labors." *Souls*, vol. 18, no. 1, 2016, pp. 166–73.

Hartman, Saidiya V. "The Plot of Her Undoing." *Notes on Feminisms 2*. Commissioned by the Feminist Art Coalition, 29 Sept. 2019, pp. 1–6. https://feministartcoalition.org/essays-list/saidiya-hartman. Accessed 6 Nov. 2019.

Hartman, Saidiya V. *Wayward Lives, Beautiful Experiments: Intimate Histories of Social Upheaval*. Norton, 2019. https://sample-9309c861f2cebb. Accessed 14 Apr. 2020.

Hartwiger, Alexander Greer. "The Postcolonial Flâneur: *Open City* and the Urban Palimpsest." *Postcolonial Text*, vol. 11, no. 1, 2016, pp. 1–17.

Havard, John C. "Slavery and the Emergence of the African American Novel." *The Cambridge Companion to Slavery in American Literature*, edited by Ezra F. Tawil. Cambridge UP, 2016, pp. 86–99.

Heiner, Brady T. "Foucault and the Black Panthers." *City*, vol. 11, no. 3, 2007, pp. 313–56.

Heinert, Jennifer L. J. *Narrative Conventions and Race in the Novels of Toni Morrison*. Routledge, 2009.

Henderson, Carol E. "Writing from No Man's Land: The Black Man's Quest for Freedom from Behind the Walls." *From the Plantation to the Prison: African-American Confinement Literature*, edited by Tara T. Green. Series Voices of the African Diaspora, edited by Chester J. Fontenot. Mercer UP, 2008, pp. 11–31.

Henderson, Mae G. "Introduction: Borders, Boundaries, and Frame(Work)s." *Borders, Boundaries, and Frames: Essays in Cultural Criticism and Cultural Studies*, edited by Mae G. Henderson. Routledge, 1995, pp. 1–30.

Henry, Natasha L. "Black Enslavement in Canada." *The Canadian Encyclopedia*, Historica Canada, 7 Sept. 2018. https://www.thecanadianencyclopedia.ca/en/article/black-enslavement. Accessed 8 Mar. 2019.

Henry, Natasha L. "Fugitive Slave Act of 1850." *The Canadian Encyclopedia*, Historica Canada, 7 Sept. 2018. https://www.thecanadianencyclopedia.ca/en/article/fugitive-slave-act-of-1850. Accessed 8 Mar. 2019.

Henry, Natasha L. "Harriet Tubman." *The Canadian Encyclopedia*, Historica Canada, 5 Feb. 2014. https://www.thecanadianencyclopedia.ca/en/article/harriet-tubman. Accessed 10 Jan. 2020.

Henry, Natasha L. "Underground Railroad." *The Canadian Encyclopedia*, Historica Canada, 15 Mar. 2018. https://www.thecanadianencyclopedia.ca/en/article/underground-railroad. Accessed 8 Mar. 2019.

Henson, Josiah. *The Life of Josiah Henson: Formerly a Slave, Now an Inhabitant of Canada as Narrated by Himself*, edited by A. D. Phelps. Bolles and Houghton, Cambridge, 1849. *HathiTrust* http://hdl.handle.net/2027/hvd.32044011936473. Accessed 5 Jan. 2020.

Henson, Josiah. *Truth Stranger than Fiction: Father Henson's Story of His Own Life*. With an introd. by Harriet Beecher Stowe, edited by Henry P. B. Jewett. John P. Jewett and Company, Boston, 1858. *HathiTrust* http://hdl.handle.net/2027/hvd.32044023298060. Accessed 5 Jan. 2020.

Henson, Josiah. *Uncle Tom's Story of His Life: An Autobiography of the Rev. Josiah Henson (1789–1876)*. With a pref. by Harriet Beecher Stowe and an introd. note by George Sturge, edited by S. Morley. 2nd ed. with a new introd. by C. Duncan Rice. Cass, 1971.

Herndon, Angelo. *Let Me Live*. Random House, c. 1937. *HathiTrust* https://hdl.handle.net/2027/mdp.39015064833224. Accessed 12 Jan. 2020.

Hill, Lawrence. "Blackness in Canada and the Accidents of Geography: Mining Fiction and Essays to Meditate on Black History, Identity, and Intersectionality." Intersectionality: Theories, Policies, Practices. 40th annual conference of the Association for Canadian Studies in German-Speaking Countries, Grainau, Germany, 15 Feb. 2019.

Hill, Lawrence. *Any Known Blood*. HarperCollins, 1997.

Hill, Lawrence. *The Book of Negroes*. Black Swan, 2010.

Himes, Chester. *If He Hollers Let Him Go*. Chatham Bookseller, 1973.

Himes, Chester. *My Life of Absurdity: The Autobiography of Chester Himes, Volume 2*. Thunder's Mouth Press, 1998.

Himes, Chester. *The Quality of Hurt: The Autobiography of Chester Himes, Volume 1*. Paragon House, 1990.

Hughes, Langston. *Autobiography: I Wonder as I Wander*, edited by Joseph McLaren. U of Missouri P, 2003.

Hughes, Langston. *Autobiography: The Big Sea*, edited by Joseph McLaren. U of Missouri P, 2002.

Hull, Gloria T., Patricia B. Scott, and Barbara Smith, editors. *All the Women Are White, All the Blacks Are Men, but Some of Us Are Brave*. Feminist Press, 1982.

Humez, Jean M. *Harriet Tubman: The Life and the Life Stories*. U of Wisconsin P, 2003.

Iasiello, Stephanie. "Photographing *a Subtlety* or the Marvelous Sugar Baby: Kara Walker's Take on the Neo-Slave Narrative." *Callaloo*, vol. 40, no. 4, 2017, pp. 14–29.

Iromuanya, Julie. "Humor as Deconstructive Apparatus in Bernardine Evaristo's *Blonde Roots*." *Callaloo*, vol. 40, no. 4, 2017, pp. 174–82.

Irr, Caren. *Toward the Geopolitical Novel: US Fiction in the Twenty-First Century*. Columbia UP, 2014. *EBSCOhost*, http://search.ebscohost.com/login.aspx?direct=true&db=nlebk&AN=677397&site=ehost-live. Accessed 22 Mar. 2020.

Jackson, George. *Soledad Brother: The Prison Letters of George Jackson*. Coward-McCann, 1970.

Jackson, George. *Blood in My Eye*. Random House, 1972.

Jacobs, Harriet A. *Incidents in the Life of a Slave Girl: Contexts, Criticism*, edited by Nellie Y. McKay and Frances Smith Foster. Norton, 2001, pp. 1–158.

Jacobs, Karen. "Teju Cole's Photographic Afterimages." *Image and Narrative*, vol. 15, no. 2, 2014, pp. 87–105.

James, Joy, editor. *Imprisoned Intellectuals: America's Political Prisoners Write on Life, Liberation, and Rebellion*. Rowman & Littlefield, 2003. *EBSCOhost*, http://search.ebsco

host.com/login.aspx?direct=true&db=nlebk&AN=83397&site=ehost-live. Accessed 15 Sept. 2018.

James, Joy, editor. *The New Abolitionists: (Neo) Slave Narratives and Contemporary Prison Writings*. State U of New York P, 2005.

James, Joy, editor. *Warfare in the American Homeland: Policing and Prison in a Penal Democracy*. Duke UP, 2007.

James, Joy. "Afrarealism and the Black Matrix." *The Black Scholar*, vol. 43, no. 4, 2013, pp. 124–31.

James, Joy. "Black Revolutionary Icons and 'Neoslave' Narratives." *Social Identities*, vol. 5, no. 2, 1999, pp. 135–59.

James, Joy. "Framing the Panther: Assata Shakur and Black Female Agency." *Want to Start a Revolution? Radical Women in the Black Freedom Struggle*, edited by Dayo F. Gore, Jeanne Theoharis, and Komozi Woodard. New York UP, 2009, pp. 138–60.

James, Joy. "Introduction." *The Angela Y. Davis Reader*, edited by Joy James. Blackwell, 1998, pp. 1–25.

James, Joy. "Introduction: Democracy and Captivity." *The New Abolitionists: (Neo) Slave Narratives and Contemporary Prison Writings*, edited by Joy James. State U of New York P, 2005, pp. xiii–xiv.

James, Joy. "Introduction: Violations." *Warfare in the American Homeland: Policing and Prison in a Penal Democracy*, edited by Joy James. Duke UP, 2007, pp. 3–16.

James, Joy. "Toni Morrison, an Activist." *Boston Review*, sec. Arts in Society, 7 Aug. 2019. https://bostonreview.net/arts-society/joy-james-toni-morrison-pieces-i-am?utm_source=Boston+Review+Email+Subscribers&utm_campaign=428%E2%80%A6. Accessed 7 Aug. 2019.

JanMohamed, Abdul R. *The Death-Bound-Subject: Richard Wright's Archaeology of Death*. Duke UP, 2005.

Jay, Paul. *Global Matters: The Transnational Turn in Literary Studies*. Cornell UP, 2010.

Johansen, Emily. "History in Place: Territorialized Cosmopolitanism in Teju Cole's *Open City*." *Diaspora: A Journal of Transnational Studies*, vol. 20, no. 1, 2011, pp. 20–39.

Johnson, Charles R. *Middle Passage*. Atheneum, 1990.

Johnson, Charles R. *Oxherding Tale*. Scribner, 2005.

Johnson, Yvonne. *The Voices of African American Women: The Use of Narrative and Authorial Voice in the Works of Harriet Jacobs, Zora Neale Hurston, and Alice Walker*. Peter Lang, 1999.

Jones, Edward P. *The Known World*. Amistad, 2003.

Jones, Gayl. *Corregidora*. Camden Press, 1988.

Jones, Gayl. *Mosquito*. Beacon Press, 2016.

Jordan, Jennifer. "Ideological Tension: Cultural Nationalism and Multiculturalism in the Novels of Ishmael Reed." *Contemporary African American Fiction: New Critical Essays*, edited by Dana A. Williams. Ohio State UP, 2009, pp. 37–61.

Joseph, Peniel E. "Historians and the Black Power Movement." *OAH Magazine of History*, vol. 22, no. 3, 2008, pp. 8–15.

Juan-Navarro, Santiago. "Self-Reflexivity and Historical Revisionism in Ishmael Reed's Neo-Hoodoo Aesthetics." *The Grove: Working Papers on English Studies*, vol. 17, 2010, pp. 77–100.

Kaplan, Caren. "Resisting Autobiography: Out-Law Genres and Transnational Feminist Subjects." *Women, Autobiography, Theory: A Reader*, edited by Sidonie Smith and Julia Watson. U of Wisconsin P, 1998, pp. 208–16.
Keita, Michelle Nzadi, and James Jones. "Murray-Douglass, Anna (1813–1882)." *The Frederick Douglass Encyclopedia*, edited by Julius E. Thompson, James L. Conyers, Jr., and Nancy J. Dawson. Greenwood Press, 2010, pp. 124–25.
Keizer, Arlene R. *Black Subjects: Identity Formation in the Contemporary Narrative of Slavery*. Cornell UP, 2018. https://www.degruyter.com/view/product/525987. 10 Dec. 2019.
Kelley, Robin D. "Black Study, Black Struggle." *Boston Review*, sec. Forum, 7 Mar. 2016. http://bostonreview.net/forum/robin-d-g-kelley-black-study-black-struggle. Accessed 9 Sept. 2019.
Khan-Cullors, Patrisse, and asha bandele. *When They Call You a Terrorist: A Black Lives Matter Memoir*. St. Martin's Press, 2018.
King, Martin Luther Jr. "I Have a Dream." *American Rhetoric: Top 100 Speeches*. American Rhetoric. https://www.americanrhetoric.com/speeches/mlkihaveadream.html. Accessed 10 Jan. 2015.
King, Martin Luther Jr. *The Autobiography of Martin Luther King, Jr.*, edited by Clayborne Carson. Abacus, London, 2000.
King, Mary. *Freedom Song: A Personal Story of the 1960s Civil Rights Movement*. Morrow, 1987.
King, Tiffany Lethabo, Jenell Navarro, and Andrea Smith, editors. *Otherwise Worlds: Against Settler Colonialism and Anti-Blackness*. Duke UP, 2020.
King, Tiffany Lethabo. *The Black Shoals: Offshore Formations of Black and Native Studies*. Duke UP, 2019.
Knopf, Kerstin. "The Gendered Prison: Female Bodies and the Carceral Space in American Women's Prison Literature." *We the People? The United States and the Question of Rights*, edited by Irina Brittner, Sabine N. Meyer, and Peter Schneck. Winter, 2020, pp. 211–29.
Koerner, Michelle. "Line of Escape: Gilles Deleuze's Encounter with George Jackson." *Genre*, vol. 44, no. 2, 2011, pp. 157–80.
Krampe, Christian J. "Inserting Trauma into the Canadian Collective Memory: Lawrence Hill's The Book of Negroes and Selected African-Canadian Poetry." *Zeitschrift für Kanada-Studien*, vol. 29, no. 1, 2009, pp. 62–83.
Krishnan, Madhu. "Postcoloniality, Spatiality, and Cosmopolitanism in the Open City." *Textual Practice*, vol. 29, no. 4, 2015, pp. 675–96.
Kristeva, Julia. *Powers of Horror: An Essay on Abjection*. Translated by Leon S. Roudiez. Columbia UP, 2010.
Kubitschek, Missy D. *Toni Morrison: Critical Companion*. Critical Companions to Popular Contemporary Writers. Greenwood Press, 1998.
Larsen, Nella. *Quicksand*. Collier-Macmillan, 1971.
Larsen, Nella. *Passing*. Collier Books, 1971.
Latner, Teishan A. "'Assata Shakur is Welcome here': Havana, Black Freedom Struggle, and US-Cuba Relations." *Souls*, vol. 19, no. 4, 2017, pp. 455–77.
Lejeune, Philippe. "The Autobiographical Pact." *The Routledge Auto/Biography Studies Reader*, edited by Ricia Anne Chansky and Emily Hipchen. Routledge literature readers series. Routledge, 2016, pp. 34–48.

Levecq, Christine. "Nation, Race, and Postmodern Gestures in Ishmael Reed's *Flight to Canada*." *NOVEL: A Forum on Fiction*, vol. 35, no. 2, 2002, pp. 281–98.
Levine, Robert S. "The Slave Narrative and the Revolutionary Tradition of American Autobiography." *The Cambridge Companion to the African American Slave Narrative*, edited by Audrey A. Fisch. Cambridge UP, 2007, pp. 99–114.
Levy-Hussen, Aida. *How to Read African American Literature: Post-Civil Rights Fiction and the Task of Interpretation*. New York UP, 2016.
Levy-Hussen, Aida. "Trauma and the Historical Turn in Black Literary Discourse." *The Psychic Hold of Slavery: Legacies in American Expressive Culture*, edited by Soyica D. Colbert, Robert J. Patterson, and Aida Levy-Hussen. Rutgers UP, 2016, pp. 195–211.
Li, Stephanie. *Pan-African American Literature: Signifyin(g) Immigrants in the Twenty-First Century*. Rutgers UP, 2018.
Lindroth, James. "Images of Subversion: Ishmael Reed and the Hoodoo Trickster." *African American Review*, vol. 30, no. 2, 1996, pp. 185–96.
Lock, Helen. "'A Man's Story Is His Gris-Gris': Ishmael Reed's Neo-HooDoo Aesthetic and the African-American Tradition." *South Central Review*, vol. 10, no. 1, 1993, pp. 67–77.
Loeb, Jeff. "Fath's Fickle Covenant: African-American Captivity Narratives from the Vietnam War." *From the Plantation to the Prison: African-American Confinement Literature*, edited by Tara T. Green. Series Voices of the African Diaspora, edited by Chester J. Fontenot. Mercer UP, 2008, pp. 154–76.
Lovejoy, Paul E. "Autobiography and Memory: Gustavus Vassa, Alias Olaudah Equiano, the African." *Slavery & Abolition*, vol. 27, no. 3, 2006, pp. 317–47.
Lovejoy, Paul E. "'Freedom Narratives' of Transatlantic Slavery." *Slavery & Abolition*, vol. 32, no. 1, 2011, pp. 91–107.
MacDonald, Joyce Green. "Border Crossings: Women, Race, and Othello in Gayl Jones's *Mosquito*." *Tulsa Studies in Women's Literature*, vol. 28, no. 2, 2009, pp. 315–36.
Maddox, John. "The Black Atlantic Revisited: Ana Maria Gonçalves's *Um defeito de cor*." *Callaloo*, vol. 40, no. 4, 2017, pp. 155–73.
Malaklou, M. Shadee, and Tiffany Willoughby-Herard, editors. "Afro-Pessimism and Black Feminism." Special issue, *Theory & Event*, vol. 21, no. 1, 2018, pp. 1–332.
Malaklou, M. Shadee, and Tiffany Willoughby-Herard. "Notes from the Kitchen, the Crossroads, and Everywhere Else, too: Ruptures of Thought, Word, and Deed from the 'Arbiters of Blackness Itself.'" *Theory & Event*, vol. 21, no. 1, 2018, pp. 2–67.
Manzanas Calvo, Ana María, editor. *Border Transits: Literature and Culture across the Line*. Rodopi, 2007.
Marriott, David. "Judging Fanon." *Rhizomes*, vol. 29, 2016. https://doi.org/10.20415/rhiz/029. e03. Accessed 13 Apr. 2016.
Martinot, Steve, and Jared Sexton. "The Avant-garde of White Supremacy." *Social Identities*, vol. 9, no. 2, 2003, pp. 169–81.
Maynard, Robyn. "Black Life and Death across the US-Canada Border: Border Violence, Black Fugitive Belonging, and a Turtle Island View of Black Liberation." *Critical Ethnic Studies*, vol. 5, no. 1–2, 2019, pp. 124–51.
Mbembe, Achille. "Afropolitanism." *Africa Remix: Contemporary Art of a Continent*, edited by Simon Njami and Lucy Durán. Jacana Media, 2007, pp. 26–30.
McCormick, Stacie Selmon. *Staging Black Fugitivity*. Ohio State UP, 2019.

McCormick, Stacie Selmon. "'Stories that Never Stop': Fugitivity and Neo-Slave Performance." *Modern Drama*, vol. 62, no. 4, 2019, pp. 517–38.

McCoy, Beth A. "Flights of Principled Fancy Dress: Steve Prince's *Katrina Suite* and the Neo-Slave Narrative." *Callaloo*, vol. 40, no. 4, 2017, pp. 183–200.

McDougall, Taija. "Left Out: Notes on Absence, Nothingness and the Black Prisoner Theorist." *Anthurium*, vol. 15, no. 2, 2019, p. 8. http://doi.org/10.33596/anth.391. Accessed 26 Sept. 2019.

McDowell, Deborah E. "In the First Place: Making Frederick Douglass and the Afro-American Narrative Tradition." *Narrative of the Life of Frederick Douglass, an American Slave, Written by Himself: Authoritative Text, Contexts, Criticism*, edited by William L. Andrews and William S. McFeely. Norton, 1997, pp. 172–83.

McDowell, Deborah E. "Telling Slavery in 'Freedom's' Time: Post-Reconstruction and the Harlem Renaissance." *The Cambridge Companion to the African American Slave Narrative*, edited by Audrey A. Fisch. Cambridge UP, 2007, pp. 150–67.

McKay, Claude. *A Long Way from Home*, edited by Gene A. Jarrett. Rutgers UP, 2007.

McKay, Claude. *My Green Hills of Jamaica and Five Jamaican Short Stories*, edited by Mervyn Morris. Heinemann Educational Book, 1979.

McKittrick, Katherine. *Demonic Grounds: Black Women and the Cartographies of Struggle*. U of Minnesota P, 2006.

McKittrick, Katherine. "Commentary: Worn out." *Southeastern Geographer*, vol. 57, no. 1, 2017, pp. 96–100.

McKittrick, Katherine. "On Plantations, Prisons, and a Black Sense of Place." *Social & Cultural Geography*, vol. 12, no. 8, 2011, pp. 947–63.

McKittrick, Katherine. "Plantation Futures." *Small Axe*, vol. 17, no. 3 (42), 2013, pp. 1–15.

McKittrick, Katherine. "Rebellion/Invention/Groove." *Small Axe*, vol. 20, no. 1 (49), 2016, pp. 79–91.

McKittrick, Katherine, editor. *Sylvia Wynter: On Being Human as Praxis*. Duke UP, 2015.

McKnight, Reginald. *I Get on the Bus*. Little, Brown, 1992.

Medovarski, Andrea. "Currency and Cultural Consumption: Lawrence Hill's *The Book of Negroes*." *Studies in Canadian Literature / Études En Littérature Canadienne*, vol. 38, no. 2, 2013, pp. 9–30.

Mengestu, Dinaw. *The Beautiful Things that Heaven Bears*. Riverhead Books, 2007.

Michaelsen, Scott, and David E. Johnson. "Border Secrets: An Introduction." *Border Theory: The Limits of Cultural Politics*, edited by Scott Michaelsen and David E. Johnson. U of Minnesota P, 1997, pp. 1–39.

Middleton, David L., editor. *Toni Morrison's Fiction: Contemporary Criticism*. Garland, 1997.

Mignolo, Walter D. "DELINKING: The Rhetoric of Modernity, the Logic of Coloniality, and the Grammar of De-Coloniality." *Cultural Studies*, vol. 21, no. 2, 2007, pp. 449–514.

Mignolo, Walter D. *Local Histories/Global Designs: Coloniality, Subaltern Knowledges, and Border Thinking*. Princeton UP, 2000.

Mielke, Laura L. "'The Saga of Third World Belle': Resurrecting the Ethnic Woman in Ishmael Reed's *Flight to Canada*." *Melus: Multi-Ethnic Literature of the US*, vol. 32, no. 1, 2007, pp. 3–27.

Mihăilă, Rodica. "Healing the Nation, Memorializing Trauma: Ground Zero and the Critique of Exceptionalism in the Recent American Novel." *Mapping Generations of Traumatic Memory in American Narratives: US Fiction in the Twenty-First Century*, edited by Dana

Mihăilescu, Roxana Oltean, and Mihaela Precup. Cambridge Scholars, 2014, pp. 286–99.

Mikics, David. "Postmodernism, Ethnicity and Underground Revisionism in Ishmael Reed." *Postmodern Culture*, vol. 1, no. 3, 1991. *Project MUSE*, http://doi.org/10.1353/pmc.1991.0018. Accessed 28 Jan. 2020.

Miller, Stephen. *Walking New York: Reflections of American Writers from Walt Whitman to Teju Cole*. Fordham UP, 2014.

Misrahi-Barak, Judith. "Post-*Beloved* Writing: Review, Revitalize, Recalculate." *Black Studies Papers*, vol. 1, no. 1, 2014, pp. 37–55. http://nbn-resolving.de/urn:nbn:de:gbv:46-00103775-17. Accessed 7 Nov. 2016.

Mitchell, Angelyn. *The Freedom to Remember: Narrative, Slavery, and Gender in Contemporary Black Women's Fiction*. Rutgers UP, 2002.

Montgomery, Maxine L., editor. *Contested Boundaries: New Critical Essays on the Fiction of Toni Morrison*. Cambridge Scholars Publishing, 2013. *EBSCOhost*, http://search.ebscohost.com/login.aspx?direct=true&db=nlebk&AN=649588&site=ehost-live. Accessed 2 Mar. 2020.

Moore, Darnell. *No Ashes in the Fire: Coming of Age Black and Free in America*. Nation Books, 2018.

Moraru, Christian. "'Dancing to the Typewriter': Rewriting and Cultural Appropriation in *Flight to Canada*." *Critique: Studies in Contemporary Fiction*, vol. 41, no. 2, 2000, pp. 99–113.

Morrison, Toni. *Beloved*. Vintage, 2007.

Moten, Fred. "Blackness and Nothingness (Mysticism in the Flesh)." *The South Atlantic Quarterly*, vol. 112, no. 4, 2013, pp. 737–80.

Moten, Fred. *In the Break: The Aesthetics of the Black Radical Tradition*. U of Minnesota P, 2003.

Moten, Fred. "The Case of Blackness." *Criticism*, vol. 50, no. 2, 2008, pp. 177–218.

Moten, Fred. "The Subprime and the Beautiful." *African Identities*, vol. 11, no. 2, 2013, pp. 237–45.

Mullen, Harryette. "Runaway Tongue: Resistant Orality in *Uncle Tom's Cabin, Our Nig, Incidents in the Life of a Slave Girl*, and *Beloved*." *Incidents in the Life of a Slave Girl: Contexts, Criticism*, edited by Nellie Y. McKay and Frances Smith Foster. Norton, 2001, pp. 253–78.

Murphy, Laura T. *Metaphor and the Slave Trade in West African Literature*. Ohio UP, 2012.

Murphy, Laura T. *Survivors of Slavery: Modern-Day Slave Narratives*. Columbia UP, 2014.

Murphy, Laura T. *The New Slave Narrative: The Battle Over Representations of Contemporary Slavery*. Columbia UP, 2019.

Murray, Hannah-Rose. "'My name is not Tom': Josiah Henson's Fight to Reclaim his Identity in Britain, 1876–1877." *Memory and Postcolonial Studies: Synergies and New Directions*, edited by Dirk Göttsche. Cultural Memories Series. Peter Lang, 2019, pp. 169–86. https://doi.org/10.3726/b14024/19. Accessed 28 Feb. 2020.

Mvuyekure, Pierre-Damien. "American Neo-HooDooism." *The Cambridge Companion to the African American Novel*, edited by Maryemma Graham. Cambridge UP, 2004, pp. 203–20.

"Narratable." *Merriam-Webster.com Dictionary*, Merriam-Webster, https://www.merriam-webster.com/dictionary/narratable. Accessed 10 Aug. 2021.

Nazer, Mende, and Damien Lewis. *Slave: The True Story of a Girl's Lost Childhood and Her Fight for Survival*. Virago, 2004.
Nehl, Markus. *Transnational Black Dialogues: Re-Imagining Slavery in the Twenty-First Century*. Transcript, 2016.
Neumann, Birgit, and Ansgar Nünning, editors. *Travelling Concepts for the Study of Culture*. De Gruyter, 2012.
Neumann, Birgit, and Yvonne Kappel. "Music and Latency in Teju Cole's *Open City*: Presences of the Past." *ariel: A Review of International English Literature*, vol. 50, no. 1, 2019, pp. 31–62.
Nünning, Ansgar. "Towards Transnational Approaches to the Study of Culture: From Cultural Studies and Kulturwissenschaften to a Transnational Study of Culture." *The Trans/National Study of Culture*, edited by Doris Bachmann-Medick. De Gruyter, 2014, pp. 23–49.
Nünning, Vera, and Ansgar Nünning. *An Introduction to the Study of English and American Literature*. 4th ed. Klett Lerntraining, 2018.
Nyong'o, Tavia. "Habeas Ficta: Fictive Ethnicity, Affecting Representations, and Slaves on Screen." *Migrating the Body: The African Diaspora and Visual Culture*, edited by Leigh Reiford and Heike Raphael-Hernandez. U of Washington P, 2017, pp. 287–305.
Nyong'o, Tavia. "Unburdening Representation." *The Black Scholar*, vol. 44, no. 2, 2014, pp. 70–80.
Obenland, Frank, Nele Sawallisch, and Elizabeth J. West. "Transnational Black Politics and Resistance: From Enslavement to Obama: Through the Prism of 1619." *Journal of Transnational American Studies*, vol. 10, no. 1, 2019, pp. 223–35. https://escholarship.org/uc/item/54r8w4jz. Accessed 19 Aug. 2019.
Oduwobi, Oluyomi, Harry Sewlall, and Femi Abodunrin. "The Postcolonial Female 'Bildung' in Lawrence Hill's *The Book of Negroes*." *Journal of Black Studies*, vol. 47, no. 5, 2016, pp. 383–401.
O'Gorman, Daniel. *Fictions of the War on Terror: Difference and the Transnational 9/11 Novel*. Palgrave Macmillan, 2015.
Olney, James. "'I was Born': Slave Narratives, their Status as Autobiography and as Literature." *Callaloo*, no. 20, 1984, pp. 46–73.
O'Neale, Sondra A. "Ishmael Reed's Fitful Flight to Canada: Liberation for Some, Good Reading for All." *Callaloo*, no. 4, 1978, pp. 174–77.
Oniwe, Bernard Ayo. "Cosmopolitan Conversation and Challenge in Teju Cole's *Open City*." *Ufahamu*, vol. 39, no. 1, 2016, pp. 43–65.
Osofsky, Gilbert, editor. *Puttin' on Ole Massa: The Slave Narrative of Henry Bibb, William Wells Brown, and Solomon Northrup*. Harper, Row, 1969.
Owusu, Portia. *Spectres from the Past: Slavery and the Politics of History in West African and African-American Literature*. Routledge Research in American Literature and Culture series. Routledge, 2020.
Pabst, Naomi. "'Mama, I'm Walking to Canada': Black Geopolitics and Invisible Empires." *Macmillan Center for African Studies*, vol. 7, 2005, pp. 25–45.
Paris, William M. "Assata Shakur, Mamphela Ramphele, and the Developing of Resistant Imaginations." *Critical Philosophy of Race*, vol. 4, no. 2, 2016, pp. 205–20.
Park, Linette, and Frank B. Wilderson III. "Afropessimism and Futures of …: A Conversation with Frank Wilderson." *The Black Scholar*, vol. 50, no. 3, 2020, pp. 29–41.

Patterson, Orlando. *Slavery and Social Death: A Comparative Study*. Harvard UP, 1982.
Patton, Venetria K. "Black Subjects Re-Forming the Past through the Neo-Slave Narrative Tradition." *Modern Fiction Studies*, vol. 54, no. 4, 2008, pp. 877–83.
Patton, Venetria K. *Women in Chains: The Legacy of Slavery in Black Women's Fiction*. State U of New York P, 2000.
Paul, Heike. "Remembering the Fugitive as a Foundational Figure? The (Black) Canadian Narrative Revisited." *Pirates, Drifters, Fugitives: Figures of Mobility in the US and Beyond*, edited by Alexandra Ganser, Katharina Gerund, and Heike Paul. Winter, 2012, pp. 259–78.
Paul, Heike. *The Myths that Made America: An Introduction to American Studies*. Transcript, 2014.
Péréz-Fernandéz, Irene. "Breaking Historical Silence: Emotional Wealth in Joan Anim-Addo's 'Daughter and His Housekeeper' and Andrea Levy' The Long Song." *Callaloo*, vol. 40, no. 4, 2017, pp. 113–26.
Perkins, Margo V. *Autobiography as Activism: Three Black Women of the Sixties*. UP of Mississippi, 2000. *ProQuest Ebook Central*, http://ebookcentral.proquest.com/lib/suub/detail.action?docID=866925. Accessed 13 Sept. 2018.
Perry, Pamela. "White." *Keywords for American Cultural Studies*. 2nd ed., edited by Bruce Burgett and Glenn Hendler. New York UP, 2014. http://hdl.handle.net/2333.1/7sqv9tzq. Accessed 12 Dec. 2019.
Petry, Ann. *The Street*. World Publ. Co, 1947.
Pettit, Eber M. *Sketches in the History of the Underground Railroad, Comprising Many Thrilling Incidents of the Escape of Fugitives from Slavery, and the Perils of those Who Aided them*. With an Introd. by W. McKinstry. W. McKinstry and Son, Fredonia, NY, 1879. *HathiTrust* http://hdl.handle.net/2027/loc.ark:/13960/t3125zz71. Accessed 19 Feb. 2020.
Pierce, Yolanda. "Redeeming Bondage: The Captivity Narrative and the Spiritual Autobiography in the African American Slave Narrative Tradition." *The Cambridge Companion to the African American Slave Narrative*, edited by Audrey A. Fisch. Cambridge UP, 2007, pp. 83–98.
Pierre, Catherine. "Terrible Journey, Beautiful Tale." *John Hopkins Magazine*, vol. 60, no. 2, 2008. http://pages.jh.edu/~jhumag/0408web/hill.html. Accessed 15 May 2018.
Polk, Olivia R. "Review of *Wayward Lives, Beautiful Experiments: Intimate Histories of Social Upheaval*, by Saidiya Hartman." *Women's Studies*, vol. 48, no. 6, 2019, pp. 652–55.
Pratt, Mary Louise. "Arts of the Contact Zone." *Profession*, 1991, pp. 33–40.
Pratt, Mary Louise. *Imperial Eyes: Travel Writing and Transculturation*. Routledge, 2008.
Prince, Gerald. "On a Postcolonial Narratology." *A Companion to Narrative Theory*, edited by James Phelan and Peter J. Rabinowitz. Blackwell, 2008.
Prince, Mary. *The History of Mary Prince, a West Indian Slave: Related by Herself*. University of North Carolina at Chapel Hill Library, 2017. *EBSCOhost*, http://search.ebscohost.com/login.aspx?direct=true&scope=site&db=nlebk&AN=1520292. Accessed 8 June 2020.
Puertas, Lucia Llano. "Touching the Past: The Inscription of Trauma and Affect in Francophone Neo-Slave Narratives." *Callaloo*, vol. 40, no. 4, 2017, pp. 78–93.
Quijano, Aníbal. "Coloniality and Modernity/Rationality." *Cultural Studies*, vol. 21, no. 2, 2007, pp. 168–78.
Ramsey, Joseph G. "Revolutionary Relatability: *Assata: An Autobiography* as a Site of Radical Teaching and Learning." *Socialism and Democracy*, vol. 28, no. 3, 2014, pp. 118–39.

Ransby, Barbara. *Making All Black Lives Matter: Reimagining Freedom in the Twenty-First Century*. U of California P, 2018.
Reed, Ishmael, and Reginald Martin. "A Conversation with Ishmael Reed." *Review of Contemporary Fiction*, vol. 4, no. 2, 1984. https://www.dalkeyarchive.com/a-conversation-with-ishmael-reed-by-reginald-martin/. Accessed 6 Feb. 2019.
Reed, Ishmael. *Flight to Canada*. Scribner, 1998.
Reed, Ishmael. *Mumbo Jumbo*. Simon & Schuster, 1996.
Reese, Sam, and Alexandra Kingston-Reese. "Teju Cole and Ralph Ellison's Aesthetics of Invisibility." *Mosaic: an interdisciplinary critical journal*, vol. 50, no. 4, 2017, pp. 103–19.
Roberts, Neil. *Freedom as Marronage*. U of Chicago P, 2015.
Roberts, Sam. "Influx of African Immigrants Shifting National and New York Demographics." *New York Times*, sec. New York Region, 1 Sept. 2014. https://nyti.ms/1x3rTds. Accessed 30 Mar. 2020.
Rodríguez, Dylan. *Forced Passages: Imprisoned Radical Intellectuals and the US Prison Regime*. U of Minnesota P, 2006.
Rodríguez, Dylan. "Forced Passages." *Warfare in the American Homeland: Policing and Prison in a Penal Democracy*, edited by Joy James. Duke UP, 2007, pp. 35–57.
Rodríguez, Dylan. "The Black Presidential Non-Slave: Genocide and the Present Tense of Racial Slavery." Rethinking Obama. *Political Power and Social Theory*, vol. 22, 2011, pp. 17–50.
Rodríguez, Dylan. "'Nuance' as Carceral Worldmaking: A Response to Darren Walker." Google Documents, 30 Sept. 2019. https://docs.google.com/document/d/1vFghXDUR5VRv0F0gREl33pgxF33yEi_TRhU_WmM75xU/edit?fbclid=IwAR2tX8Zz_EehC2PL4PcUie0B_uopYDahqw-aItnbwfFnLBDTtyh17mIugGY. Accessed 1 Oct. 2019.
Rodríguez, Richard T. "The Locations of Chicano/a and Latino/a Studies." *A Concise Companion to American Studies*, edited by John Carlos Rowe. Wiley-Blackwell, 2010, pp. 190–209.
Rody, Caroline. *The Daughter's Return: African-American and Caribbean Women's Fictions of History*. Oxford UP, 2001.
Rolle, Dominick D. "Marronage and Re-Creation in *Assata*." *CLA Journal*, vol. 61, no. 3, 2018, pp. 155–70.
Rothberg, Michael. *Multidirectional Memory: Remembering the Holocaust in the Age of Decolonization*. Stanford UP, 2009.
Roynon, Tessa. *The Cambridge Introduction to Toni Morrison*. Cambridge UP, 2013.
Rushdy, Ashraf H. A. *Neo-Slave Narratives: Studies in the Social Logic of a Literary Form*. Oxford UP, 1999.
Rushdy, Ashraf H. A. *Remembering Generations: Race and Family in Contemporary African American Fiction*. U of North Carolina P, 2001.
Rushdy, Ashraf H. A. "Slavery and Historical Memory in Late-Twentieth-Century Fiction." *The Cambridge Companion to Slavery in American Literature*, edited by Ezra Tawil. Cambridge UP, 2016, pp. 236–49.
Rushdy, Ashraf H. A. "The Neo-Slave Narrative." *The Cambridge Companion to the African American Novel*, edited by Maryemma Graham. Cambridge UP, 2004, pp. 87–105.
Ryan, Tim A. *Calls and Responses: The American Novel of Slavery since 'Gone with the Wind.'* Louisiana State UP, 2008.

Sadowski-Smith, Claudia. *Border Fictions: Globalization, Empire, and Writing at the Boundaries of the United States.* U of Virginia P, 2008.

Sagawa, Jessie, and Lawrence Hill. "Projecting History Honestly: An Interview with Lawrence Hill." *Studies in Canadian Literature / Études En Littérature Canadienne*, vol. 33, no. 1, 2008, pp. 307–22.

Sagawa, Jessie, and Wendy Robbins. "Resister and Rebel Storytellers: Slave Narratives and Neo-Slave Novels by and/or about Women Connected to Canada." *Postcolonial Text*, vol. 6, no. 4, 2011, pp. 1–21.

Said, Edward. "Traveling Theory." *The Edward Said Reader*, edited by Mustafa Bayoumi and Andrew Rubin. Granta Books, 2001, pp. 195–217.

Saint, Lily "From a Distance: Teju Cole, World Literature, and the Limits of Connection." *NOVEL: A Forum on Fiction*, vol. 51, no. 2, 2018, pp. 322–38.

Saldívar, José David. *Border Matters: Remapping American Cultural Studies.* U of California P, 1997.

Saldívar-Hull, Sonia. *Feminism on the Border: Chicana Gender Politics and Literature.* U of California P, 2000.

Santamarina, Xiomara. "Black Womanhood in North American Women's Slave Narratives." *The Cambridge Companion to the African American Slave Narrative*, edited by Audrey A. Fisch. Cambridge UP, 2007, pp. 232–45.

Sawallisch, Nele. *Fugitive Borders: Black Canadian Cross-Border Literature at Mid-Nineteenth Century.* Transcript, 2018.

Scacchi, Anna. "'*If You Go There* – You Who Was Never *There*:' On Contemporary Uses of The Memory of Slavery." *Iperstoria*, vol. 8, 2016, pp. 4–15. http://www.iperstoria.it/joomla/images/PDF/Numero_8/monografica_8/Anna%20Scacchi_intestato.pdf. Accessed 15 Dec. 2016.

Schwalm, Helga. "Autobiography." *The Living Handbook of Narratology*, edited by Peter Hühn et al. Hamburg University, 2014. http://www.lhn.uni-hamburg.de/article/autobiography. Accessed 12 Feb. 2019.

Scott, David, and Sylvia Wynter. "The Re-Enchantment of Humanism: An Interview with Sylvia Wynter." *Small Axe*, vol. 8, 2000, pp. 119–207.

Selasi, Taiye. "Bye-Bye Babar." *LIP Magazine*, Features LIP#5 Africa, 3 Mar. 2005. http://thelip.robertsharp.co.uk/?p=76. Accessed 15 Oct. 2017.

Selasi, Taiye. *Ghana Must Go.* Penguin Press, 2013.

Sexton, Jared. "African American Studies." *A Concise Companion to American Studies*, edited by John Carlos Rowe. Wiley-Blackwell, 2010, pp. 210–28.

Sexton, Jared. "Afro-Pessimism: The Unclear Word." *Rhizomes*, vol. 29, 2016. https://doi.org/10.20415/rhiz/029.e02. Accessed 13 Apr. 2016.

Sexton, Jared. "Ante-Anti-Blackness: Afterthoughts." *Lateral*, vol. 1, 2012. https://doi.org/10.25158/L1.1.16. Accessed 14 Oct. 2013.

Sexton, Jared. "People-of-Color-Blindness: Notes on the Afterlife of Slavery." *Social Text*, vol. 28, no. 2, 2010, pp. 31–56.

Sexton, Jared. "Racial Profiling and the Societies of Control." *Warfare in the American Homeland: Policing and Prison in a Penal Democracy*, edited by Joy James. Duke UP, 2007, pp. 197–218.

Sexton, Jared. "The Social Life of Social Death: On Afro-Pessimism and Black Optimism." *InTensions Journal*, vol. 5, 2011. http://www.yorku.ca/intent/issue5/articles/jaredsexton.php. Accessed 2 Jan. 2013.

Sexton, Jared. "Unbearable Blackness." *Cultural Critique*, vol. 90, 2015, pp. 159–78.

Shakur, Assata. *Assata: An Autobiography*, edited by Lennox S. Hinds. Zed Books, 1987.

Shakur, Assata. "An Open Letter from Assata Shakur." *World History Archives*. Hartford Web Publishing, 1 Apr. 1998. http://www.hartford-hwp.com/archives/45a/103.html. Accessed 3 July 2019.

Shakur, Assata. "Women in Prison: How We Are." *The New Abolitionists: (Neo) Slave Narratives and Contemporary Prison Writings*, edited by Joy James. State U of New York P, 2005, pp. 79–89.

Sharpe, Christina E. *In the Wake: On Blackness and Being*. Duke UP, 2016.

Sharpe, Christina E. "'And to Survive.'" *Small Axe*, vol. 57, no. 3 (57), 2018, pp. 171–80.

Sharpe, Christina E. "Black Life, Annotated." *The New Inquiry*, sec. Essays and Reviews, 8 Aug. 2014. https://thenewinquiry.com/black-life-annotated/. Accessed 18 Aug. 2014.

Sharpe, Christina E. "Black Studies: In the Wake." *The Black Scholar*, vol. 44, no. 2, 2014, pp. 59–69.

Sharpe, Christina E. *Monstrous Intimacies: Making Post-Slavery Subjects*. Duke UP, 2010.

Siemerling, Wilfried. "Slave Narratives and Hemispheric Studies." *The Oxford Handbook of the African American Slave Narrative*, edited by John Ernest. Oxford UP, 2014, pp. 344–61.

Siemerling, Wilfried. *The Black Atlantic Reconsidered: Black Canadian Writing, Cultural History, and the Presence of the Past*. MQUP, 2015. EBSCOhost, http://search.ebscohost.com/login.aspx?direct=true&db=nlebk&AN=978067&site=ehost-live. Accessed 3 Feb. 2020.

Simpson, Audra. *Mohawk Interruptus: Political Life across the Borders of Settler States*. Duke UP, 2014.

Simson, Rennie. "Josiah Henson (1789–1883)." *African American Autobiographers: A Sourcebook*, edited by Emmanuel S. Nelson. Greenwood, 2002, pp. 179–84. EBSCOhost, http://search.ebscohost.com/login.aspx?direct=true&db=nlebk&AN=86611&site=ehost-live. 28 Feb. 2020.

Sinanan, Kerry. "The Slave Narrative and the Literature of Abolition." *The Cambridge Companion to the African American Slave Narrative*, edited by Audrey A. Fisch. Cambridge UP, 2007, pp. 61–80.

Smith, Valerie. "Form and Ideology in Three Slave Narratives." *Incidents in the Life of a Slave Girl: Contexts, Criticism*, edited by Nellie Y. McKay and Frances Smith Foster. Norton, 2001, pp. 222–36.

Smith, Valerie. "'Loopholes of Retreat': Architecture and Ideology in Harriet Jacobs's *Incidents in the Life of a Slave Girl*." *Reading Black, Reading Feminist: A Critical Anthology*, edited by Henry Luis Gates, Jr. Meridian, 1990, pp. 212–26.

Smith, Valerie. "Neo-Slave Narratives." *The Cambridge Companion to the African American Slave Narrative*, edited by Audrey A. Fisch. Cambridge UP, 2007, pp. 168–85.

Smith, Sidonie, and Julia Watson. *Reading Autobiography: A Guide for Interpreting Life Narratives*. 2nd ed. U of Minnesota P, 2010.

Smith, Sidonie. *Where I'm Bound: Patterns of Slavery and Freedom in Black American Autobiography*. Greenwood Press, 1974.

Smith, Stephanie A. "Harriet Jacobs: A Case History of Authentication." *The Cambridge Companion to the African American Slave Narrative*, edited by Audrey A. Fisch. Cambridge UP, 2007, pp. 189–200.

Sollors, Werner. "Cosmopolitan Curiosity in an Open City: Notes on Reading Teju Cole by way of Kwame Anthony Appiah." *New Literary History*, vol. 49, no. 2, 2018, pp. 227–48.

Somerville, Siobhan B. "Queer." *Keywords for American Cultural Studies*. 2nd ed., edited by Bruce Burgett and Glenn Hendler. New York UP, 2014. https://keywords.nyupress.org/american-cultural-studies/essay/queer/. Accessed 29 Oct. 2019.

Soto, Isabel. "'Idea I'a need' or, Enough Said: The Poetics of Reticence in Teju Cole's *Every Day Is for the Thief* and *Open City*." *Atlantic Studies: Global Currents*, vol. 18, no. 3, 2021, pp. 368–86.

Spaulding, A. Timothy. *Re-Forming the Past: History, the Fantastic, and the Postmodern Slave Narrative*. Ohio State UP, 2005.

Spillers, Hortense J. "Mama's Baby, Papa's Maybe: An American Grammar Book." *Diacritics*, vol. 17, no. 2, 1987, pp. 65–81.

Spillers, Hortense J. "Changing the Letter: The Yokes, the Jokes of Discourse, or, Mrs. Stowe, Mr. Reed." *Black, White, and in Color: Essays on American Literature and Culture*. U of Chicago P, 2003, pp. 176–202.

Stauffer, John. "Frederick Douglass's Self-fashioning and the Making of a Representative American Man." *The Cambridge Companion to the African American Slave Narrative*, edited by Audrey A. Fisch. Cambridge UP, 2007, pp. 201–17.

Stender, Fay. "The Need to Abolish Corrections." *Santa Clara Law Review*, vol. 14, no. 4, 1974, pp. 793–809.

Steward, Austin. *Twenty-Two Years a Slave, and Forty Years a Freeman; Embracing a Correspondence of several Years, while President of Wilberforce Colony, London, Canada West, by Austin Steward*. William Elling, Rochester, NY, 1857. https://hdl.handle.net/2027/loc.ark:/13960/t4wh2p944. Accessed 6 June 2020.

Stowe, Harriet Beecher. *Uncle Tom's Cabin*. Bantam Books, 1981.

Sudbury, Julia. "Celling Black Bodies: Black Women in the Global Prison Industrial Complex." *Feminist Review*, vol. 80, no. 1, 2005, pp. 162–79.

Suhr-Sytsma, Nathan. "The Extroverted African Novel and Literary Publishing in the Twenty-first Century." *Journal of African Cultural Studies*, vol. 30, no. 3, pp. 339–55.

Syedullah, Jasmine. "Loopholes in the Fugitive Heart of Harriet Jacobs." Home/Not Home: Centering American Studies Where We Are. American Studies Association Annual Convention, Colorado Convention Center, Denver, 20 Nov. 2016.

Tally, Justine, editor. *The Cambridge Companion to Toni Morrison*. Cambridge UP, 2007.

Tally, Justine. *Toni Morrison's Beloved: Origins*. Routledge Transnational Perspectives on American Literature. Routledge, 2008. https://doi.org/10.4324/9780203884706. Accessed 2 Mar. 2020.

Tawil, Ezra F. "Introduction." *The Cambridge Companion to Slavery in American Literature*, edited by Ezra F. Tawil. Cambridge UP, 2016, pp. 1–15.

Tawil, Ezra F., editor. *The Cambridge Companion to Slavery in American Literature*. Cambridge UP, 2016.

Teller, Katalina. "Mieke Bals Ansatz zur kulturellen Analyse und ihre Theorie der wandernden Begriffe im Lichte einer möglichen 'interkulturellen Narratologie.'" *Narratologie*

Interkulturell: Entwicklungen – Theorien, edited by Magdolna Orosz and Jörg Schönert. Peter Lang, 2004, pp. 167–77.

Terrefe, Selamawit D. "Speaking the Hieroglyph." *Theory & Event*, vol. 21, no. 1, 2018, pp. 124–47.

Terrefe, Selamawit D., and Christina Sharpe E. "What Exceeds the Hold? An Interview with Christina Sharpe." *Rhizomes*, vol. 29, 2016. https://doi.org/10.20415/rhiz/029.e06. Accessed 13 Apr. 2016.

Teshome, Tezeru, and K. Wayne Yang. "Not Child but Meager: Sexualization and Negation of Black Childhood." *Small Axe*, vol. 22, no. 3 (57), 2018, pp. 160–70.

Thomas, Greg, and Sylvia Wynter. "Proud Flesh Inter/Views: Sylvia Wynter." *ProudFlesh: New Afrikan Journal of Culture, Politics, and Consciousness*, vol. 4, 2006, pp. 1–36.

Thomas, Greg. "Afro-Blue Notes: The Death of Afro-pessimism (2.0)?." *Theory & Event*, vol. 21, no. 1, 2018, pp. 282–317.

Thomas, Greg. "'Neo-Slave Narratives' and Literacies of Maroonage: Rereading Morrison with Dionne Brand, George Jackson, and Assata Shakur." *Toni Morrison: Au-Delà Du Visible Ordinaire / Beyond the Ordinary Visible*, edited by Andrée-Anne Kekeh-Dika, Maryemma Graham, and Janis A. Mayes. Presses universitaires de Vincennes, Saint-Denis, 2015, pp. 195–230.

Thomas, Rhondda Robinson. "Locating Slave Narratives." *The Oxford Handbook of the African American Slave Narrative*, edited by John Ernest. Oxford UP, 2014, pp. 328–43.

Tompkins, Jane. *Sensational Designs: The Cultural Work of American Fiction, 1790–1870*. Oxford UP, 1985.

Trans-Atlantic Slave Trade Database. "Estimates Only Embarked." http://www.slavevoyages.org/estimates/zJvBmm7n. Accessed 30 Mar. 2020.

Trans-Atlantic Slave Trade Database. "Estimates Only Disembarked." http://www.slavevoyages.org/estimates/pidDSOIe. Accessed 30 Mar. 2020.

Tremblay-McGaw, Robin. "Enclosure and Run: The Fugitive Recyclopedia of Harryette Mullen's Writing." *Melus: Multi-Ethnic Literature of the US*, vol. 35, no. 2, 2010, pp. 71–94.

Troiano, Edna M. *Uncle Tom's Journey from Maryland to Canada: The Life of Josiah Henson*. Arcadia Publishing, 2019.

Truth, Sojourner with Olive Gilbert. *Narrative of Sojourner Truth*. Project Gutenberg, Champaign, Ill. 2000. *EBSCOhost*, http://search.ebscohost.com/login.aspx?direct=true&db=nlebk&AN=2009734&site=ehost-live. Accessed 12 Nov. 2019.

Tucker, Jeffrey Allen. "Beyond the Borders of the Neo-Slave Narrative: Science Fiction and Fantasy." *The Cambridge Companion to Slavery in American Literature*, edited by Ezra F. Tawil. Cambridge UP, 2016, pp. 250–64.

Umoja, Akinyele K. "Maroon: Kuwasi Balagoon and the Evolution of Revolutionary New Afrikan Anarchism." *Science and Society*, vol. 79, no. 2, 2015, pp. 196–220.

Varvogli, Aliki. "Urban Mobility and Race: Dinaw Mengestu's *The Beautiful Things That Heaven Bears* and Teju Cole's *Open City*." *Studies in American Fiction*, vol. 44, no. 2, 2017, pp. 235–57.

Velikova, Roumiana. "Flight to Canada via Buffalo: Ishmael Reed's Parody of Local History." *ANQ*, vol. 17, no. 4, 2004, pp. 37–44.

Vermeulen, Pieter. "Flights of Memory: Teju Cole's *Open City* and the Limits of Aesthetic Cosmopolitanism." *Journal of Modern Literature*, vol. 37, no. 1, 2013, pp. 40–57.

Vint, Sherryl. "'Only by Experience': Embodiment and the Limitations of Realism in Neo-Slave Narratives." *Science Fiction Studies*, vol. 34, no. 2, 2007, pp. 241–61.
Walcott, Rinaldo. *Black Like Who? Writing Black Canada*. Insomniac Press, 2003. *EBSCOhost*, http://search.ebscohost.com/login.aspx?direct=true&scope=site&db=nlebk&AN=391218. Accessed 28 Jan. 2020.
Walcott, Rinaldo. "Freedom Now Suite: Black Feminist Turns of Voice." *Small Axe*, vol. 22, no. 3 (57), 2018, pp. 151–59.
Walcott, Rinaldo, Phanuel Antwi, and David Chariandy. "The Ethics of Criticism." *Transition*, no. 124, 2017, pp. 51–61.
Wald, Gayle. *Crossing the Line: Racial Passing in Twentieth-Century US Literature and Culture*. Duke UP, 2000.
Walker, Alice. *In Search of our Mothers' Gardens: Womanist Prose*. Women's Press, 1985.
Walker, Margaret. *Jubilee*. Houghton Mifflin, 1966.
Wallace, Michelle. "Female Troubles: Ishmael Reed's Tunnel Vision." *The Critical Response to Ishmael Reed*, edited by Bruce Allen Dick with assistance of Pavel Zemliansky. Critical Responses in Arts and Letters. Greenwood Press, 1999, pp. 183–91.
Walsh, Richard. "'A Man's Story Is His Gris-Gris': Cultural Slavery, Literary Emancipation and Ishmael Reed's *Flight to Canada*." *Journal of American Studies*, vol. 27, no. 1, 1993, pp. 57–71.
Warren, Calvin. "Black Care." *liquid blackness*, vol. 3, no. 6, 2016, pp. 36–47.
Warren, Calvin. "Black Time: Slavery, Metaphysics, and the Logic of Wellness." *The Psychic Hold of Slavery: Legacies in American Expressive Culture*, edited by Soyica D. Colbert, Robert J. Patterson, and Aida Levy-Hussen. Rutgers UP, 2016, pp. 55–68.
Warren, Kenneth Wayne. "'Blackness' and the Sclerosis of African American Cultural Criticism." *Nonsite.org*, vol. 28, 10 May 2019. https://nonsite.org/article/blackness-and-the-sclerosis-of-african-american-cultural-criticism#. Accessed 5 May 2020.
Warren, Kenneth Wayne. *What Was African American Literature?*. Harvard UP, 2011.
Weheliye, Alexander G. *Habeas Viscus: Racializing Assemblages, Biopolitics, and Black Feminist Theories of the Human*. Duke UP, 2014.
Weier, Sebastian. "Consider Afro-Pessimism." *Amerikastudien / American Studies*, vol. 59, no. 3, 2014, pp. 419–33.
Weinstein, Cindy. "The Slave Narrative and Sentimental Literature." *The Cambridge Companion to the African American Slave Narrative*, edited by Audrey A. Fisch. Cambridge UP, 2007, pp. 115–34.
White, Dan. "Toni Morrison and Angela Davis on Friendship and Creativity." *Newscenter*, UC Santa Cruz, 29 Oct. 2014. https://news.ucsc.edu/2014/10/morrison-davis-q-a.html. Accessed 7 Aug. 2019.
Whitehead, Colson. *The Underground Railroad: A Novel*. Doubleday, 2016.
Wilderson, Frank B. III. *Afropessimism*. Liveright, 2020.
Wilderson, Frank B. III. "Conversation." *Poetry Center Digital Archives*, San Francisco State College, 14 Feb. 2019. https://diva.sfsu.edu/collections/poetrycenter/bundles/238243. Accessed 1 Sept. 2019.
Wilderson, Frank B. III. "Grammar and Ghosts: The Performative Limits of African Freedom." *Theatre Survey*, vol. 50, no. 1, 2009, pp. 119–25.
Wilderson, Frank B. III. "Gramsci's Black Marx: Whither the Slave in Civil Society?." *Social Identities*, vol. 9, no. 2, 2003, pp. 225–40.

Wilderson, Frank B. III. *Incognegro: A Memoir of Exile and Apartheid*. South End Press, 2008.

Wilderson, Frank B. III. *Red, White, and Black: Cinema and the Structure of US Antagonisms*. Duke UP, 2010.

Wilderson, Frank B. III. "The Black Liberation Army and the Paradox of Political Engagement." *Postcoloniality – Decoloniality – Black Critique: Joints and Fissures*, edited by Sabine Broeck and Carsten Junker. Campus, 2014, pp. 175–207.

Wilderson, Frank B. III. "The Prison Slave as Hegemony's (Silent) Scandal." *Warfare in the American Homeland: Policing and Prison in a Penal Democracy*, edited Joy James. Duke UP, 2007, pp. 23–34.

Wilderson, Frank B. III. "The Vengeance of Vertigo: Aphasia and Abjection in the Political Trials of Black Insurgents." *InTensions Journal*, vol. 5, 2011. http://www.yorku.ca/intent/issue5/articles/frankbwildersoniii.php. Accessed 1 July 2015.

Wilderson, Frank B. III, and Saidiya V. Hartman. "The Position of the Unthought." *Qui Parle*, vol. 13, no. 2, 2003, pp. 183–201.

Williams, Dana A. "Lessons before Dying: The Contemporary Confined Character-in-Process." *From the Plantation to the Prison: African-American Confinement Literature*, edited by Tara T. Green. Series Voices of the African Diaspora, edited by Chester J. Fontenot. Mercer UP, 2008, pp. 32–57.

Williams, Evelyn. *Inadmissible Evidence: The Story of the African-American Trial Lawyer Who Defended the Black Liberation Army*. IUniverse.com, 2000.

Williams, Sherley A. *Dessa Rose*. Morrow, 1986.

Willoughby, Christopher D. E. "Black New Yorkers and the Fight for the African Burial Ground from the Colonial Period to the Present." Blog, Lapidus Center for the Historical Analysis of Transatlantic Slavery. https://www.lapiduscenter.org/black-new-yorkers-and-the-fight-for-the-african-burial-ground-from-the-colonial-period-to-the-present/?fbclid=IwAR1%E2%80%A6. Accessed 1 Mar. 2019.

Winks, Robin W. "The Making of a Fugitive Slave: Josiah Henson and Uncle Tom – A Case Study." *The Slave's Narrative*, edited by Henry Louis Gates, Jr., and Charles T. Davis. Oxford UP, 1990, pp. 112–46.

Womack, Autumn. "Contraband Flesh: On Zora Neale Hurston's *Barracoon*." *The Paris Review*, sec. Arts and Culture, 7 May 2018. https://www.theparisreview.org/blog/2018/05/07/contraband-flesh-on-zora-neale-hurstons-barracoon/. Accessed 3 June 2020.

Wooden, Isaiah Matthew. "Review of *Wayward Lives, Beautiful Experiments: Intimate Histories of Social Upheaval*, by Saidiya Hartman." *Theatre Journal*, vol. 71, no. 4, 2019, pp. 537–38.

Woodfox, Albert. *Solitary*. Grove, 2019.

Woolf, Virginia. *A Room of One's Own*. Penguin Books, 1945.

Wright, Michelle M. *Physics of Blackness: Beyond the Middle Passage Epistemology*. U of Minnesota P, 2015.

Wright, Richard. *Black Boy: A Record of Childhood and Youth*. Vintage Books, 2000.

Wright, Richard. *Native Son*. HarperCollins, 1998.

Wynter, Sylvia. "'No Humans Involved': An Open Letter to My Colleagues." *Forum N. H. I.: Knowledge for the 21st Century*, vol. 1, no. 1, 1994, pp. 42–73.

X, Malcolm, with Alex Haley. *The Autobiography of Malcolm X as told to Alex Haley*. Ballantine Books, 2015.

Yancy, George. "Whiteness as Ambush and the Transformative Power of Vigilance." *Black Bodies, White Gazes: The Continuing Significance of Race*. 2nd ed., edited by George Yancy. Rowman & Littlefield, 2017, pp. 217–42.

Yellin, Jean Fagan. "Written by Herself: Harriet Jacobs' Slave Narrative." *Incidents in the Life of a Slave Girl: Contexts, Criticism*, edited by Nellie Y. McKay and Frances Smith Foster. Norton, 2001, pp. 203–09.

Yorke, Stephanie. "The Slave Narrative Tradition in Lawrence Hill's *The Book of Negroes*." *Studies in Canadian Literature / Études En Littérature Canadienne*, vol. 35, no. 2, 2010, pp. 129–44.

Zafar, Rafia. "Introduction: Over-Exposed, Under-Exposed: Harriet Jacobs and *Incidents in the Life of a Slave Girl*." *Harriet Jacobs and Incidents in the Life of a Slave Girl: New Critical Essays*, edited by Deborah M. Garfield and Rafia Zafar. Cambridge UP, 1996, pp. 1–10.

Acknowledgements

The research for this monograph was generously funded through a *Brückenstipendium* by the University of Bremen (Apr.–Dec. 2014) and a dissertation fellowship by the Evangelisches Studienwerk e.V. (Jan. 2015–Mar. 2017). I also received financial support for research stays and conference travels from the Evangelisches Studienwerk, the Central Research Development Fund of the University of Bremen, the Bremen Institute of Canada and Québec Studies, the Postgraduate Forum of the German Association for American Studies, and the Emerging Scholars Forum of the Association for Canadian Studies in German-Speaking Countries. I thank the series editors Prof. Dr. Carsten Junker, Prof. Dr. Julia Roth und Prof. Dr. Darieck Scott for accepting my manuscript for their American Frictions series.

First ideas and different versions of parts of the introduction and chapters 1.1.–1.4 as well as chapters 2.2. and 2.3. were published in

von Gleich, Paula. "African American Narratives of Captivity and Fugitivity: Developing Post-Slavery Questions for *Angela Davis: An Autobiography*." *Current Objectives of Postgraduate American Studies*, vol. 16, no. 1, 2015. http://dx.doi.org/10.5283/copas.221. Accessed 23 June 2015.
von Gleich, Paula. "Fugitivity against the Border: Afro-pessimism, Fugitivity, and the Border to Social Death." *Borders, Borderthinking, Borderlands: Developing a Critical Epistemology of Global Politics*, edited by Sebastian Weier and Marc Woons. E-International Relations, 2017, pp. 203–15. https://www.e-ir.info/2017/06/27/afro-pessimism-fugitivity-and-the-border-to-social-death/. Accessed 24 Nov. 2021.
von Gleich, Paula. "How Black is the Border? Border Concepts Travelling North American Knowledge Landscapes." *Knowledge Landscapes North America*, edited by Christian Klöckner, Simone Knewitz, and Sabine Sielke. Winter, 2016, pp. 191–210. https://www.winter-verlag.de/en/detail/978-3-8253-6627-8/Kloeckner_ea_Eds_Knowledge_Landscapes/. Accessed 24 Nov. 2021.
von Gleich, Paula. "Markus Nehl, *Transnational Black Dialogues: Re-Imagining Slavery in the Twenty-First Century* (Bielefeld: transcript, 2016), 212 pp." *Amerikastudien / American Studies*, vol. 62, no. 4, 2018. https://dgfa.de/wp-content/uploads/von-Gleich_62.4.pdf. Accessed 22 Dec. 2018.
von Gleich, Paula. "The 'Fugitive Notes' of Teju Cole's *Open City*." *Atlantic Studies: Global Currents*, 2021. 10.1080/14788810.2020.1870399.

I also presented work in progress at several conferences: the Postgraduate Forum of the German Association for American Studies, 10–12 Oct. 2014, Johannes Gutenberg Universität Mainz; FORUM INPUTS, 15 Jan. 2015, Institute for Postcolonial and Transcultural Studies, University of Bremen; Summer Institute 'Borders,

Borderthinking, Borderlands,' 15–25 May 2015, University of Bremen; 62nd annual conference of the German Association for American Studies, 28–31 May 2015, Rheinische Friedrich-Wilhelms-Universität Bonn; 11th biennial conference of the Collegium for African American Research "Mobilising Memory: Creating African Atlantic Identities," 24–28 June 2015, Liverpool Hope University; annual meeting of the American Studies Association, 17–20 Nov. 2016, Denver, Colorado; international conference "Die un-sichtbare Stadt: Perspektiven – Räume – Randfiguren in Literatur und Film," 7–9 Feb. 2018, University of Bremen; annual conference of the Emerging Scholars Forum of the Association for Canadian Studies in German-Speaking Countries, 29 June–1 July 2018, University of Bern; 2nd International Conference of the Black Americas Network, "Black Power: Movements, Cultures, and Resistance in the Black Americas," 18–19 Oct. 2018, University of Bielefeld; 66th annual conference of the German Association for American Studies, 13–15 June 2019, University of Hamburg; lecture series Gender – Culture – Feminism, 8 Jan. 2020, University of Bremen; 41st annual conference of the Association for Canadian Studies in German-Speaking Countries, 14–16 Feb. 2020, Grainau; and the BAAS digital conference #BAAS2021 of the British Association for American Studies, online, 5–11 Apr. 2021.

I thank all co-panelists, conference participants, editors, and anonymous reviewers for their valuable feedback. I also thank my co-editors at *COPAS*, my companions at the Dobben office, and the members of the doctoral network Perspectives in Cultural Analysis: Decoloniality, Black Diaspora, and Transnationalism at the University of Bremen (esp. Dr. Cedric Essi, Dr. Marius Henderson, Dr. Mariya Nikolova, Dr. Ina Schenker, and Dr. Samira Spatzek) for their encouragement, exchange, and friendship. For support that helped me secure funding at the beginning of this project, I thank Prof. Dr. Elisabeth Arend and Prof. Dr. Norbert Schaffeld. I am indebted to Prof. Dr. Frank Wilderson and Prof. Dr. Jaye Austin Williams for feedback on this project in its early stages. I also thank Prof. Dr. Tina Campt for welcoming me as a visiting researcher to the Barnard Center for Research on Women and the Institute for Research on Women, Gender, and Sexuality at Columbia University in the City of New York in 2016. Special thanks go out to my supervisor and mentor Prof. Dr. Sabine Broeck, my referees Prof. Dr. Carsten Junker and Prof. Dr. Annika McPherson as well as Dr. Karin Esders, Prof. Dr. Arnim von Gleich and Cecilie Eckler-von Gleich, Prof. Dr. Kerstin Knopf, Vinnie Paul Melendez, and Prof. Dr. Isabel Soto.

Index

9/11 terrorist attacks 223, 233

Abani, Chris 222f.
abjection 7f., 11, 37, 72, 114, 141f., 198, 244
abolitionism 103, 125, 140, 172, 206, 255
– and Black Power autobiographies 113, 118, 138
– and neo-slave narratives 162, 167f., 180, 182, 186, 205f., 219
– and slave narratives 67–69, 71, 75, 79–81, 174f.
ab ovo 113f., 139, 168, 173, 208
Abu-Jamal, Mumia 95f., 106, 135, 153
Acoli, Sundiata 135, 150, 153
Adichie, Chimamanda Ngozi 222, 224, 227
African Burial Ground Monument, New York City 239–41
African diaspora, see also Black diaspora 40, 60, 155, 214, 217f., 221, 223, 226, 231, 244
African literature, see also West African literature 159, 222
Afrofabulation, see also critical fabulation 36, 39
Afropessimism (the monograph), see also Wilderson, Frank B. III 7, 48
Afro-pessimism 2–8, 10–14, 19–26, 28–39, 42f., 46, 48, 50–55, 60–63f., 247f., 253–57
– and Black Power autobiographies 133f., 141f., 149, 151
– and slave narratives 67, 77
Afropolitanism 10, 159, 222f., 235, 243
afterlife of slavery, see also Hartman, Saidiya 5–7, 9, 11, 27, 33, 44f., 51, 56–58, 64, 91, 167, 253, 255
– and *Angela Davis: An Autobiography* 125, 127
– and neo-slave narratives 162, 165f., 204, 210, 216
– and *Open City* 159, 226f., 236, 241, 243, 247

– and time 46–48
agency 7, 36, 38, 41–43, 52f., 66f., 81, 131, 137, 162, 198, 210f., 255f.
– and border concepts 17, 52
– narrative agency 70, 183
Ahmed, Sara 62
Alexander, Michelle 1, 4, 20, 100, 102, 156
A Mercy, see also Morrison, Toni 165
Americanah, see also Adichie, Chimamanda Ngozi 222, 227
American Civil War 55, 80, 140, 144, 157, 167, 189f., 199, 204, 221, 254
American Revolutionary War 144, 206, 240
American studies 4, 14f., 18, 29, 62, 189, 248, 257
anagrammatical, see also Sharpe, Christina 50, 74, 233
Angela Davis: An Autobiography, see also Davis, Angela 9, 93, 120–35, 146, 153
Angelou, Maja 43
anti-blackness 6–8, 14, 23, 27, 31, 33–35, 39, 42f., 47–50, 56, 62f., 66, 103, 189, 254f.
– and Black Power autobiographies 106, 108, 112f., 120, 133, 136, 143
– and gender 24, 48, 111, 120, 143, 151, 153, 250
– and neo-slave narratives 156, 185f., 214, 221–23, 227, 232
– and settler colonialism 1, 256
– and slave narratives 88, 175
anti-black violence 1, 7, 25, 52, 72f., 78, 121, 136
– and gender 24, 65, 74, 79, 88, 137, 143, 147, 152, 230, 243f., 246
– in neo-slave narratives 51, 157, 200f., 213f., 227, 233, 236, 240, 243f., 247, 252
Any Known Blood, see also Hill, Lawrence 205
Anzaldúa, Gloria 15–19, 29f.

appropriation 1, 17, 153, 155, 158, 166–68, 171, 182 f., 246
Aquash, Anna Mae 98
Arbery, Ahmaud 3
aspiration, *see also* Campt, Tina; Sharpe, Christina 40, 42, 52
Assata: An Autobiography, *see also* Shakur, Assata 9, 59, 92 f., 105, 127, 135–51, 153, 179, 194, 216, 250 f., 253
assimilation 158, 167, 181
assumptive logic 19, 25, 60 f., 256
astro-travel 9, 93, 146, 251
A Taste of Power 98
autobiographical I 140, 142, 253, 256
Autobiography of Miss Jane Pittman, *see also* Gaines, Ernest J. 160

baby ghost 187, 201 f., 215, 251
Baker, Ella 98
Balagoon, Kuwasi 97, 136
Baldwin, James 205
Bal, Mieke, *see also* travelling concept, cultural analysis 3, 7, 10, 12–15, 19, 30, 41, 58 f., 61, 248
Bambara, Toni Cade 121
Baraka, Amiri 110, 165
Baroque music 224 f.
Beloved (the fictional character) 187, 190 f., 195 f., 201–03, 213, 249, 252
Beloved (the novel), *see also* rememory 8 f., 49, 51, 57 f., 136, 155–59, 161–63, 165 f., 182 f., 187–204, 208 f., 215, 219, 245 f., 249, 251 f., 254
– and *Open City* 200, 229, 241, 245, 253
– and *Someone Knows My Name* 158 f., 187, 208–10, 213, 215, 218 f., 221
Best, Stephen 31, 37, 45, 155 f., 163, 187 f.
bildungsroman, *see also* coming-of-age-story 138, 206, 245
Black Atlantic 10, 43, 155, 206, 215, 223, 226, 240, 242, 245
Black Book 188
Black border 6, 11 f., 19, 27–33, 36, 39, 52 f., 62, 176, 248, 254 f., 257
– and Black Power autobiographies 115, 120, 123, 127, 136 f.

– and neo-slave narratives 159, 176, 192, 198, 221, 232, 243 f., 247
– and slave narratives 66, 77, 91
Black Codes 1, 23, 100
Black diaspora, *see also* African diaspora 2, 5, 40, 48 f., 52, 163 f., 179, 212, 222, 226, 243 f., 246
Black Europe 40 f.
Black feminism 3, 24, 33 f., 71, 79, 92, 107, 111, 152 f., 167, 206
Black feminist fugitive thought 7 f., 11 f., 27, 31, 33, 35, 38, 48, 53 f., 60 f., 67, 108, 134, 243, 255 f.
Black Feminist Fugitivity 7, 22, 33, 52, 247
Black feminist theory 5–8, 10, 21 f., 34, 38, 42–45, 62, 248, 257
Black geographies 39 f., 40, 46, 67, 87
Black Liberation Army 55, 106, 135–37, 139, 144, 150
Black liberation movements 8, 55, 253
Black Lives Matter movement 3–5, 9, 55, 96 f., 106, 152 f., 247 f.
Black loyalists 205, 207
Black Optimism 35–38
Black Panther Party 95, 98, 106, 119–21, 136, 144
Black Power movement 5, 8, 27, 55, 69, 91, 93–99, 111 f., 122 f., 152 f.
– and neo-slave narratives 154, 158, 160, 166, 181, 186, 246
Black Thunder, *see also* Bontemps, Arna 160
Black time 46, 48, 107, 120, 253
Bland, Sandra 3
Bloch, Ernst 13, 140
bloodhounds, trope of 124 f., 173, 178 f., 195, 198
Blood in My Eye, *see also* Jackson, George 92, 109, 121, 240
"blood-stained gate," trope of the 73 f., 83
Bok, Francis 65
Bontemps, Arna 160, 205, 255
Book of Negroes (the historical document) 205, 212
Book of Negroes (the novel), *see also* *Someone Knows My Name* 10, 204

border, *see also* Black border, Canadian-US border, US-Mexican border   1, 12–21, 26–33, 38, 68, 73, 172, 179, 190, 196, 222, 247, 249
border concepts   6, 12, 14f., 17–21, 26, 28–31, 32, 223, 256
borderlands   15–18, 28–30
Border studies   6, 12, 15f., 60, 256
border thinking   7, 12, 17f., 20, 28, 30
Bradley, David   228
Brand, Dionne   162, 187, 205
Brent, Linda, *see also* Jacobs, Harriet   80, 82f., 86, 89
Broeck, Sabine   1, 7f., 23, 43–46, 60, 77, 141, 156, 187, 190, 203
Brotherhood (political/solidary)   197, 232, 238
Brown, Elaine   98f.
Browne, Simone   38, 114, 205
Brown, John   89
Brown, Michael   3f., 152
Brown, William Wells   65, 77, 158, 168, 170, 179f.
Brussels, Belgium   224f., 228–32, 235, 241, 244, 250
Bukhari, Safiya   106, 115, 136
Bulawayo, NoViolet   222
burial, trope of the, *see also* grave, tomb   218, 239–41
Butler, Octavia, *see also* Kindred   44, 155, 182

Cameroon   225
camouflage, *see also* disguise   81, 92, 123, 145, 180, 185, 253
Campt, Tina M.   6f., 12, 27, 33, 37f., 40–43, 48–54, 87f., 118, 133f., 139, 256
Canada   1, 5, 10f., 39, 65, 67f., 90, 140, 152, 176, 204–07, 212, 249f.
– as mythical place of freedom   1, 150, 158, 167f., 171–77, 181–87, 207, 210, 220f.
Canadian-US border   1, 15, 158, 176, 187
captive   1, 5, 7, 25, 44, 47, 64, 68, 94, 255
– and Black Power autobiographies   92, 104, 106, 109, 111, 113–15, 123, 129f., 132–34, 137, 144, 146–48, 153
– and neo-slave narratives   159, 194, 197f., 200, 203, 210f., 215f., 219, 232, 241
– and slave narratives   82
captivity, *see also* confinement   1, 2, 5, 8–11, 31f., 34, 39f., 44, 46, 49–56, 58f., 62–64, 66, 79, 91, 94f., 101, 103, 160, 166, 248–57
– and Black Power autobiographies   92, 97, 107, 109, 111, 113–15, 118–20, 122f., 127–29, 131, 133f., 136f., 140, 142, 144, 146f., 153
– and neo-slave narratives   156–59, 167, 186, 190, 192, 196–200, 203, 207, 211, 215f., 223, 236f., 243, 245
– and slave narratives   69, 84f., 87
carceral   10, 49, 105–08, 132, 196, 203, 234
carcerality   94, 120
care   9, 48f., 60, 88, 93, 120, 147, 152f., 205f., 221, 251–55
– in *Beloved*   50f., 194, 197f., 200–04
– in *Open City*   236, 240, 244, 246
Caribbean   3, 16f., 38–40, 56f., 65, 67, 163f., 223, 244
Castile, Philando   3
chain gang   104, 131, 190, 196f., 221, 249f., 252
Chicanx   6, 12, 15–17, 20, 28, 30
Child, Lydia Maria   79f., 82, 89f.
Childs, Dennis   57, 96, 104f., 128, 146, 156, 164, 189f., 196–99, 203
Christmas, William A.   93, 129
Civil Rights movement   8, 27, 55, 91, 94f., 98f., 107, 117, 119, 120
Clarke, George Elliot   207
Clark, Septima   98
class   18, 20, 24, 39, 61, 92f., 103, 116, 119, 133f., 139, 143, 225, 231, 233f., 250
Cleaver, Eldridge   95f., 110
Clutchette, John   93, 113, 122, 129
coffle   114, 196, 211, 215–17
Cole, Teju, *see also* *Every Day Is For the Thief*, *Open City*   8, 10, 221–35, 240f., 244f.
colonialism   5f., 16–19, 21, 24, 28, 30, 37, 46, 159, 222

colonial captivity narrative 68
color line, *see also* Du Bois, W. E. B. 20, 23
Columbus, Christopher 233
coming-of-age story, *see also* bildungsroman 110, 113, 118, 122f., 138f., 142, 152
Commander, Michelle D. 2, 43f., 50f.
communist 92, 104, 109, 120, 123, 128, 153
comparative freedom 75–78, 250
confinement, *see also* captivity, solitary confinement 8–10, 13, 21, 32, 40, 55f., 58f., 64, 95, 97, 248–52, 254–56
– and Black Power autobiographies 91f., 94, 114f., 120f., 127f., 134, 136, 147f.
– and neo-slave narratives 157, 186, 188–91, 201–04, 215f., 232, 237
– and slave narratives 66, 73, 81, 83, 85–88, 90, 94, 211
Congo 225
contact zone 16–18, 28–30, 52
convict, *see also* prisoner 91, 96
convict leasing 1, 77, 100, 106
Cooper, Afua 205
Corregidora, *see also* Jones, Gayl 155, 161
cosmopolitanism 159, 222, 224–26, 231, 243f., 247, 250
Counter Intelligence Program (COINTELPRO) 99, 135
Craft, William and Ellen 71
crime fiction 56, 79, 178f.
critical ethnic studies 30, 54, 101, 103
critical fabulation, *see also* Afrofabulation 7, 36, 47f.
critical prison studies 94, 100, 107, 154
Cuba 38, 60, 93, 98, 135, 137, 149–51
cultural analysis, *see also* Bal, Mieke 3, 12–14, 41, 54, 61, 248

Davis, Angela 50, 78, 102, 109, 119, 152f., 161
– and her autobiographical writing 8f., 92–95, 97–99, 104, 106, 108, 113, 120–36, 138, 140f., 145, 149, 157f., 179, 195, 249f.
– and her scholarly writing 34, 71, 100f., 104, 106f., 111, 115, 187

death 8f., 45, 48–51, 67, 93, 95, 105f., 252–54, 256
– and Black Power autobiographies 109–11, 115f., 119f., 133, 138f., 145f.
– and neo-slave narratives 158f., 173, 177, 183, 187, 191, 193, 195, 200–05, 211, 215–21, 237, 240f.
death-boundedness 117, 252f.
Deleuze, Gilles 117–19
Dessa Rose 107, 136, 154f., 161, 188, 194
deus ex machina 149, 195
Dillon, Stephen 45f., 101, 106–08, 115, 117, 127, 147
disequilibrium 60, 79, 137
disguise 67, 75, 81, 83f., 87, 91f., 123–26, 144–46, 169, 178–81
Douglass, Frederick 9, 50, 59, 64–68, 70–82, 90f., 94, 98, 100, 105
– and Black Power autobiographies 112f., 125, 137, 140f., 208, 251, 254
– and neo-slave narratives 158, 160, 168, 170f., 173–76, 178–80, 189, 206, 218
Douglass, Patrice D. 4, 25, 48, 60, 150
Draft Riots, New York City 236f.
drama 39, 58, 79, 163f., 169
Drumgo, Fleeta 93, 113, 129
Du Bois, W. E. B. 20, 77f., 100

Edugyan, Esi 5, 205
Ellis Island, New York City 231f.
Ellison, Ralph 55f., 97, 180
emancipation (of enslaved people) 1, 4, 8, 27, 66, 69, 86, 177
end of the world 27, 32, 53, 119
Engels, Friedrich 153
enslavism 1, 23, 45f., 107, 178
– and Black Power autobiographies 91, 125, 132, 135
– and neo-slave narratives 180, 185f., 193, 200, 215, 243, 247
– and slave narratives 66f., 71, 77, 81, 83f., 90f., 176f., 185, 252
Equiano, Olaudah 65, 206, 208
equilibrium 60, 79, 137
escape, *see also* flight, fugitive, fugitivity, inescapability 1f., 6, 8, 10, 35–37, 39–42, 47, 52, 55, 58, 62, 154, 249–54

– and Black Power autobiographies   93, 106, 108f., 112, 115–20, 127f., 130, 135f., 139f., 142–51, 153f.
– and neo-slave narratives   158f., 163, 167f., 177–80, 182, 185, 187f., 190–200, 202–05, 216, 219f., 225, 230, 232
– and slave narratives   65–67, 69–71, 74f., 78, 81–88, 90f., 112, 147, 173f., 176
ethnicity   18, 20, 23f., 28, 36, 218, 225
Europe   4, 17, 21, 40, 59, 77, 158, 231f., 241, 247
evasion, as narrative technique   10, 64, 159, 200, 227, 237, 241f., 245, 253
Every Day Is for the Thief, *see also* Cole, Teju   243
exile   9, 59, 94, 108, 137, 150f., 249, 253

Fanon, Frantz   22, 27, 35, 37, 77f.
Federal Bureau of Investigation (FBI)   98f., 120f., 124, 135, 139, 144, 150, 179, 249
flânerie   159, 224f., 244f.
flight, *see also* escape, fugitive, fugitivity   1–3, 5, 7–12, 33–35, 38f., 41f., 44, 46–49, 56–58, 61, 63f., 248–56
– and Black Power autobiographies   91–94, 108, 111, 115–21, 123, 125–27, 130, 133–36, 140–43, 145–47, 149–51, 156
– and neo-slave narratives   156–59, 166f., 169, 171, 177–79, 181–84, 186, 188, 190–93, 195, 197, 199f., 202, 204, 209, 211, 219–21, 223–27, 230f., 234–37, 242–45
– and slave narratives   65–69, 71, 75f., 79–82, 84, 86, 90, 156, 173f., 177
flight from   11, 31, 52, 55f., 68, 75, 77, 93, 121, 126, 130, 149, 154, 192, 220, 233–35, 237, 242f.
flight to   37, 66, 79, 171, 186, 202
Flight to Canada, *see also* Reed, Ishmael   9–11, 57f., 154, 165–90, 205, 220f., 251, 253f.
Flint, Dr. (Norcom, James)   65, 82–89, 185, 252
Floyd, George   3

forward-haunting, *see also* haunting   190, 204
Foucault, Michel   119
Franklin, Benjamin   141
Franklin, Howard Bruce   11, 56, 64, 70, 72, 79, 94–97, 101, 109, 149
freedom, *see also* freedom as marronage, comparative freedom, freedom dream, unconfined freedom   1, 3, 8f., 11, 27, 32, 38, 44, 46, 52, 58, 60, 154, 251–53, 255
– and Black Power autobiographies   91f., 94, 99, 102f., 107f., 116, 124–28, 130, 133f., 136, 140, 145–49, 151f.
– and neo-slave narratives   158, 163f., 166–68, 173f., 178, 180f., 183–86, 189, 194, 197, 199, 204, 206, 210, 213–20, 235, 246
– and slave narratives   65–68, 70f., 74–79, 81–91, 140, 148
freedom as marronage   38, 75, 77, 102
freedom dream   27, 44, 47f., 52f., 55, 61, 63, 255
– in Black Power autobiographies   109, 119, 128, 130, 134, 136, 153
– in neo-slave narratives   183, 186, 193, 221, 232
– in slave narratives   74, 87f., 90, 181
Free North   1, 59, 66, 84, 89, 187, 193, 196, 198, 239, 241, 252
fugitive, *see also* escape, flight, fugitivity   1f., 5, 7, 9, 11f., 31, 34f., 37, 39, 41–44, 50–54, 57f., 91, 250, 252–55
– and Black Power autobiographies   60, 91–95, 101, 104, 106–08, 111, 116–21, 123–27, 132f., 135, 141, 144f., 148–51, 153, 156, 251
– and neo-slave narratives   136, 156–59, 161, 167–69, 171, 177–83, 185–88, 191, 193–96, 198–200, 202, 204, 207, 210, 218f., 221, 223–27, 229–31, 240–44, 246
– and slave narratives   64–69, 73f., 76f., 79f., 88, 90, 172, 174–77
fugitive legacy   43f., 46, 52, 68, 87, 125, 150, 153, 232

Fugitive Slave Act   1, 73, 76, 88, 172, 176, 193
fugitivity, *see also* escape, flight, fugitive   2f., 5–9, 11–14, 32–36, 38–44, 46, 50–56, 58f., 62, 160, 248, 250–57
– and Black Power autobiographies   93, 95, 106, 108, 116, 118–20, 122–28, 134–36, 145, 147
– and neo-slave narratives   156f., 159, 166, 179, 183, 186f., 190, 199, 203, 207, 221, 223, 225f., 245
– and slave narratives   64, 66, 81, 88, 136, 172
fungibility   25, 34
futurity   9, 40, 42f., 52, 93, 108, 119f., 134, 190, 193f., 196, 220

Gaines, Ernest J.   160f.
Garner, Eric   3
Garner, Margret   58, 188, 202
garret   83–88, 90, 144, 185, 201, 252
Garrison, William Lloyd   68
Garza, Alicia   3
Gates, Henry Luis Jr.   58, 67, 135
gender   3, 18, 20, 23f., 28, 48, 77, 100, 106f., 247, 250, 253
– and Black Power autobiographies   103, 111, 120, 125, 136f., 139, 142–44, 147, 151–53
– and neo-slave narratives   161–63, 225–27, 230, 243f.
– and slave narratives   65, 67, 71, 73f., 79, 84, 88
Genet, Jean   111
genocide   10, 94, 101, 159, 171, 222, 233, 235
genre (literary)   7, 11, 43, 52, 57–59, 63, 148, 255
– and Black Power autobiographies   93–95, 98, 103, 108, 121, 123, 132, 136f., 141f.
– and neo-slave narratives   163–66, 208, 222f., 245, 253
– and slave narratives   65, 67, 69, 76
genre of Man, *see also* Wynter, Sylvia   6, 20f., 26, 32, 48, 59, 63

genre of the human, *see also* Wynter, Sylvia   22, 28, 30
George Jackson Brigade   106
Germany   40, 120f., 231, 237
Ghana   43f.
Ghana Must Go   222
ghost, *see also* baby ghost   154, 157, 169, 189, 195, 208, 219, 221, 236f., 245
Gilroy, Paul, *see also* Black Atlantic   10, 43, 155, 206, 223
Gómez, Esteban   233
Gordon, Lewis   4, 6, 23, 26, 31f., 35, 37, 53, 58
gothic   9, 80, 157, 162, 165, 168, 189, 201, 204, 221, 253
Goyal, Yogita   2, 11, 45, 65, 222f., 227, 234f., 242–45
grammar of suffering   6, 25, 34, 233
gratuitous violence   24–26, 61, 72, 79, 90, 137, 139, 146f., 233
grave, trope of the, *see also* burial   85, 115, 197, 203, 218, 239f.
Gray, Freddie   3
Great Migration   1
Green, Tara   97, 115, 136
Gumbs, Alexis Pauline   7, 33f.
Gyasi, Yaa   4f.

hair   145, 198f., 213, 235
Haiti   38, 170, 235, 249
Haitian Revolution   235
Haley, Alex   43, 95
Hamer, Fannie Lou   98
Hames-García, Michael   4, 31, 60, 69, 78, 80–82, 89f., 96, 101f., 136, 139–41, 145–50, 194
happy ending   89, 150, 158f., 174, 209f., 212, 220
Harlem, New York City   47, 232, 236, 239
Harlem Renaissance   55, 128, 151, 160
Hartman, Saidiya, *see also* afterlife of slavery, *Lose Your Mother*, *Scenes of Subjection*, "Venus in Two Acts"   6–8, 12, 22, 25, 27, 33f., 36–38, 42–52, 54, 61, 64, 154, 253
– and Black Power autobiographies   107, 120, 125, 134, 136, 146

– and neo-slave narratives   155, 159, 165f., 192, 208f., 211, 214, 227, 232, 237
– and slave narratives   56, 71–73, 78f., 83f., 87f., 208
haunting, *see also* forward-haunting, ghost   90, 107, 154
– and neo-slave narratives   161, 185, 187, 189, 197, 201f., 213, 215, 226, 236–38, 243
Henson, Josiah   8f., 57f., 65, 72, 78f., 158, 167–79, 181f., 189, 191, 198, 254f.
Herndon, Angelo   56f., 104f., 128, 160, 164
Hill, Lawrence, *see also Someone Knows My Name*   8f., 166, 172, 183, 202, 240, 255
Himes, Chester   56, 79
Hinds, Lennox S.   138
historical novel   9f., 107, 154, 157–59, 162, 164, 166, 169, 189, 205
History of Mary Prince, A West Indian Slave   65, 71
hold   36, 47–51, 74, 77, 88, 107, 114, 132, 196f., 203, 211, 216, 221, 250f.
Holocaust   226, 241, 256
home   40, 86, 89, 142f., 149, 151, 178, 250
– and neo-slave narratives   180, 186f., 191, 201, 204, 210, 213–21
Homegoing   4
homemaking   10, 40, 159, 176f., 210, 214f., 217, 219–21, 248
hope   7, 10, 28, 30, 42, 84, 90, 134, 147
– and neo-slave narratives   159, 173, 175, 184, 194, 196f., 210f., 213f., 216–19
Hughes, Langston   151, 205
hybridity   18f., 28f., 222, 227
hypervisibility   180, 199

If They Come in the Morning: Voices of Resistance   122
immigration, *see also* migration   207, 223, 232, 249
imperialism   109f., 131, 149, 185
imprisonment, *see also* incarceration   1, 6, 9, 23, 51, 194, 249f., 253f.
– and Black Power autobiographies   92, 99–105, 107, 110, 113f., 117f., 121–23, 127–29, 133, 135, 138, 146, 151, 156

– and neo-slave narratives   159, 196f., 222, 227
– and slave narratives   78, 87, 89, 96, 252
incarceration, *see also* imprisonment, mass incarceration, racialized mass incarceration   4, 8, 10, 21, 31, 45, 50, 82, 95, 97, 100, 152 159, 249
– and Black Power autobiographies   92f., 97, 99, 101, 103–05, 118, 129f., 133–38, 148
– and *Open City*   234, 236f., 243
Incidents in the Life of a Slave Girl: Written by Herself, *see also* Jacobs, Harriet   56, 64, 79–90, 94, 112, 126, 136, 140, 211, 249, 253
Incognegro, *see also* Wilderson, Frank B. III   7, 48
Indigeneity   1, 11, 18, 23–25, 28, 42, 101, 167, 170f., 199, 256
inescapability   9f., 93, 114, 159, 189
infanticide   58, 200–03, 209, 221, 249, 251
in medias res   99, 122f., 138, 168, 208, 228
intertextuality   8–10, 54, 58, 66, 69, 93f., 136, 140, 152, 248
– in neo-slave narratives   156–58, 160, 167, 170, 179, 186, 206, 208, 224, 229, 240, 254
"in the wake," *see also* wake work, Sharpe, Christina   9, 49, 93, 200, 204, 218, 240, 246, 251, 255
In the Wake, *see also* Sharpe, Christina   5, 7, 21f., 42, 45, 47–51, 59, 61, 88, 93, 114, 197, 201, 211, 233, 236f., 240, 250, 252
intramural   44, 49, 51, 197, 251, 255
in ultimas res   168, 208
invisibility   67, 131, 164, 180

Jackson, George, *see also Soledad Brother*   8f., 11, 57, 92–95, 98f., 102, 104–06, 108–22, 129–34, 136f., 140–42, 144–47, 153, 158, 178, 199, 240, 250–53
Jackson, Jonathan   93, 110f., 119, 121, 129, 240, 252

Jacobs, Harriet, *see also Incidents in the Life of a Slave Girl* 8–11, 55–57, 59, 64–66, 69, 71, 78–91, 175 f., 249 f., 252
– and Black Power autobiographies 94, 98, 105, 112 f., 126, 136, 140 f., 147–49, 153
– and neo-slave narratives 157, 161, 171, 178, 185, 188, 194, 201, 207 f., 211
jail 31, 93, 95, 111, 116, 121–23, 125, 127–32, 134 f., 138 f., 146 f., 152, 200 f., 213, 249, 251
Jamaica 38, 151
James, Joy 23, 77, 81, 84, 91–93, 95–98, 100–04, 106–09, 111, 121, 125, 135, 137, 144 f., 153 f., 156–58, 166, 240
Japanese American internment 229, 256
Jewishness 11, 171, 180, 206, 226, 256
Jim Crow, *see also* segregation 1, 20, 23, 83, 100 f., 106, 236
Johnson, Charles 154 f., 170
Jones, Edward 155, 165
Jones, Gayl, *see also Corregidora* 68, 121, 155
Jordan, June 108, 121
Jordan, Robert Charles 108, 129 f.
Jubilee, *see also* Walker, Margaret 160 f.
junior partners of civil society 24 f., 29 f.

Khan-Cullors, Patrice 3–5, 9, 93, 106, 152 f.
Kindred 44, 155, 163, 182, 189 f.
King, Martin Luther Jr. 47, 78, 95 f., 117, 215
King, Mary 99
King, Rodney 21
King, Tiffany Lethabo 1, 7, 24, 255 f.
Koerner, Michelle 117–19, 251
Kristeva, Julia 7

Lagos, Nigeria 226, 237 f.
Larsen, Nella 55 f.
Latin America, *see also* South America 3, 17, 38
Latinx 6, 12, 15, 20 f., 28, 30, 62, 102, 247
Lebron, Lolita 98
Lejeune, Philippe 59

Let Me Live, *see also* Herndon, Angelo 56, 104
Liberia 230, 234, 249
Life and Times of Frederick Douglass 65
literacy 58, 65, 68, 70 f., 74, 79, 174, 206, 209
literatures of migration 18, 159
Local Histories/Global Designs: Coloniality, Subaltern Knowledges, and Border Thinking, *see also* Mignolo, Walter 17–19, 29
London, UK 158, 205–07, 211, 219 f.
Look for me in the Whirlwind: The Collective Autobiography of the New York 21 97, 136
"loophole of retreat," *see also* Jacobs, Harriet 81, 84–88, 90 f., 132, 148, 153, 185 f., 215, 252
Lorde, Audre 108
Lose Your Mother, *see also* afterlife of slavery; Hartman, Saidiya 7, 21, 38, 43–48, 52, 125, 154 f., 159, 165, 214, 220, 233, 237
lynching 100, 104, 135, 214, 236

Magee, Ruchell 121 f., 128 f.
magical realism 162, 195
Mali 205, 234
Manhattan, New York City 120, 143 f., 225 f., 228, 233, 235 f., 238 f.
Marable, Manning 135
marronage/maroonage, *see also* "freedom as marronage" 38 f., 57 f., 143, 151
Martin, Trayvon 3 f.
Marxism 23, 60, 120
Marx, Karl 153
mask 66, 145, 170, 230
Mason-Dixon Line 1
mass incarceration, *see also* racialized mass incarceration 100, 105
Mbembe, Achille 222
McClain, James 93, 129
McDougall, Taija 113 f., 117, 119 f., 142
McKay, Claude 151
McKittrick, Katherine 20–22, 38–40, 42, 46, 49, 61, 67, 85–88, 156
McKnight, Reginald 44

mediation
- and Black Power autobiographies   112
- and neo-slave narratives   168, 171, 178, 180, 182f., 190, 195, 204, 209, 221, 229, 234, 242, 244–46, 253f.
- and slave narratives   68, 77, 175, 177
melancholia/melancholy   49, 155f.
melodrama   209f.
Mengestu, Dinaw   222f.
meta-fiction, see also self-referentiality   162, 184
Mexico   1, 3, 15f., 18f., 38, 68
Middle Passage   1, 8, 28, 51, 78, 102, 223, 250
- and Black Power autobiographies   114f., 119, 123
- and neo-slave narratives   190–93, 196, 203–05, 208f., 211, 215, 218, 221, 232f., 237f., 241, 244, 253
"Middle Passage Epistemology"   46
"Middle Passage carceral model"   196, 203
midwifery   195, 206
Mignolo, Walter, see also border thinking, Local Histories/Global Designs   7, 12, 17–20, 29f.
migration, see also immigration   1f., 5, 8, 10, 68, 159, 210, 212f., 215, 219f., 250, 254
- and Open City   222, 227f., 234, 237, 243, 247
militancy   98, 180, 251
mise en abyme   168, 205
Monstrous Intimacies, see also Sharpe, Christina   21f., 46, 73f., 76f., 83
Morrison, Toni, see also Beloved   58, 121, 157f., 165f., 187–94, 197–204, 210
Morocco   229, 230, 234
Mosquito, see also Jones, Gayl   68
Moten, Fred   35–41, 49f., 139
motherhood   9, 64, 70, 80f., 85, 92, 114, 116f., 120, 136, 147f., 161, 163f., 187f., 194
MOVE   95
Moynihan Report   110f., 161
multiculturalism   3, 207
Mumbo Jumbo, see also Reed, Ishmael   170
Murray, Anne   75

music   7, 34, 39, 52, 164, 224f., 228
My Bondage and My Freedom, see also Douglass, Frederick   65, 73–78, 81, 250

narratibility   61, 158f., 191, 203, 210, 221, 227, 244f., 247, 253
narrative closure   9, 47, 61, 90, 154, 157, 159, 169, 204, 209f., 220, 254
Narrative of the Life and Adventures of Henry Bibb, An American Slave, Written by Himself   65
Narrative of the Life of Frederick Douglass, an American Slave, Written by Himself, see also Douglass, Frederick   64f., 67f., 70, 72–76, 78, 80f., 94, 112
Narrative of William Wells Brown, an American Slave, see also Brown, William Wells   65
narrativity   58f.
Nazer, Mende   65
neo-hoodooism, see also Reed, Ishmael   169f.
neoliberalism   26, 106–08
neo-slave narrative, see also novel of slavery   5, 8–10, 43f., 56–59, 66, 80, 94, 154–67, 205, 249, 253–55
- and Beloved   187, 189f., 202f., 251
- and Black Power autobiographies   104, 107, 125, 136
- and Flight to Canada   168, 170, 172, 182f., 185–87
- and Open City   221–23, 228, 236f., 239, 244–47
- and Someone Knows My Name   204–09, 252
neo-slavery   10, 57, 92, 105, 111, 147, 150, 154, 156, 189, 250, 252, 254
New Jim Crow, see also Alexander, Michelle   4, 20, 100
Newton, Huey P.   98, 121f., 150, 158
New York City, New York   10, 56, 74f., 88, 106, 128, 143f., 151, 159, 205f., 212f., 218
- and Open City   220–23, 225, 231–33, 235–41, 243–45, 247, 249f.
Nigeria   59, 159, 208, 222–27, 230, 237, 241–44, 250

Nixon, Richard 100f.
North Star 67, 173, 191f., 198, 251
Nova Scotia 205f., 211–15, 221
novel of slavery, *see also* neo-slave narrative 8f., 79, 107, 154, 159, 164, 166, 171, 221
Nyong'o, Tavia 35f., 39

Obama, Barack 3, 47, 55, 247
Ohio River 1, 187, 193, 199
Olney, James 67, 70f.
Olugbala, Rema 150
Open City, see also Cole, Teju 4, 10, 34, 57–59, 159, 166f., 200, 221–49, 253–56
origin (cultural, ethnic) 67, 207, 216, 218, 225, 233f.

Parks, Rosa 98
partus sequitur ventrem 70, 114, 193
Patterson, William 128
Patterson, Orlando 1, 24f., 37
people-of-color-blindness 4
Perkins, Margo 93, 98f., 108, 110–12, 114, 121, 130, 140–43, 246
"permutations," *see also* Sexton, Jared 27, 32, 106, 114, 120, 134, 247
"perpetual migration" 211, 215f., 219f., 244, 247, 250
Petry, Ann 56, 255
Pettit, Eber 182
Phillips, Wendell 68
photography 7, 34, 40–42, 49, 52, 224, 337
Poe, Edgar Allan 168–70
poetry 11, 15, 58, 95, 111, 138f., 163f., 168–70, 177–79, 181f., 185
political prisoner 66, 95f., 98, 104–06, 121–23, 129–30, 132, 135, 251
Poma, Guáman 16
Portugal 205, 233f.
postcolonial 19, 28, 46, 54, 157, 206, 224
postcolonial studies 15, 17, 28, 54, 141, 223
postmodern 9, 157f., 162–65, 167, 170, 186
post-racialism 3, 23, 47, 247
post-slavery 38, 46, 87, 92, 156

Pratt, Mary Louise 16–19, 28–30, 52
Prince, Mary 65, 71
prison, *see also* imprisonment, jail, prison industrial complex 1f., 9, 45, 56, 59, 92, 154, 249–51, 253f.
– and Black Power autobiographies 93, 95–97, 99–101, 103f., 106–10, 112–16, 119f., 122, 127–36, 138f., 145–51, 153
– and neo-slave narratives 184, 196f., 215, 237
– and slave narratives 79, 82, 85f., 92
prison as plantation, trope of the 92, 111, 132, 254
prison abolitionism 100–04, 106, 122, 125, 153, 166
prisoner, *see also* captive, convict, political prisoner 9, 51, 80, 92–98, 101–04, 107f., 115, 117, 120, 123, 125, 127, 129, 130–32, 146, 157, 196f., 203, 214, 250–52
prison industrial complex 23, 97, 99f., 105, 107, 190, 197
prison labor 51, 92, 197, 252
prison letters 9, 61, 92, 95, 109–20, 122, 140, 153, 240, 251
prison narrative 8, 101, 105, 166
prison writing 94–97, 99, 101–03, 105
"prison-slave-in-waiting," *see also* Wilderson, Frank B. III 32, 120, 134
prison songs, *see also* work songs 96
Promised Land 1
Prosser, Gabriel 160
psychoanalytic approaches 61

queer 3, 9, 34, 46, 100, 106–08, 115, 153

race 3, 14, 16, 20f., 23, 45, 54, 56, 62, 100f., 106, 117, 142, 150
– and Black Power autobiographies 99, 103, 125, 139
– and neo-slave narratives 155, 161, 180, 185, 194, 197, 207f., 222f., 225–27, 231, 241, 244f.
– and slave narratives 76, 78, 87, 91
racialized mass incarceration 9, 50, 97, 101f., 105f., 115, 122, 133, 154

racial profiling   101, 113
racism   3f., 6, 23, 35, 39, 42, 62, 66
– and Black Power autobiographies   92f., 99–101, 103f., 107, 110, 112, 114, 118, 132–34, 141f., 144f., 149f.
– and neo-slave narratives   155, 161, 164, 169, 171, 183, 231
– and slave narratives   69, 76f., 90, 92
radical prison praxis   102, 108, 132
rape, *see also* sexualized violence, sexual abuse/assault   51, 60, 70f., 73, 82–84, 95, 135, 143, 196f., 201, 252
– in *Open City*   226, 228, 230, 237f., 241f., 244, 246
Reagan, Ronald   100
realism   10, 58f., 158–60, 163, 170f., 207, 222, 245
Reconstruction   101, 189
Reed, Ishmael, *see also Flight to Canada*   156, 158f., 163, 167–87, 195, 198, 201, 209–11, 229, 250
Red, White, and Black: Cinema and the Structure of US Antagonisms, *see also* Wilderson, Frank B. III   2, 5f., 11, 23–28, 32, 34, 45, 48, 51, 55, 60f., 79, 179, 211, 233, 250, 255
refuge   2, 9, 40f., 44, 64, 68, 249f., 252
– and Black Power autobiographies   133, 136, 143, 150f., 153
– and neo-slave narratives   156, 169, 180f., 185, 190f., 199f., 204, 210f., 213, 215f., 220, 225, 227
– and slave narratives   81, 84, 87f., 90, 94, 173, 176
refugee   220, 225f., 232, 234f., 246
refusal   1, 4f., 7, 9, 12, 24, 33–35, 37f., 40–43, 47, 50–53, 66f., 148, 151, 251f., 255f.
– and Black Power autobiographies   92f., 111, 118–20, 127–34, 136, 146, 153
– and neo-slave narratives   157, 159, 177f., 181, 186f., 192, 202, 204, 219, 233, 240, 244
– and slave narratives   78, 88f.
rememory
– in *Beloved*   187, 189, 190, 192, 194–96, 199f., 202, 221, 253

– in *Open City*   235, 238, 243
resistance   5–7, 16, 33–38, 40–43, 50–52, 56f., 60, 255
– and Black Power autobiographies   100, 109, 112, 118, 120, 127f., 131, 133–35, 139f., 151
– and neo-slave narratives   158, 162, 167, 169, 181, 202, 209, 233
– and slave narratives   74, 77, 86f.
Rice, Tamir   3
Roberts, Neil   38, 57, 69, 74–78, 102, 125f.
Rodríguez, Dylan   100–04, 107f., 123, 130–32, 149–51, 153, 156
romance   9, 80, 87, 157, 162, 207, 245
runaway   9, 87, 93, 138, 143f., 147, 192
ruse of analogy   20, 25, 33, 50
Rushdy, Ashraf H. A.   69, 94, 103, 142, 154f., 162–66, 169f., 177f., 181, 185, 202, 209, 228, 245f.
Rwanda   225f., 230, 235

Sadiki, Kamau   147
Sands, Mr.   82–84, 88f., 252
satire   9, 158, 162, 170f., 209, 254
Scenes of Subjection, *see also* Hartman, Saidiya   21f., 25, 43, 45, 48, 60f., 64, 71–73, 78f., 83, 88, 120, 192, 227
Scott, Walter   3
Second World War   226, 229, 237, 256
segregation   1, 8, 20, 100, 120, 123, 136, 214
Selasi, Taiye   222
self-fashioning   42f., 77, 125
self-referentiality, *see also* meta-fiction   158, 245
senior partners of civil society   24, 30f., 52
sentimentalism   10, 65, 79–81, 83, 87, 90f., 158, 162, 171, 179, 189, 207, 253
settler colonialism, *see also* colonialism   1, 10, 24, 42, 94, 101, 198, 233, 256
Sexton, Jared   4, 6, 22, 27, 30–33, 35–37, 45f., 53, 61, 103, 106, 222
sexual abuse/assault, *see also* rape, sexualized violence   71, 82, 85, 95, 136, 143, 193
sexuality, *see also* gender, queer   15, 20, 69, 83, 87, 98, 106, 185, 225, 231, 237

sexualized Black body   180, 251
sexualized violence, *see also* rape, sexual abuse/assault   73, 82, 84, 100, 185, 200f., 237, 243, 246
Shakur, Assata, *see also Assata: An Autobiography*   8f., 11, 34, 37, 57, 60, 78, 249–51, 253, 256
– and Black Power autobiographies   92–95, 97–99, 104–06, 108, 111, 113, 115, 120f., 126f., 130, 132, 135–53
– and neo-slave narratives   158, 161, 179, 188, 194, 246
Shakur, Kakuya Amala Olugbala   147–49
Shakur, Zayd Malik   135, 149
Sharpe, Christina, *see also* anagrammatical, aspiration, "in the wake," *In the Wake*, *Monstrous Intimacies*, wake work   6f., 12, 21f., 24, 27, 33, 38, 42, 45, 47–52, 54, 62, 73f., 76f., 88, 107, 114f., 197, 200, 236f., 240, 250, 252, 257
Sierra Leone   206, 211, 213, 215, 230
Simpson, Audra   1, 42
slave narrative   5, 8–11, 53, 56–59, 64–92, 171–77, 249–55
– and Black Power autobiographies   94, 96, 98, 103, 105, 108, 111–14, 123, 125, 127, 136, 138–41, 145, 148
– and neo-slave narratives   154, 156–62, 164–68, 170f., 177–79, 181, 185, 188–91, 195, 198, 205–10, 235, 244–47
slavery   1f., 4f., 8–10, 16, 19–21, 23, 28f., 37–39, 43, 45–49, 52, 56–58, 154, 172–77, 249–51, 253f., 256
– and Black Power autobiographies   92, 94f., 99–103, 105–07, 110f., 113–15, 119, 124f., 127f., 133, 136, 143f., 149, 151
– and neo-slave narratives   154–67, 170f., 177–87, 189–93, 195f., 199, 201, 203–14, 216f., 219–21, 223f., 226f., 230–32, 235–47
– and slave narratives   64–91
Smith, Sidonie   56, 65–68, 90, 96, 110, 145, 151
Smith, Sidonie, and Watson, Julia   58–60, 110, 132, 138, 142

Smith, Valerie   65, 68, 71, 80, 83f., 86f., 160, 163, 170, 188f., 203f., 228
social death   1f., 5, 7, 10f., 13, 19, 24–30, 32–35, 37f., 43, 47, 51–54, 56, 148, 247f., 254
– and Black Power autobiographies   118, 120, 130, 133f., 139, 142
– and neo-slave narratives   159, 192f., 197f., 204, 209–11, 221, 252
– and slave narratives 66, 78f., 176
socialism   92, 109, 123, 214
Soledad Brother, *see also* Jackson, George   9, 92f., 109–20, 130–32, 136, 141f., 146, 153
Soledad Brothers, *see also* Jackson, George   93, 120, 129
solidarity   102, 129–31, 133, 232, 240, 243f.
solitary confinement   128f., 131
Sollors, Werner   224, 226–28, 231, 235, 241
Someone Knows My Name, *see also* Hill, Lawrence   4, 10f., 57f., 61, 158f., 165f., 168, 186f., 204–21, 244–46, 250, 252, 254
South America, *see also* Latin America   15, 57, 65, 160, 223
speculative fiction   59, 155, 162f., 165
Spillers, Hortense   8, 21f., 34, 37, 51, 77, 87, 106, 144, 155, 169f., 186, 197, 214, 216
spirituals   96
Statue of Liberty, New York City   231f., 238
Stender, Fay   93, 111, 115–17, 119, 129
Sterling, Alton   3
Steward, Austin   198
storytelling   59, 196, 206, 208f., 212, 222, 230, 244f.
Stowe, Harriet Beecher, *see also Uncle Tom's Cabin*   78, 80, 140, 158, 168–70, 175f., 179, 185
Student Nonviolent Coordinating Committee (SNCC)   99
Sudbury, Judy   97, 100
suicide   200, 202
surveillance   2, 4, 31, 50, 71, 74, 86, 97, 114f., 117, 120, 135, 143, 152, 179, 244

Taylor, Breonna   3
Taylor, Moses   239
Terrefe, Selamawit   4, 49, 60, 62, 74, 197
The Beautiful Things That Heaven Bears   222
The Black Book   188
The Chaneysville Incident   228
The Life of Josiah Henson, Formerly a Slave, Now an Inhabitant of Canada, as Narrated by Himself, see also Henson, Josiah   167
Thomas, Greg   205
– on Afro-pessimism   4, 6, 22, 31, 37, 136, 151
– on *Beloved*   188, 190–93, 199–204, 251
– on Black Power writing   109, 114, 116, 147, 149, 154, 157f., 166, 240
– on maroonage   57, 116, 130, 188, 199, 201f., 204
Thompson Patterson, Louise   128
tomb, trope of the   105, 114, 197, 215
Tometi, Opal   3
transatlantic slave trade, see also Middle Passage   1, 4, 8, 16, 23, 25, 27, 32, 44f., 67, 106, 172, 214, 222f., 248
– and neo-slave narratives   211, 223, 226f., 231–33, 236, 238f., 241, 243f., 247
transnational   10, 15, 18, 28, 43, 46, 159, 165, 210, 216f., 222–24, 235f.
trauma   72, 139, 157, 162, 190, 196, 206, 208f., 212, 221, 226, 230, 237f., 243, 245
travelling concept, see also Bal, Mieke   12–15, 19, 30–32, 39, 59
trickster-ism   145, 168, 170
Trump, Donald J.   4, 152
Truth, Sojourner   34, 206
Truth Stranger Than Fiction: Father Henson's Story of His Own Life, see also Henson, Josiah   168, 174
Tubman, Harriet   98, 140
Turner, Hayes and Mary   104f.
Turner, Nat   55, 155, 181, 184
Twelve Years a Slave: Narrative of Solomon Northup, a Citizen of New-York, Kidnapped in Washington City in 1841…   65

Uncle Tom's Cabin, see also Stowe, Harriet Beecher   80, 140, 158, 167f., 170, 189
Uncle Tom   78, 168f., 175f., 185
Uncle Tom's Story of His Life: An Autobiography of the Rev. Josiah Henson, see also Henson, Josiah   158, 173–76, 198
unconfined freedom   5–7, 9, 31–33, 38, 44, 47, 248, 250, 253
– and Black Power autobiographies   107, 109, 140, 150
– and neo-slave narratives   167, 171, 176, 178, 198, 204, 209, 221, 252
– and slave narratives   66, 84, 90f.
underground (in hiding)   93, 106–08, 121, 126–28, 134, 139, 144, 147, 241, 247, 249, 251
underground (subterranean)   56, 236f., 239, 241, 245f., 253
Underground Railroad   67, 124, 140, 172, 174, 176, 191–94, 205
Underground Railroad (the novel), see also Whitehead, Colson   4, 155
United Kingdom   68, 160, 164, 175f., 229, 250
US-Mexican border   15f.
utopia   29f., 140, 184

"Venus in Two Acts," see also Hartman, Saidiya   36, 43, 45, 47f., 61, 136, 146, 208f.
Verrazano, Giovanni da   233
Vietnam War   94
Virginia, United States   160, 168, 178f., 183, 185, 196
Virgin of Flames   222

wake work   7, 48–51, 256
Walcott, Rinaldo   38, 49f., 207
Walker, Alice   34, 135
Walker, Margaret, see also *Jubilee*   160f.
war, see also American Revolutionary War, Civil War, Second World War, Vietnam War   10, 104, 159, 222, 234–36, 249
'war on crime'   154
'war on drugs'   100, 106
'war on gangs'   106
'war on terror'   3, 224

Warren, Calvin, *see also* Black time 46, 51, 61, 107, 200, 251, 253
Warren, Kenneth W. 31, 155
Wells-Barnett, Ida B. 100
We Need New Names, *see also* Bulawayo, NoViolet 222
West Africa 10, 44, 47, 158f., 206, 211, 213, 215, 217, 220, 222, 234, 250
West, Cornel 135
When They Call You a Terrorist: A Black Lives Matter Memoir, *see also* Khan-Cullors, Patrice 93, 152f.
white feminism 23, 60
Whitehead, Colson, *see also* Underground Railroad 4f., 155
white 6, 8, 19, 23f., 28, 48f., 52, 55, 77f., 90, 98f., 101, 121, 135, 141, 151, 155, 170, 178, 182, 195, 201, 231
whiteness 4, 62, 234
white savior complex 194
white supremacy 4, 39, 41f., 62, 87, 105, 107f., 118, 133, 247

wig 123, 145, 178
Wilderson, Frank B. III, *see also* Afropessimism; Incognegro; *Red, White, and Black* 2, 4–7, 19f., 22–37, 39, 42f., 45f., 48–52, 55, 60f., 72, 79, 103, 202, 211, 250
– and Black Power autobiographies 107, 119f., 127, 134, 136f., 141, 149f.
Williams, Evelyn A. 135f.
Williams, Sherley Anne, *see also* Dessa Rose 107, 136, 154f.
working class 57, 106, 133
work songs, *see also* prison songs 131, 196
Wright, Michelle 46
Wright, Richard 55f., 97, 117
Wynter, Sylvia 6, 20–22, 24, 26–28, 30, 38–40, 48, 53, 59

X, Malcolm 79 95–97, 103

www.ingramcontent.com/pod-product-compliance
Lightning Source LLC
Chambersburg PA
CBHW020222170426
43201CB00007B/289